Universitext

Universitext is a series of textbooks that presents material from a wide variety of mathematical disciplines at master's level and beyond. The books, often well class-tested by their author, may have an informal, personal even experimental approach to their subject matter. Some of the most successful and established books in the series have evolved through several editions, always following the evolution of teaching curricula, into very polished texts.

Thus as research topics trickle down into graduate-level teaching, first textbooks written for new, cutting-edge courses may make their way into *Universitext*.

More information about this series at http://www.springer.com/series/223

Norbert A'Campo

Topological, Differential and Conformal Geometry of Surfaces

 Springer

Norbert A'Campo
Mathematisches Institut
Universität Basel
Basel, Switzerland

ISSN 0172-5939   ISSN 2191-6675   (electronic)
Universitext
ISBN 978-3-030-89031-5   ISBN 978-3-030-89032-2   (eBook)
https://doi.org/10.1007/978-3-030-89032-2

Mathematics Subject Classification (2020): 57-02, 53-02, 55-02, 30Fxx, 53Cxx

This Springer imprint is published by the registered company Springer Nature Switzerland AG
The registered company address is: Gewerbestrasse 11, 6330 Cham, Switzerland

Preface

This book paves a road with some detours towards the analytic theory of Riemann surfaces. Aimed at the undergraduate/graduate level, it comprises extended lecture notes based on courses given at the University of Basel over many years, a series of lectures that started at 昆明理工大学, the Center of Engineering Mathematics, Kunming University of Science and Technology, December 2014, and continued at the Erwin Schrödinger International Institute for Mathematics and Physics, Vienna, the International Centre for Theoretical Sciences of the Tata Institute of Fundamental Research, Bangalore, and the CIMPA School at Banaras Hindu University, Varanasi. The aim is a study of surfaces from the topological, differential, metrical and conformal viewpoint.

The first sections introduce basic concepts such as the differential of a function and also various fields: vector fields; 1- and 2-differential forms; the path integral of 1-forms; the exterior differential of forms; the almost complex structure given by J-fields; the Poincaré Lemma for 1-forms; the Morse Lemma; Morse functions on manifolds; the Frobenius Integrability Criterion; the classification of compact connected oriented surfaces by genus or Euler number; Stokes' Theorem and de Rham cohomology; the Brouwer, Kakutani, Banach and Schauder Fixed Point Theorems; coordinates and basic smooth local Rigidity Theorems; and the abstract field of complex numbers $\mathbb{C}$ versus the $\mathbb{R}$-algebra of complex numbers $\mathbb{C}$.

Two interpretations $\mathbb{H}_I$, $\mathbb{H}_J$ of the hyperbolic plane $\mathbb{H}$ as a space of geometric objects are given: Starting from quadratic equations, as in high-school, we arrive at $\mathbb{H}_I$, using surjective ring-homomorphisms from $\mathbb{R}[X]$ to a field (up to automorphism of the target field). The second interpretation $\mathbb{H}_J$ is based on oriented linear complex structures on the real plane. Hyperbolic trigonometry is introduced, mainly using the interpretation $\mathbb{H}_I$.

The next sections follow less basic concepts: Riemannian metrics on manifolds; the isometric torsion-free connection on vector fields and the curvature; using the Ansatz of Koszul; the Gauss–Bonnet Theorem for surfaces; conformal structure on surfaces via J-fields; local rigidity of J-fields and integrability to complex local coordinates on J-surfaces; and J-fields in higher dimensions and the obstruction to integrability by the Nijenhuis tensor, a specialization and simplification due to

Nijenhuis and Woolf of the Nijenhuis tensor in the case of families of J-surfaces over a smooth complex base.

The introduced concepts are then used for the study of Riemann surfaces: definition of Riemann surfaces; examples of Riemann surfaces with a hyperbolic metric; hyperbolic metrics on compact conformal surfaces of genus ≥ 2 and the Uniformization Theorem for compact Riemann surfaces by a perhaps new continuity method; Riemann's Theorem on holomorphic 1-forms and the Existence Theorem; reconstruction of Riemann surfaces from the field of meromorphic functions; holomorphic line bundles; Čech and Dolbeault cohomology; the Riemann–Roch Theorem; Teichmüller spaces as leafs of a foliation on the space of J-fields, together with a construction of the Weil–Petersson Kähler structure; holomorphic embeddings of Riemann surfaces in projective spaces; the theorem of 周炜良 (Wei-Liang Chow) with an "elementary" proof and Riemann surfaces as projective varieties; the Teichmüller curve and its universal functorial property; the Jacobian of a Riemann surface; the Bergman metric and stratification of Teichmüller space; the energy of canonical embeddings.

Acknowledgements

After my teaching time at the University of Basel I had the opportunity to participate in many conferences on Low Dimensional Topology, especially on Singularities, Teichmüller Theory and Symplectic Topology. This participation and the resulting discussions were extremely stimulating for me and they are also reflected, unfortunately only partially, in the present book. I thank all the organizers of these events: Alejandro Melle, Amine Bahayou, András Némethi, Anna Wienhard, Athanase Papadopoulos, Bankteshwar Tiwari, Bob Penner, Duco van Straten, Francisco-Jesus Castro-Jimenez, François Laudenbach, François Loeser, Georgios Tsapogas, Gianluca Bande, Javier Fernandez de Bobadilla, Ken'itchi Oshika, Krishnendu Gongopadhyay, Lizhen Ji, Makoto Sakuma, Michel Boileau, Renzo Caddeo, Shing-Tung Yau, Sumio Yamada, Tsichlias Charalampos, Vincent Colin, and Yasha Eliashberg.

I thank Annette A'Campo-Neuen, Athanase Papadopoulos, Himalaya Senapati, Patrick Popescu Pampu and Sumio Yamada for their useful suggestions.

Witterswil, 4 June 2021   *Norbert A'Campo*

Contents

Chapter 1
Basic Differential Geometry

1.1 Fields on Open Sets in Real Vector Spaces

Let V be a finite-dimensional real vector space of finite dimension equipped with a norm function $\| \cdot \|_V : V \to \mathbb{R}$. A norm function allows estimations of vectors, defines convergence of sequences in V, and provides a topology on V. Remember, a real-valued function $\| \cdot \|_V : V \to \mathbb{R}$ on a real (or complex) vector space is a *norm* if the following properties for $\lambda \in \mathbb{R}$, $u, v \in V$ or $\lambda \in \mathbb{C}$, $u, v \in V$ hold:

(i) $\|u - v\|_V > 0$ if and only if $u \neq v$,   (*Detects equality*)
(ii) $\|\lambda u\|_V = |\lambda| \, \|u\|_V$,   (($\lambda, |\lambda|$)-*homogeneous*)
(iii) $\|u + v\|_V \leq \|u\|_V + \|v\|_V$.   (*Sub-additivity*)

Associated to a norm $\| \cdot \|_V$ on V is the distance function $\mathrm{Dist} : V \times V \to \mathbb{R}$ defined by $\mathrm{Dist}(u, v) = \|u - v\|_V$.

More generally, a *distance function* on a set X is a two-point function $\mathrm{Dist} : X \times X \to \mathbb{R}$ satisfying for $a, b, c \in X$

(i) $\mathrm{Dist}(a, b) \geq 0$ and $\mathrm{Dist}(a, b) = 0$ if and only if $a = b$,   (*Detects equality*)
(ii) $\mathrm{Dist}(a, b) = \mathrm{Dist}(b, a)$,   (*Symmetry*)
(iii) $\mathrm{Dist}(a, c) \leq \mathrm{Dist}(a, b) + \mathrm{Dist}(b, c)$.   (*Triangle inequality*)

A distance function on a set X provides a *topology* on X. A *topology* T on a set X is a subset $T \subset \mathcal{P}(X)$ of the power set $\mathcal{P}(X)$ of the set X. A set X endowed with a topology $T \subset \mathcal{P}(X)$ is called the *topological space* (X, T). Elements of T are called *open* subsets of (X, T) or in X if the topology is not explicitly mentioned. In order to be a topology on the set X the subset $T \subset P(X)$ must satisfy the axioms:

(i) $\emptyset, X \in T$,
(ii) if $(A_i)_{i \in I}$ is a finite family with $A_i \in T$ then $\cap_{i \in I} A_i \in T$,
(iii) if $(A_i)_{i \in I}$ is any family with $A_i \in T$ then $\cup_{i \in I} A_i \in T$.

In words, the above properties say that a union of open subsets in a topological space is again an open subset, and that a finite intersection of open subsets in a topological space is again an open subset. The provided topology T on a *metric space* (X, Dist) is constructed in three steps: first declare the so-called *open balls*,

N. A'Campo, *Topological, Differential and Conformal Geometry of Surfaces*, Universitext,
https://doi.org/10.1007/978-3-030-89032-2_1

i.e. subsets $b_{\mathrm{Dist}}(a, r) := \{b \in X \mid \mathrm{Dist}(a, b) < r\}$, $a \in X, r \in \mathbb{R}$, to be elements of the set T; second declare unions of open balls to be elements of T; and finally (please) do the third step by verifying that T is indeed a topology on X.

A distance function Dist on a vector space V is constructed using a norm on V if and only if the following properties for $u, h \in V$ and $\lambda \in \mathbb{R}$ hold:

(i) $\mathrm{Dist}(u, u + h)$ only depends upon h,
(ii) $\mathrm{Dist}(u, u + \lambda h) = |\lambda| \mathrm{Dist}(u, u + h)$.

The first property says that distances are invariant under translation, the second says that distances are positively homogeneous.

In the case where the distance function on a vector space V is constructed from a norm $\| \cdot \|_V$, the open ball in V with center $p \in V$ and radius $r \geq 0$ is the subset $b_V(p, r) = \{q \in V \mid \|q - p\|_V < r\}$. A subset U of V is *open* if it is a union of open balls. The collection of these open subsets in V is the topology on V induced by the given norm. If the real or complex vector space is of finite dimension, two different norms provide equal topologies.

Let U be an open subset of the vector space V. Of central importance is the notion of a field on U. There are many kinds of fields, so we proceed by giving the first basic examples.

A *vector field* on U is a map $X: U \to V$. For a vector field we denote the value of the map X at the point $p \in U$ by $X_p \in V$.

Let V^* be the vector space of all linear maps $L: V \to \mathbb{R}$. An element of V^* is called a *linear form* on V. Sometimes we more briefly refer to a *form* on V. The vector space V^* is called the *dual vector space* of the vector space V.

A 1-*differential form* on U is a field given by a map $\omega: U \to V^*$. Again we denote by $\omega_p \in V^*$ its value at the point $p \in U$.

A real-valued function $f: U \to \mathbb{R}$ or a complex-valued function $f: U \to \mathbb{C}$ with numerical values can also be seen as a "field of numbers" on U.

Let $f: U \to \mathbb{R}$ be a function. Let $p \in U$ be a point. We wish to study the behavior of the function f near the point $p \in U$. In order to do so we consider a number $r > 0$ such that the ball $b_V(p, r) := \{q \in V \mid \|p - q\|_V < r\}$ with center p and radius r is contained in U. Such a positive radius r exists since we have assumed U to be an open subset in V. The behavior of the function f near p, we say also the *local behavior* of f at p, is encoded in the map $p + h \in b_V(p, r) \mapsto f(p + h) \in \mathbb{R}$. The following notion of *differentiability* of f at p is very important.

We say that a function f is *differentiable* at p if for $p + h \in b_V(p, r)$ a three-term expansion of $f(p + h)$ exists

$$f(p + h) = A + L(h) + R(h)$$

that satisfies the following properties for each term (and as we will see, the terms are uniquely determined by these properties):

(i) The first term A is independent of h, so $A \in \mathbb{R}$ is a number.
(ii) The second term $L(h)$ is the value at $h \in V$ of a linear form $L \in V^*$.
(iii) The third term $R(h)$ is the value at $h \in V$ of a map $R: V \to \mathbb{R}$ with the property

$$R(h) = 0, \quad \lim_{h \to 0, \, \|h\|_V \neq 0} |R(h)| / \|h\|_V = 0.$$

Remember that the dual vector space V^* of the real vector space V is the vector space of all linear maps $L: V \to \mathbb{R}$. So, for $h \in V$ and $L \in V^*$, $L(h) \in \mathbb{R}$ denotes the value of L at h.

The property of the term $R(h)$ is expressed in words by saying that $R(h)$ is arbitrarily small relative to h. This condition can be expressed using quantifiers on positive numbers ϵ, δ in the following way: For every $\epsilon > 0$ there exists a $\delta > 0$ such that for all $h \in V$ with $\|h\|_V < \delta$ the inequality $|R(h)| < \epsilon\|h\|_V$ holds.

The first observation is that the first term A is the value $f(p)$ of the function f at the point p. Indeed, evaluating the three-term expansion for $h = 0$, one gets $f(p) = f(p + 0) = A + 0 + 0$ since $L(0) = R(0) = 0$.

The next observation is that the linear form L that appears in the second term of the three-term expansion is independent of the expansion. Indeed, let

$$f(p + h) = A + L(h) + R(h), f(p + h) = A + L'(h) + R'(h)$$

be two expansions satisfying the above typical properties. It follows that the difference

$$L(h) - L'(h) = (L - L')(h) = R'(h) - R(h)$$

is linear in h and also arbitrarily small relative to $\|h\|_V$, hence $L = L'$ since $L - L'$ is a linear form on V. Indeed, suppose the contrary: let $u \in V$ be such that $L(u) \neq L'(u)$, then we would have

$$\lim_{s \to 0, s > 0} |(L - L')(su)|/\|su\|_V = |(L - L')(u)|/\|u\|_V \neq 0$$

contradicting that $R'(h) - R(h) = L(h) - L'(h)$ is arbitrarily small relative to $\|h\|_V$.

Since now the linear form of the second term is well defined, it deserves a name and a notation. We call the linear form $L \in V^*$ the *differential* of the function f at the point p. We denote the differential of f at the point p by $(Df)_p \in V^*$.

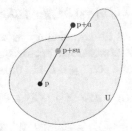

Fig. 1.1 $(Df)_p(u) = \lim_{s \to 0, s > 0} \frac{1}{s}(f(p + su) - f(p))$

The value $(Df)_p(u)$ is also called the derivative at p of f in the direction of $u \in V$ and can be computed by the limit

$$(Df)_p(u) = \lim_{s \to 0, s > 0} \frac{1}{s}(f(p + su) - f(p)).$$

A norm function on V admits at $0 \in V$ a directional derivative for all $u \in V$, but is not differentiable at $0 \in V$.

Let the function $f: U \to \mathbb{R}$ be differentiable at every point of U. In short, we say that f is differentiable. From the differentiable field of numbers f we get a field of linear forms. Indeed, the map

$$p \in U \mapsto (\mathrm{D}f)_p \in V^*$$

is a 1-differentiable form that we denote by $\mathrm{d}f$. Therefore $\mathrm{d}f$ is a field defined on $U \subset V$ with values in V^*.

From a function f we get by differentiation the field $\mathrm{d}f: U \to V^*$. The value of $\mathrm{d}f$ at p is the linear form $(\mathrm{D}f)_p$. It seems to be self-explanatory to call $\mathrm{d}f$ the *differential* of the function f, but be aware that not all 1-differential forms appear in this way as differentials of functions. Even locally, as we shall see, not all differential 1-forms appear as differentials of functions.

We will see later that there are many ways to make a field out of two given fields. Here is an example: let $\omega: U \to V^*$ be a 1-differentiable form and let $X: U \to V$ be a vector field. We get a field of numbers, a function $f := i_X \omega$, by putting $f(p) = (i_X \omega)_p := \omega_p(X_p)$. The operation i_X is called the *insertion* of the vector field X. Let $f: U \to \mathbb{R}$ be a differentiable function and let $X: U \to V$ be a vector field. We get a third field Xf of numbers, so a function, by putting $(Xf)(p) := (\mathrm{d}f)_p(X_p)$. Note that $Xf = i_X(\mathrm{d}f)$. The function Xf is called the derivative of f in the direction of X. There are other common notations for this, for instance $\mathrm{d}_X f$.

The action of vector fields on functions $f \mapsto Xf$ satisfies the Leibniz rule[1]

$$X(f \cdot g) = (Xf) \cdot g + f \cdot (Xg),$$

where the dot $\cdot$ stands for the multiplication of functions. Multiplication of functions will however mostly be written without this dot. The Leibniz rule without the dot $\cdot$ looks like

$$X(fg) = (Xf)g + f(Xg).$$

We also want the notion of differentiability of vector-valued functions. So, let W be a second normed finite-dimensional vector space with norm $\| \cdot \|_W$. A map $f: U \to W$ is differentiable at a point $p \in U$ if a three-term expansion

$$f(p + h) = A + L(h) + R(h)$$

exists with the properties: $A \in W$, $L \in \mathrm{Hom}(V, W)$ is a linear map $L: V \to W$ and finally $R: V \to W$ is a map such that $\|R(h)\|_W$ is arbitrarily small relative to $\|h\|_V$. Again, A has to be the value $f(p)$ of f at p and the linear map L is well defined independently of the expansion and is denoted by $(\mathrm{D}f)_p$. Now $\mathrm{d}f$ is a field on U with values in $\mathrm{Hom}(V, W)$, the vector space of linear maps from V to W.

As an example: From a vector field X on $U \subset V$ we get by differentiation the field

---

[1] Named after German mathematician and philosopher Gottfried Wilhelm Leibniz (1646–1716).

$$dX: U \to \mathrm{End}(V), \ p \mapsto (u \in V \mapsto (DX)_p(u) \in V),$$

which is a field of endomorphisms of the ambient vector space V.

A map $f: U \to W$ is twice differentiable if f is differentiable and the map $df: U \to \mathrm{Hom}(V, W)$ is also differentiable. Inductively on $k = 1, 2, \ldots$ we define a map $f: U \to W$ to be k-times differentiable if f is differentiable and the map $df: U \to \mathrm{Hom}(V, W)$ is $(k-1)$-times differentiable. We call a map $f: U \to W$ smooth if f is k-times differentiable for all $k \in \mathbb{N}$.

In this text, unless explicitly stated to the contrary, we consider only smooth fields, functions, maps,...

Let $(x_1, x_2, \ldots, x_n)$ be a basis for the vector space V^* of linear forms on V. The restrictions of the functions $x_i: V \to \mathbb{R}$ to the open subset $U \subset V$ are functions on U that we again denote by the symbols x_i. A number $a \in \mathbb{R}$ can be considered as the constant function on U that takes only the value a. As an example, for $p \in U$ the expression $x_1 - x_1(p)$ denotes a function on U that vanishes at p.

Theorem 1.1. *Let $U \subset V$ be an open subset of a real vector space V of finite dimension n. Let $(x_1, x_2, \ldots, x_n)$ be a basis of V^*. Let $p \in U$. Let f be a smooth function on U with $f(p) = 0$. Then there exist smooth functions $g_1, g_2, \ldots, g_n$ on U such that*

$$f = (x_1 - x_1(p))g_1 + (x_2 - x_2(p))g_2 + \cdots + (x_n - x_n(p))g_n.$$

If, moreover, $(Df)_p = 0$ holds, the function g_i in the above expansion satisfies $g_i(p) = 0$, $i = 1, \ldots, n$.

Proof. We assume that the open set U is star-shaped with p as center. This means that for all $q \in U$ the segment $\{tq + (1-t)p \mid t \in [0, 1]\}$ lies in U. Moreover, we assume that the coordinate functions are centred at p, i.e. $x_i(p) = 0$. Let $(e_1, e_2, \ldots, e_n)$ be the basis of V with $x_j(e_i) = \delta_i^j$ and $q = q - p = \sum_i x_i(q)e_i$. With this notation, for $q \in U$, remembering $f(p) = 0$, one expands

$$f(q) = f(q) - f(p) = \int_0^1 (Df)_{tq+(1-t)p}(q-p)dt$$

$$= (Df)_p(q-p) + \int_0^1 ((Df)_{tq+(1-t)p} - (Df)_p)(q-p)dt$$

$$= \sum_i x_i(q)((Df)_p(e_i) + \int_0^1 ((Df)_{tq+(1-t)p} - (Df)_p)(e_i)dt)$$

$$= \sum_i x_i(q)g_i(q)$$

with

$$g_i(q) = (Df)_p(e_i) + \int_0^1 ((Df)_{tq+(1-t)p} - (Df)_p)(e_i)dt.$$

Hence $f = \sum_i x_i g_i$. Clearly $g_i(p) = 0$ if $(Df)_p = 0$ holds. The assumption that U is star-shaped is removed by using a smooth partition of unity. $\qquad\square$

Before going on, please consider the following exercises.

Exercise 1.2. Let X be a set having three elements and let $D: X \times X \to \mathbb{R}$ be a distance on X. Show that there exists an isometric embedding of the metric space (X, D) into the Euclidean plane. Let $Y = \{a, b, c, d\}$ be a set having four elements. Let $D: Y \times Y \to \mathbb{R}$ be the distance function given by $D(d, a) = D(d, b) = D(d, c) = 1, D(a, b) = D(b, c) = D(c, a) = 2$. Show that no isometric embedding of (Y, D) into a Euclidean space of any dimension exists. Find a norm on a two-dimensional vector space into which (Y, D) isometrically embeds.

Exercise 1.3. Let V, W be finite-dimensional vector spaces with norms $\| \cdot \|_V$ and $\| \cdot \|_W$. On $\mathrm{Hom}(V, W)$ define the operator norm

$$\|A\|_{\mathrm{Operator}} := \mathrm{Max}\{\|A(v)\|_W / \|v\|_V \mid v \in V, v \neq 0\}.$$

Show that this maximum is achieved.

The following exercise was proposed, with a $*$, for the freshmen of 1959 at the University of Utrecht.[2]

Exercise 1.4. Let $f: \mathbb{R} \to \mathbb{R}$ be a function that is continuous at $0 \in \mathbb{R}$. Assume that $\lim_{h \to 0, h \neq 0} \frac{f(2h) - f(h)}{h}$ exists. Show that f is differentiable at 0.

Exercise 1.5. Let $\mathrm{sq}: \mathrm{Hom}(V, V) \to \mathrm{Hom}(V, V)$ be the map defined by $\mathrm{sq}(A) := A \circ A$. Compute $(\mathrm{D}\,\mathrm{sq})_A(H)$, $A, H \in \mathrm{Hom}(V, V)$. For $V = \mathbb{R}^2$ with the standard Euclidean norm and $A = \left(\begin{smallmatrix} 0 & 1 \\ 1 & 1 \end{smallmatrix}\right)$, compute the operator norms of A and of $(\mathrm{D}\,\mathrm{sq})_A$.

Exercise 1.6. Let $\exp: \mathrm{Hom}(V, V) \to \mathrm{Hom}(V, V)$ be the exponential mapping

$$\exp(A) := \sum_{n=0}^{\infty} \frac{1}{n!} A^n.$$

Show that the exponential mapping is differentiable. Do not try to compute its differential, but compute instead $(\mathrm{D}\exp)_A(H)$ for $A, H \in \mathrm{Hom}(V, V)$, assuming $A \circ H = H \circ A$ or $A \circ H = -H \circ A$.

Exercise 1.7. For a finite-dimensional vector space V over the field of real numbers, let $\mathrm{Det}: \mathrm{Hom}(V, V) \to \mathbb{R}$ be the determinant function. Show that $(\mathrm{D}\,\mathrm{Det})_{\mathrm{Id}}(H)$ is equal to $\mathrm{Trace}(H)$. Hint: Study first the cases of $\mathrm{Dim}(V) = 1, 2, 3$.

Exercise 1.8. For $H \in \mathrm{Hom}(V, V)$ let H also denote the constant vector field on $\mathrm{Hom}(V, V)$ with value H. Compute the function $i_H \mathrm{d}\,\mathrm{Det}$.

Exercise 1.9. Let V be the vector space $\mathbb{R}^2$. Let $A \in \mathrm{Hom}(V, V)$ have $0, 1, 2$ or ∞ many 1-dimensional eigenspaces and let $v \in V$ be a non-eigenvector, if possible. Sketch the curves $t \in \mathbb{R} \mapsto \exp(tA)(v) \in V$. Consider $A: V \to V$ as a vector field on V. The corresponding differential equation for parametrized curves $\gamma: \mathbb{R} \to V$ is $\dot{\gamma}(t) = A(\gamma(t))$, $\gamma(0) = v$. Extend the list of the above sketched curves in order to get a complete list of all solutions of differential equations of the form $\dot{\gamma}(t) = A(\gamma(t))$, $\gamma(0) = v$.

---

[2] By Dutch mathematician Professor Frederik van der Blij (1923–2018).

Exercise 1.10. Let X, Y be vector fields on U. Show that the additive commutator of the actions of the vector fields X and Y on smooth functions again satisfies the Leibniz rule. More precisely, we call $[X, Y]f := X(Yf) - Y(Xf)$ the additive commutator and check the rule $[X, Y](f \cdot g) = [X, Y]f \cdot g + f \cdot [X, Y]g$.

Exercise 1.11. Let X, Y be vector fields on U. Show the existence of a uniquely defined vector field Z with $Zf = [X, Y]f$ for all smooth functions f on U. We denote this vector field Z by $[X, Y]_{\text{Lie}}$. The operation $(X, Y) \mapsto [X, Y]_{\text{Lie}}$ is called the Lie bracket[3] of vector fields.

Exercise 1.12. (*Exercise 1.11 continued*) For a smooth function f and vector fields X, Y on U, show the properties:

$$[fX, Y]_{\text{Lie}} = f[X, Y]_{\text{Lie}} - (Yf)X,$$

$$[X, fY]_{\text{Lie}} = f[X, Y]_{\text{Lie}} + (Xf)Y.$$

Conclude that $([X, Y]_{\text{Lie}})_p$ does not only depend on the values X_p, Y_p of the vector fields X, Y.

Hint: If $X_p \neq 0$, choose a function f with $f(p) = 1$ and $(Df)_p(Y_p) \neq 0$.

Exercise 1.13. Let X, Y, Z be vector fields on U. Show the Jacobi identity:[4]

$$[X, [Y, Z]_{\text{Lie}}]_{\text{Lie}} + [Z, [X, Y]_{\text{Lie}}]_{\text{Lie}} + [Y, [Z, X]_{\text{Lie}}]_{\text{Lie}} = 0.$$

Exercise 1.14. For a function f denote by m_f the operation on the space of functions defined by multiplication by f. So, $m_f(g) = fg$. We say that a vector field F compensates m_f if the commutator $[F, m_f]$ is the identity operator on functions. Which functions are compensated by a vector field?

Exercise 1.15. Let $f : U \to \mathbb{R}$ be a smooth function on an open subset in a real vector space V of finite dimension n. Let the function f and the higher order differentials up to order k vanish at $p \in U$. Show that f admits an expansion $f = \sum_e (x - x(p))^e g_e$, where $e = (e_1, e_2, \ldots, e_n)$ are the integral multi-exponents with $0 \leq e_i$, $|e| = \sum_i e_i = k$, the g_e are smooth functions on U, $x_1, x_2, \ldots, x_n$ form a basis for V^*, and $(x - x(p))^e$ denotes the product $\prod_i (x_i - x_i(p))^{e_i}$.

Hint: First expand $f = \sum (x_i - x_i(p))g_i$ and apply the statement inductively to the functions g_i.

Exercise 1.16. Keeping the notation from the previous exercise, let $M_k(p)$ be the ideal in the ring of smooth real functions $C^\infty(U, \mathbb{R})$ on U consisting of those functions f such that f and all higher differentials of f, up to order k, vanish at p. Show that $m_p := M_0(p)$ is a maximal ideal. Show $M_k(p) = m_p^{k+1}$. Compute the dimensions of the vector spaces $M_{k-1}(p)/M_k(p)$. Identify $M_0(p)/M_1(p)$ with V^*.

We have seen that a vector field X defines an operation of derivation on functions, the operation of insertion i_X and also defines a differential equation. Solutions of the latter are parametrized curves. What can we do with a 1-differential form?

---

[3] Named after Norwegian mathematician Marius Sophus Lie (1842–1899).

[4] Named after German mathematician Carl Gustav Jacob Jacobi (1804–1851).

We have already seen that a 1-differentiable form ω together with a vector field X yields a function: $i_X\omega$. What more can we do?

Let $\omega: U \to V^*$ be a 1-differentiable form. Let $\gamma: [a,b] \to U$ be a continuously differentiable curve in U. The *path integral* of ω over γ is defined by:

$$\int_\gamma \omega := \int_a^b \omega_{\gamma(t)}(\dot\gamma(t))dt,$$

which is a number. Here $\dot\gamma(t) \in V$ is the velocity vector of γ at instant t:

$$\dot\gamma(t) := (D\gamma)_t(1),$$

i.e. the differential $(D\gamma)_t: \mathbb{R} \to V$ of γ at the point $t \in [a,b]$ evaluated at the vector $1 \in \mathbb{R}$.

The path integral $\int_\gamma \omega$ does not depend on the parametrization, as long as the re-parametrization respects the orientation.

The path integral is additive with respect to catenation of paths. So, the path integral is also well defined for paths that are piece-wise continuously differentiable.

The path integral $\int_\gamma df$, where the 1-form is the differential of a function, evaluates to the difference of the values of the function at the endpoints:

$$\int_\gamma df = \int_a^b (df)_{\gamma(t)}(\dot\gamma(t))dt$$
$$= \int_a^b \frac{d}{dt}(f \circ \gamma(t))dt = [f \circ \gamma]_a^b = f(\gamma(b)) - f(\gamma(a)).$$

In particular for a closed path, i.e. a path with $\gamma(a) = \gamma(b)$, one has $\int_\gamma df = 0$.

In general, the path integral over a closed path of a 1-differential form does not evaluate to 0.

In order to give examples, we consider on $\mathbb{R}^2$ the coordinate functions

$$x, y: \mathbb{R}^2 \to \mathbb{R}.$$

The functions x, y are defined as follows. Let $e_1 = (1,0)$, $e_2 = (0,1) \in \mathbb{R}^2$ be the two standard basis vectors. Every vector $u \in \mathbb{R}^2$ can be written in a unique way as a linear combination $u = ae_1 + be_2$, $a, b \in \mathbb{R}$. We define the functions x, y on $\mathbb{R}^2$ by $x(u) = a$, $y(u) = b$. The differentials of the functions x, y are the 1-differential forms dx, dy. The function $y: \mathbb{R}^2 \to \mathbb{R}$ has at $p \in \mathbb{R}^2$ the typical expansion $y(p+h) = y(p) + y(h) + 0$, so we get $(dy)_p(h) = y(h)$.

Example 1.17. Let ω be the 1-differential form on $\mathbb{R}^2$ given by $\omega = xdy$. This very short and dense notation expands by evaluating at a point p and vector h the number $\omega_p(h)$, $p, h \in \mathbb{R}^2$, given by

$$\omega_p(h) = x(p)((dy)_p(h)) = x(p)y(h).$$

Evaluating the path integral of ω on the closed path γ defined by

$$\gamma(t) := \cos(t)e_1 + \sin(t)e_2, \; t \in [0, 2\pi],$$

with $\dot{\gamma}(t) = -\sin(t)e_1 + \cos(t)e_2$, one gets

$$\int_\gamma \omega = \int_0^{2\pi} x(\gamma(t))((dy)_{\gamma(t)})(\dot{\gamma}(t))dt = \int_0^{2\pi} \cos(t)^2 dt = \pi \neq 0. \qquad (1.1)$$

If we denote by $\partial B_p(r)$ the counterclockwise oriented boundary of the closed Euclidean disk with center $p \in \mathbb{R}^2$ and radius $r \in \mathbb{R}_{>0}$, we have

$$\int_{\partial B_p(r)} \omega = \pi r^2 \neq 0. \qquad (1.2)$$

We conclude from the first computation (1.1) that the 1-differential form ω is not the differential of a function on $\mathbb{R}^2$.

We conclude by the second computation (1.2) that for no non-empty open subset U in $\mathbb{R}^2$ is the restriction ω_U of the form ω to U the differential of a function on U.

In short, the 1-differential form ω is at no point locally the differential of a function. Even more briefly, we say, the 1-differential form ω is locally not the differential of a function.

Example 1.18. Let U be the punctured plane $U := \mathbb{R}^2 \setminus \{0\}$. Let $\rho : U \to \mathbb{R}$ be the radius function $\rho := (x^2 + y^2)^{1/2}$. Put $\alpha := \frac{x}{\rho}dy - \frac{y}{\rho}dx$. For $p \in U$, let B_p be the open Euclidean ball with center p and radius $\rho(p)$. We have the inclusion $B_p \subset U$. Let $\text{angle}_p : B_p \to [-\pi/2, \pi/2]$ be the function where $\text{angle}_p(q)$ equals the counterclockwise oriented Euclidean angle in radians between the rays emanating from the origin through p and q. We notice, that its differential $d\,\text{angle}_p$ is equal to the restriction of α to B_p. So the path-integral of α evaluates to 0 for every closed path in B_p. In short, the 1-differential form α is locally the differential of a function. Nevertheless,

$$\int_{\partial B_p(r)} \alpha = 2\pi \neq 0, \; r > \rho(p), \; p \in U$$

and α cannot be globally on U the differential of a function.

Let $U \subset V$ be an open subset of a vector space V. We say that a 1-differential form $\omega : U \to V^*$ is *exact* if there exists a function f on U with $\omega = df$. We say that a 1-differential form $\omega : U \to V^*$ is *locally exact* if for every $p \in U$ there exist an open subset U' of U containing p and a function f on U' such that we have for the restriction of $\omega_{|U'}$ to U' the equality $df = \omega_{|U'}$.

The 1-form ω of Example 1.17 is not exact, and also not locally exact. The 1-form of Example 1.18 is locally exact, but not exact.

We wish to have an infinitesimal criterion for local exactness. First we define the so-called exterior derivative $d\omega$ of a 1-differential form ω on $U \subset V$.

The exterior derivative $d\omega$ will be a field on U with values in $\Lambda^2(V)$. So first we define $\Lambda^k(V)$, $k \in \mathbb{N}$. If $k = 0, 1$ we put $\Lambda^0(V) = \mathbb{R}$ and $\Lambda^1(V) = V^*$. If $k = 2$, we define $\Lambda^2(V)$ as the vector space of all maps

$$\Lambda^2(V) := \{B : V \times V \to \mathbb{R} \mid B \text{ bi-linear and anti-symmetric}\}.$$

If $k > 2$ we define $\Lambda^k(V)$ as the vector space of all maps

$$\Lambda^k(V) := \{B : V^k \to \mathbb{R} \mid B \text{ multi-linear and alternating}\}.$$

A field on U with values in $\Lambda^k(V)$ is called a *k-differential form* on U. Such a field is a real function in the case $k = 0$.

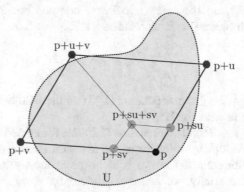

Fig. 1.2 $(d\omega)_p(u, v) = \lim_{s \to 0, s > 0} \frac{1}{s^2} \int_{p \to p+su \to p+su+sv \to p+sv \to p} \omega.$

We turn back to the definition of the exterior differential $d\omega$ of a 1-differential form ω. In order to do so, we have to define the number

$$(d\omega)_p(u, v) \in \mathbb{R}, \ p \in U, \ u, v \in V.$$

For $p \in U$, $s \in \mathbb{R}$, $s > 0$, $u, v \in V$ let $P_p(u, v, s)$ be the parallelogram in V with corners p, $p + su$, $p + su + sv$, $p + sv$. Let $\partial P_p(u, v, s)$ be its oriented boundary from p to $p + su$ to $p + su + sv$ to $p + sv$ back to p (Fig. 1.2). For s small enough we have $P_p(u, v, s) \subset U$ since U is open. We define the required number by

$$(d\omega)_p(u, v) := \lim_{s \to 0, \, s > 0} 1/s^2 \int_{\partial P_p(u, v, s)} \omega.$$

For a function f on U we have the important basic fact:

Theorem 1.19.
$$d(df) = 0.$$

Proof. Since $\partial P_s(p, u, v), s > 0$, is a closed path, we have the vanishing

$$1/s^2 \int_{\partial P_p(u, v, s)} df = 0$$

even before taking limits.   □

It is important to note that the above proof works as soon as $df : U \to V^*$ is continuous, i.e. the function f is *continuously differentiable*. But, if the function f is twice differentiable, we can deduce more:

Theorem 1.20. *The differential* $(D(df))_p : V \to V^*$ *is symmetric.*

Before giving a proof, we recall some basic facts from linear algebra. Here V is a finite-dimensional real vector space. A vector $u \in V$ can also be considered as a linear form on V^* by declaring $u(L) := L(u)$, $L \in V^*$. Hence we get a canonical map

$$u \in V \mapsto (L \in V^* \mapsto L(u) \in \mathbb{R}) \in V^{**},$$

which is a canonical inclusion $i : V \hookrightarrow V^{**}$ in general and which is an isomorphism of vector spaces in the case of finite dimension. This isomorphism i allows the identification $V^{**} = V$.

According to a theorem of Erdős[5] and Kaplansky[6] a vector space V is isomorphic to its dual V^* if and only if the dimension of V is finite. More precisely, their proof shows that if V is of infinite dimension, a strict inequality of cardinals $\mathrm{Dim}(V) < \mathrm{Dim}(V^*)$ holds. From the inequalities $\mathrm{Dim}(V) < \mathrm{Dim}(V^*) < \mathrm{Dim}(V^{**})$, that hold only if V is of infinite dimension, it follows that a vector space V and its bi-dual V^{**} are isomorphic if and only if the vector space V is finite-dimensional.

We recall that the transpose ${}^t B$ of a linear map $B : V \to W$ is the linear map

$$ {}^t B : W^* \to V^*, \ L \in W^* \mapsto L \circ B \in V^*.$$

Assume V is finite-dimensional. The natural inclusion

$$i : V \to V^{**}, \ u \in V \mapsto (L \in V^* \mapsto L(u) \in \mathbb{R}) \in V^{**}$$

is an isomorphism that allows us to identify the spaces V and V^{**}. The transpose of a linear map $A : V \to V^*$ from V to its dual V^* is again a linear map ${}^t A : V^{**} = V \to V^*$ from V to its dual V^*.

A linear map $A : V \to V^*$ defines a bilinear map $B_A : V \times V \to \mathbb{R}$ by putting $B_A(u, v) := (A(u))(v)$. We do not always have the following symmetry property: $B_A(u, v) = B_A(v, u)$, $u, v \in V$. In case it does hold, we say that B_A (and also A itself) is *symmetric*.

A linear map $A : V \to V^*$ can be decomposed in a unique way as a sum of two terms $A = A^+ + A^-$ where the first term

$$A^+ := \frac{1}{2}(A + {}^t A)$$

is symmetric and the second term

$$A^- := \frac{1}{2}(A - {}^t A)$$

is anti-symmetric.

For U, V, f, as above, the Hessian[7] $(H_f)_p$ at a point $p \in U$ is the symmetric bilinear form $(H_f)_p : V \times V \to \mathbb{R}$ defined by

---

[5] Pál Erdős (1913–1996), Hungarian mathematician.

[6] Irving Kaplansky (1917–2006), Canadian mathematician.

[7] Named after German mathematician Ludwig Otto Hesse (1811–1874).

$$(H_f)_p(u, v) := ((D(df))_p(u))(v).$$

For a twice differentiable function we have a four-term expansion:

$$f(p + h) = f(p) + (Df)_p(h) + \frac{1}{2}(H_f)_p(h, h) + R_p(h),$$

where the remainder term $R_p(h)$ is arbitrarily small relative to $\|h\|_V^2$.

Proof. Using the parallelograms $P_p(u, v, s)$ and the three-term expansion $(df)_{p+h} = (df)_p + (D(df))_p(h) + R_p(h)$ with terms in V^*, we compute the anti-symmetric part of $(D(df))_p$ as

$$(((D(df))_p)^-(u))(v) = (d(df))_p(u, v).$$

From $d(df) = 0$ it follows that the anti-symmetric part of $(D(df))_p$ vanishes, hence $(D(df))_p$ is symmetric.   □

We present some computations that explicitly use the linear structure on U. The above definition of $d\omega$ refers parallelograms, so it has used the linear structure of V. Let V be an n-dimensional vector space equipped with the basis $e_1, e_2, \ldots, e_n$. The corresponding linear coordinate functions $x_1, x_2, \ldots, x_n \colon V \to \mathbb{R}$ are implicitly defined by the unique linear combination

$$u = x_1(u)e_1 + x_2(u)e_2 + \cdots + x_n(u)e_n, \ u \in V$$

of u with respect to the basis $e_1, e_2, \ldots, e_n$.

The differentials $dx_i \colon V \to V^*$, $p \in V \mapsto (dx_i)_p \in V^*$ of the coordinate functions are constant maps. The field dx_i takes on V only one value in V^*. It follows that $D(dx_i) = 0$ in $\mathrm{Hom}(V, V^*)$. Moreover, the linear forms $(dx_i)_p \in V^*$ constitute at every point $p \in U$ a basis of V^*. A 1-differential form ω on an open subset U of V can hence be written as a linear combination

$$\omega = a_1 dx_1 + a_2 dx_2 + \cdots + a_n dx_n,$$

where the coefficients a_i are functions $a_i \colon U \to \mathbb{R}$. It follows by the Leibniz product rule that

$$(D\omega)_p = (Da_1)_p dx_1 + (Da_2)_p dx_2 + \cdots + (Da_n)_p dx_n \colon V \to V^*, p \in U.$$

By testing on the basis vectors e_i one obtains for a function $f \colon U \to \mathbb{R}$ at $p \in U$:

$$(Df)_p = (Df)_p(e_1)(dx_1)_p + (Df)_p(e_2)(dx_2)_p + \cdots + (Df)_p(e_n)(dx_n)_p,$$

hence

$$df = \frac{\partial f}{\partial x_1} dx_1 + \frac{\partial f}{\partial x_2} dx_2 + \cdots + \frac{\partial f}{\partial x_n} dx_n,$$

where we have used the traditional notation for partial derivatives

$$\frac{\partial f}{\partial x_i}(p) := (Df)_p(e_i) = i_{e_i} df.$$

We have defined for a 1-differential form ω its exterior differential $d\omega$ by

$$(d\omega)_p(u, v) := \lim_{s \to 0, s > 0} \frac{1}{s^2} \int_{\partial P_p(u,v,s)} \omega.$$

In order to evaluate this limit as $s \to 0$ of an integral depending on s, we use the three-term expansion of the integrand $\omega : U \to V^*$ at the point p,

$$\omega_{p+h} = \omega_p + (D\omega)_p(h) + R_p(h).$$

The integral expands accordingly. The first term contributes 0 since contributions from opposite sides of the parallelogram $P_s(p, u, v)$ cancel. Note that the second term can be used to define the 1-differential form α on U by putting

$$q \in U \mapsto \alpha_q := (D\omega)_p(q - p) \in V^*.$$

Observe for $q = p + h$ the identity $\alpha_{p+h} = (D\omega)_p(h)$.

It follows that the integral for $s > 0$ of the second term evaluates to

$$\frac{1}{s^2} \int_{\partial P_p(u,v,s)} \alpha$$

and

$$\lim_{s \to 0, s > 0} \frac{1}{s^2} \int_{\partial P_p(u,v,s)} \alpha = (da_1 \wedge dx_1 + da_2 \wedge dx_2 + \cdots + da_n \wedge dx_n)_p(u, v),$$

where the $\wedge$-product of 1-forms is used. Define for 1-forms α, β their wedge product $\alpha \wedge \beta$ as the 2-form with

$$(\alpha \wedge \beta)_p(u, v) = \det\begin{pmatrix} \alpha_p(u) & \alpha_p(v) \\ \beta_p(u) & \beta_p(v) \end{pmatrix}, \quad p \in U, u, v \in V.$$

This evaluation is anti-symmetric in u, v since the orientations of the boundaries of the parallelograms $P_p(u, v, s)$ and $P_p(v, u, s)$ are opposite and moreover this evaluation yields the anti-symmetric part of the bilinear form $(u, v) \in V \times V \mapsto ((D\omega)_p(u))(v) \in \mathbb{R}$.

The third term also defines a 1-form ρ on U by

$$q \in U \mapsto \rho_q := R_p(q - p) \in V^*.$$

The integral

$$\int_{\partial P_s(p,u,v)} \rho$$

is arbitrarily small relative to s^2, so its contribution vanishes after taking the limit $s \to 0$.

We conclude:

$$d\omega = da_1 \wedge dx_1 + da_2 \wedge dx_2 + \cdots + da_n \wedge dx_n.$$

In particular, for a 1-form $\omega = df$ that is the differential of a twice differentiable function f, we obtain from $d\omega = d(df) = 0$ that

$$0 = \sum_{1 \le i < j \le n} \left(\frac{\partial^2 f}{\partial x_i \partial x_j} - \frac{\partial^2 f}{\partial x_j \partial x_i} \right) dx_i \wedge dx_j$$

and

$$\frac{\partial^2 f}{\partial x_i \partial x_j} - \frac{\partial^2 f}{\partial x_j \partial x_i} = 0$$

hold. Indeed, $0 = i_{e_j}(i_{e_i} d(df)) = \dfrac{\partial^2 f}{\partial x_i \partial x_j} - \dfrac{\partial^2 f}{\partial x_j \partial x_i}$. Recall the notation $\frac{\partial}{\partial x_i}\left(\frac{\partial}{\partial x_j} f\right) = $ $\dfrac{\partial^2 f}{\partial x_i \partial x_j}$. The above proves the important result due to A.C. Clairaut[8] and K.H.A. Schwarz:[9]

Theorem 1.21. *For a twice differentiable function f that is defined on an open subset in $\mathbb{R}^n$ the operations $f \mapsto \frac{\partial f}{\partial x_i}$ of partial differentiation commute, i.e. the identities*

$$\frac{\partial^2 f}{\partial x_i \partial x_j} = \frac{\partial^2 f}{\partial x_j \partial x_i}$$

hold.   □

Example 1.22. We turn back to Example 1.17. A computation shows for the 1-form $\omega = x dy$ on $\mathbb{R}^2$

$$(d\omega)_p(u, v) = x(u)y(v) - x(v)y(u)$$
$$= \det\begin{pmatrix} x(u) & x(v) \\ y(u) & y(v) \end{pmatrix} = \pm \text{Area}(P_p(u, v, 1)).$$

As an example, taking $u = e_1, v = e_2$ yields $(d\omega)_p(e_1, e_2) = 1 \ne 0$. This example shows that no restriction to a non-empty open set of ω is exact.

Before going on, please consider the following exercises.

Exercise 1.23. For $p \in \mathbb{R}^2$ put

$$\alpha(p) = \frac{x - x(p)}{((x - x(p))^2 + (y - y(p))^2)^{1/2}} dy - \frac{y - y(p)}{((x - x(p))^2 + (y - y(p))^2)^{1/2}} dx.$$

Let $\gamma \colon [0, 1] \to \mathbb{R}^2$ be a smooth closed curve. Define on $C = \mathbb{R}^2 \setminus \gamma([0, 1])$ the function

---

[8] Alexis Claude Clairaut, French mathematician, astronomer and geophysicist (1713–1765).
[9] Karl Hermann Amandus Schwarz, German mathematician (1843–1921).

$$\text{Ind}_\gamma \colon p \in C \mapsto \int_\gamma \alpha(p) \in \mathbb{R}.$$

The value $\text{Ind}_\gamma(p)$ is called the *index* of the loop γ with respect to the point p. Assume moreover that the curve γ is simply closed. Show that the function Ind_γ takes only two values 0 and 1 or 0 and -1. Show that both $C_0 = \{p \in C \mid \text{Ind}_\gamma(p) = 0\}$ and $C_1 = C \setminus C_0$ are connected. Show that C_1 is homeomorphic to the open disk in the plane.

The above exercise is a proof of the smooth Jordan Curve Theorem. For a proof of the continuous Jordan Curve Theorem see [70]. See [47] for its history and an interesting lemma of Max Dehn[10] that proves the Jordan Curve Theorem for simple polygons by induction.

Theorem 1.24 (Smooth Jordan Curve Theorem). *Let $\gamma \colon [0, 1] \to \mathbb{R}^2$ be a smooth simply closed curve. The complement of the image of γ has two connected components.* $\square$

For a stronger extension of the continuous Jordan Curve Theorem see also Dehn's early proof in [47].

Theorem 1.25 (Jordan–Schoenflies Curve Theorem). *Let $\gamma \colon S^1 \to \mathbb{R}^2$ be a continuous simply closed curve. Then there exists a homeomorphism $\phi \colon \mathbb{R}^2 \to \mathbb{R}^2$ such that the composition $\phi \circ \gamma \colon S^1 \to \mathbb{R}^2$ is a continuous parametrization of the unit circle.* $\square$

Exercise 1.26. The choice of two points A, B in the Euclidean plane E defines on $E \setminus \{A, B\}$ the vector fields X, Y given by $X_p := A - p$, $Y_p := B - p$. Let the curve $t \in [0, +\infty) \mapsto \phi_X(t, p) \in E$ solve with initial condition $\phi_X(0, p) = p$ the differential equation of the vector field X. Analogously, let the curve $t \in [0, +\infty) \mapsto \phi_Y(t, p) \in E$ solve the differential equation of Y. For $t, s > 0$, and p not on the line through A, B, does the following hold:

$$\phi_Y(s, \phi_X(t, p)) = \phi_X(t, \phi_Y(s, p))?$$

This question models politics in the capital, or at home, where one is aiming at two different goals but still fighting over priority. Compute the bracket $[X, Y]_{\text{Lie}}$. For which points $p \in E \setminus \{A, B\}$ does $([X, Y]_{\text{Lie}})_p = 0$ hold?

Exercise 1.27. Let $\gamma \colon \mathbb{R} \to \mathbb{R}^2$ be a periodic path with $\|\dot{\gamma}(t)\|_{\text{Euclidean}} = 1$ and $T \mapsto \int_0^T x(\gamma(t)) y(\dot{\gamma}(t)) dt$ bounded. Show that the curve γ has at least two flexes per period. A flex is a point where the speed $\dot{\gamma}(t)$ and acceleration $\ddot{\gamma}(t)$ vectors become linearly dependent.

Exercise 1.28. Let $Q = (A, B, C, D)$ be an oriented quadrilateral in the Euclidean plane E with side lengths a, b, c, d. Study the positions of Q in E for which the path integral $\int_Q x dy$ takes its maximal or minimal value. Is there always a position with $\int_Q x dy = 0$? Is the space of positions connected?

---

[10] Max Dehn (1878–1952), German mathematician.

Exercise 1.29. Let $U \subset V$ be an open subset of a vector space V. Let ω be a 1-differential form on U, $p \in U$, $u, v \in V$. We wish to compute $(d\omega)_p(u, v)$. Let X, Y be vector fields on U with $X_p = u$, $Y_p = v$. First we consider the functions $f := i_X\omega$, $g := i_Y\omega$, $h := i_Z\omega$, where $Z := [X, Y]_{\text{Lie}}$. The value at p of the function $Xg - Yf - h$ is the number $(d\omega)_p(u, v)$. One can eliminate the point p and write

$$d\omega(X, Y) + \omega([X, Y]_{\text{Lie}}) = X\omega(Y) - Y\omega(X).$$

Show for a function f and vector fields X, Y:

$$d\omega(fX, Y) = f d\omega(X, Y) = d\omega(X, fY).$$

1.2 Closed Forms are Locally Exact

The infinitesimal property $d\omega = 0$, i.e. that of being closed, implies local exactness for a 1-differential form ω.

Theorem 1.30. *Let $U \subset V$ be an open subset of a finite-dimensional real vector space and $\omega : U \to V^*$ a closed 1-differential form . Let $B_p(r) \subset U$ be a ball (for a norm on V). Then there exists a function $f : B_p(r) \to \mathbb{R}$ with $df = \omega$.*

Proof. Let $f : B_p(r) \to \mathbb{R}$ be the function defined by $f(q) := \int_{[p,q]} \omega$ for $q \in B_p(r)$, where $[p, q] : [0, 1] \to B_p(r)$ is the path $[p, q](t) = p + t(q - p)$. The main difficulty is to prove $(Df)_q = \omega_q$, $q \in B_p(r)$.

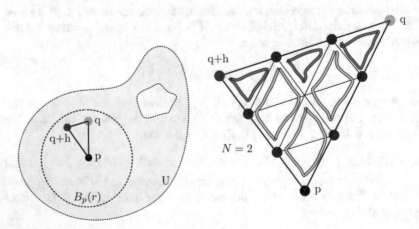

Fig. 1.3 Construction of f on $B_p(r)$. Proof of $df = \omega$.

First we prove

$$\int_{\partial\Delta(p,q,q+h)} \omega = \int_{[p,q]} \omega + \int_{[q,q+h]} \omega + \int_{[q+h,p]} \omega = 0,$$

which would follow from Stokes' Theorem, if applied to the triangle $\Delta(p, q, q + h)$ and its oriented boundary. Here we prefer to proceed differently.

We subdivide the triangle $\Delta(p, q, q + h)$ by drawing N equidistant lines parallel to each of the sides. These lines divide the triangle into $(N+1)^2$ congruent triangles, all being a scaling of $\Delta(p, q, q + h)$ by the factor $\dfrac{1}{N+1}$. We have $(N + 1)$ sub-triangles with an edge on $[q, q + h]$. We put the remaining $(N + 1)^2 - N - 1 = N^2 + N$ sub-triangles pairwise together and obtain $(N^2 + N)/2$ parallelograms, each of them a translate of the parallelogram $P_p(q - p, q + h - p, 1/(N + 1))$.

The path integral of ω over the boundary of $\Delta(p, q, q + h)$ is the sum of the path integrals over the boundaries of the $N + 1$ sub-triangles and the $N^2 + N$ sub-parallelograms. The path integral over the boundary of a sub-parallelogram is bounded in absolute value by $C_1/(N + 1)^3$ since we have $d\omega = 0$.

The path integral over the boundary of a sub-triangle

$$\Delta' = \Delta\left(p', p' + \frac{q - p}{N + 1}, p' + \frac{q + h}{N + 1}\right)$$

is in absolute value bounded by $C_2/(N + 1)^2$. Indeed,

$$\int_{\partial\Delta'} \omega = \int_{\partial\Delta'} \omega_{p'} + \int_{\partial\Delta'} \omega - \omega_{p'} = 0 + \int_{\partial\Delta'} (\omega - \omega_{p'}).$$

Here $\omega_{p'}$ is considered as the constant 1-form $q \in U \mapsto \omega_{p'} \in V^*$. So $(\omega - \omega_{p'})_q = \omega_q - \omega_{p'}$ and an operator norm estimate $\|(\omega - \omega_{p'})_q\| \leq C_3/(N + 1)$ holds for $q \in \Delta_{p'}$.

Hence the path integral of ω over the boundary of $\Delta(p, q, q + h)$ is in absolute value bounded by $\text{Maximum}\{C_1, C_2\}/(N + 1)$. This bound holds for all N, so the path integral vanishes.

We can make a three-term expansion for f at q:

$$f(q + h) = f(q) + \int_{[q, q+h]} \omega = f(q) + \omega_q(h) + R(h).$$

Now we see that $(Df)_q = \omega_q$, after having checked that the term

$$R(h) = -\omega_q(h) + \int_{[q, q+h]} \omega = \int_0^1 (\omega_{q+th} - \omega_q)(h)dt$$

is arbitrarily small relative to h. $\square$

Exercise 1.31. Let ω be a closed 1-differential form on the punctured space $\mathbb{R}^3 \setminus \{0\}$. Show that ω is exact.

Exercise 1.32. Let α be a 1-differential form on $\mathbb{R}^2$. Assume $\alpha_p \neq 0$ at a point $p \in \mathbb{R}^2$. Show that a triple (U, f, g) consisting of an open neighborhood U of p and two functions $f, g: U \to \mathbb{R}$ with $\alpha_{|U} = g df$ exists.

Exercise 1.33. Consider $\alpha = \mathrm{d}x + y\mathrm{d}z$ on $\mathbb{R}^3$. Observe that the form $\alpha \wedge \mathrm{d}\alpha$ is a volume form on $\mathbb{R}^3$. Conclude that even locally no functions f and g with $\alpha = g\mathrm{d}f$ exist.

We do not study the higher differential forms as introduced by Élie Cartan,[11] [19]. See the books of Henri Cartan[12] [23] and Paul Malliavin[13] [73]. Differential calculus provides a very powerful toolbox. As illustrations we present in the next two sections the one (punch-)line proof of the Brouwer Fixed Point Theorem, due to Nelson James Dunford and Jacob Theodore Schwartz, and a study of real commutative division algebras.

1.3 Fixed Point Theorems

The Brouwer Fixed Point Theorem[14] is a fundamental result in Topology. It has many consequences and generalizations.

Theorem 1.34 (L.E.J. Brouwer). *Every continuous mapping* $f: B^d \to B^d$ *with domain and target the closed Euclidean unit ball* $B^d \subset E^d$ *has a fixed point, i.e., there exists a* $p \in B^d$ *with* $f(p) = p$.

Proof. The following proof is by contradiction. Assume f to be a continuous self map of B^d without fixed points. The closed ball B^d being compact, the continuous function $p \in B^d \mapsto \|f(p) - p\|_{\text{Euclid}} \in \mathbb{R}$ achieves its minimal value $\epsilon > 0$. Approximate the map f up to $\epsilon/3$ by the restriction of a continuously differentiable map $g: E^d \to E^d$ with $g(B^d) \subset B^d$. Now, this restriction has no fixed point either. So, without loss of generality we may moreover assume that the map f is continuously differentiable.

Let $R: B^d \to B^d$ be the map where $R(p) \in B^d$, $p \in B^d$, is the intersection point of the half-ray from $f(p)$ that passes through p with the boundary sphere ∂B^d of B^d (Fig. 1.4). The map R is continuously differentiable, maps B^d to its boundary and satisfies $R(p) = p$ if p belongs to the boundary. The map R is a differentiable retraction of B^d to its boundary ∂B^d. Let $x_1, x_2, \ldots, x_d$ be orthonormal coordinate functions on E^d. Put $R_i(p) = x_i(R(p))$. The functions x_i and R_i coincide on the boundary of B^d.

From the identity $\|R(p)\|_{\text{Euclid}}^2 \equiv 1$ on B^d we get

$$\frac{1}{2}\mathrm{d}\|R(p)\|_{\text{Euclid}}^2 = R_1\mathrm{d}R_1 + \cdots + R_d\mathrm{d}R_d \equiv 0$$

on B^d. Now using Stokes' Theorem twice and the above linear dependency of the differentials $\mathrm{d}R_i$, we obtain the contradiction:

---

[11] Élie Cartan (1869–1951), French mathematician.

[12] Henri Cartan (1904–2008), French mathematician.

[13] Paul Malliavin (1925–2010), French mathematician.

[14] Named after Dutch mathematician Luitzen Egbertus Jan Brouwer (1881–1966).

$$0 \neq \int_{B^d} \mathrm{d}x_1 \wedge \mathrm{d}x_2 \wedge \cdots \wedge \mathrm{d}x_d = \int_{\partial B^d} x_1 \wedge \mathrm{d}x_2 \wedge \cdots \wedge \mathrm{d}x_d$$

$$= \int_{\partial B^d} R_1 \wedge \mathrm{d}R_2 \wedge \cdots \wedge \mathrm{d}R_d = \int_{B^d} \mathrm{d}R_1 \wedge \mathrm{d}R_2 \wedge \cdots \wedge \mathrm{d}R_d = 0. \qquad \square$$

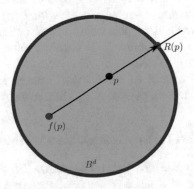

Fig. 1.4 The retraction R in the proof of Brouwer's Theorem.

See Nikolai V. Ivanov's comments about the above proof of Brouwer's Theorem in [56].

Exercise 1.35. Let $X \colon \mathbb{R}^2 \to \mathbb{R}^2$ be a smooth vector field. Assume that the differential equation $\dot{\gamma}(t) = X_{\gamma(t)}$ admits a periodic solution $\gamma \colon \mathbb{R} \to \mathbb{R}^2$. Show that the vector field vanishes at some point $p \in \mathbb{R}^2$.

Exercise 1.36. Let $K \subset \mathbb{R}^n$ be compact non-empty and convex. Then a continuous map $f \colon K \to K$ has a fixed point.
  Hint: The space K is homeomorphic to a ball in Euclidean space $\mathbb{R}^m$.

Exercise 1.37. Let B be the closed unit ball in the Banach space $C^0([0,1], \mathbb{R})$ of all continuous real functions on $[0,1]$ with the supremum norm. Construct a fixed-point-free continuous map $f \colon B \to B$.

Exercise 1.38. Let $K \subset C^0([0,1], \mathbb{R})$ be a convex compact non-empty subset and let $f \colon K \to K$ be a continuous map. Prove that the map f has a fixed point.

Exercise 1.39 (Schauder Fixed Point Theorem). [15] Let $(E, \| \cdot \|_E)$ be a normed real vector space of any finite or infinite dimension and let $K \subset E$ be compact, non-empty and convex. Then a continuous map $f \colon K \to K$ has a fixed point.
  Hint: Construct for $\epsilon = \frac{1}{n}$, $n \geq 1$, a finite subset $A_n \subset K$ such that each point in K is at distance ϵ from A_n. Let $B_n \subset K$ be the convex hull of A_n. Construct a continuous map $f_n \colon B_n \to B_n$ with $\| f_n(b) - f(b) \| \leq \frac{3}{n}$. The maps $f_n \colon B_n \to B_n$ have fixed points, hence by a subsequence argument, it will follow that the map $f \colon K \to K$ too

---

[15] Named after Polish mathematician Juliusz Schauder (1901–1943).

has a fixed point. One possible way of constructing the map $f_n: B_n \to B_n$ would be to first map $p \in B_n$ to the compact convex subset in B_n consisting of those points $q \in B_n$ that realize the minimal distance $\mathrm{Inf}_{b \in B_n} \|b - f(p)\|_E$. Then proceed as in the proof of the Kakutani Fixed Point Theorem, see Theorem 1.41 below.

Exercise 1.40 (Perron–Frobenius Theorem). [16] Let $A: \mathbb{R}^n \to \mathbb{R}^n$ be a linear map that maps the set $\{u \in \mathbb{R}^n \mid u \neq 0, x_i(u) \geq 0\}$ into the set $\{u \in \mathbb{R}^n \mid x_i(u) > 0\}$. Show that there exist a unique vector $u \in \mathbb{R}^n$ with $x_i(u) \geq 0$, $\sum_i x_i(u) = 1$, and $A(u) = \lambda u$, $\lambda \in \mathbb{R}$. Show that the eigenvalue λ is of multiplicity 1. Moreover, show that the eigenvalue λ strictly dominates the absolute values of the other eigenvalues of A.

Hint: define a self map $\bar{A}$ of the compact convex set $\{u \in \mathbb{R}^n \mid x_i(u) \geq 0, \sum_i x_i(u) = 1\}$; the use of the so-called Hilbert metric, see Section 3.1 later, is of great help.

The Perron–Frobenius Theorem is fundamental for the Google Search Engine.

Theorem 1.41 (Kakutani Fixed Point Theorem). [17] *Let $K \subset V$ be a non-empty convex compact subset in a finite-dimensional real vector space. Let $CC(K)$ be the space of non-empty, convex and compact subsets in K, the topology on $CC(K)$ being the Hausdorff topology. Let $f: K \to CC(K)$ be continuous. Then there exists a $p \in K$ satisfying $p \in f(p)$.*

The Hausdorff topology on $CC(K)$ can be defined in terms of a metric. Let D_V be a Euclidean metric on V. For $A \in CC(K)$ and $\epsilon \geq 0$ let A_ϵ be the intersection of K with the union of the closed D_V-balls $B(a, \epsilon)$, $a \in A$. For $A, B \in CC(K)$ define the Hausdorff distance $D_{\mathrm{Hausdorff}}(A, B)$ [18] as the smallest ϵ for which the inclusions $A \subset B_\epsilon$ and $B \subset A_\epsilon$ hold.

The proof will use the following geometric statement, which is in particular true for the Euclidean distance D_V on V:

Theorem 1.42 (Minimal Ball Lemma). *Let $A \subset V$ be a non-empty, convex, closed and bounded subset in a Euclidean Vector space V with Euclidean norm D_V. There exists a unique closed Euclidean ball $B(c(A), r)$ of minimal radius r that contains A. Moreover $c(A) \in A$. The map $c: CC(K) \to K$ is continuous.*

Proof. Such balls of minimal radius r exist since A is assumed to be bounded. If A is a subset of two such balls with centers c_1, c_2 one has $A \subset B(c_1, r) \cap B(c_2, r)$. But, unless $c_1 = c_2$, the inclusions

$$A \subset B(c_1, r) \cap B(c_2, r) \subset B\left(\frac{1}{2}(c_1 + c_2), r'\right), \quad r' = \sqrt{r^2 - \delta^2} < r,$$

for $\delta = \mathrm{dist}(c_1, c_2)/2$, contradict the minimality of r. The last inclusion reflects a

---

[16] Named after German mathematicians Oskar Perron (1880–1975) and Ferdinand Georg Frobenius (1849–1917).

[17] Named after Japanese-American mathematician 角谷静夫, Kakutani Shizuo (1911–2004).

[18] Named after Felix Hausdorff (1868–1942), German mathematician.

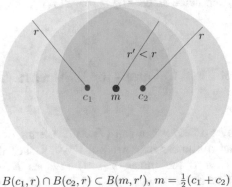

$$B(c_1,r) \cap B(c_2,r) \subset B(m,r'), \; m = \tfrac{1}{2}(c_1+c_2)$$

Fig. 1.5 See proof: Uniqueness of Euclidean balls of minimal radius that contain a compact set.

special property of the Euclidean distance due to Pythagoras' Theorem. This proves uniqueness of the center $c(A)$. Again from Pythagoras' Theorem it follows that the center $c(A)$ belongs to A. Indeed, assume $c(A) \notin A$ but $A \subset B(c(A),r)$ with minimal r. Let $c' \in A$ realize the minimal distance $t > 0$ from $c(A)$ to A. Then $A \subset B(c', \sqrt{r^2 - t^2})$, contradicting the minimality of r. Continuity of the map c is clear. $\qquad \square$

Proof (Kakutani's Theorem). The composition of $c \colon CC(K) \to K$ with the map f is a continuous map $c \circ f \colon K \to K$ that has a fixed point $p \in K$ by Brouwer's Theorem 1.34. So, $p = c(f(p)) \in f(p)$ by the previous Theorem 1.42. $\qquad \square$

The above proof of the Minimal Ball Lemma works also for compact, convex, non-empty subsets in a Hilbert space. It follows that Kakutani's Theorem holds for convex compact subsets in a Hilbert space. Kakutani's Theorem is used in Game Theory and Econometrics, showing the existence of the Nash Equilibrium.[19]

The following theorem holds for complete metric spaces and contracting maps. No compactness is assumed.

Theorem 1.43 (Banach Fixed Point Theorem). [20] *Let (X,d) be a complete non-empty metric space. Let $f \colon X \to X$ be a contracting map, i.e. there exists a $\lambda \in [0,1[$ such that for $p,q \in X$ the inequality $d(f(p),f(q)) \le \lambda d(p,q)$ holds. Then the map f has a unique fixed point.*

Proof. Let $p \in X$. The sequence $(p_k)_{k \in \mathbb{N}}$ of points in X inductively defined by putting $p_0 = p$ and $p_{k+1} = f(p_k)$ is a Cauchy sequence. The map f being Lipschitz continuous, it follows that the limit point $q = \lim_{k \to \infty} p_k$ is a fixed point since

$$f(q) = f\left(\lim_{k \to \infty} p_k\right) = \lim_{k \to \infty} f(p_k) = \lim_{k \to \infty} p_{k+1} = q.$$

---

[19] Named after American mathematician John Forbes Nash Jr. (1928–2015).
[20] Named after Polish mathematician Stefan Banach (1892–1945).

Fixed points q, q' coincide since $d(q, q') = d(f(q), f(q')) \leq \lambda d(q, q')$ only holds if $q = q'$. $\qquad\qquad\qquad\qquad\qquad\qquad\qquad\qquad\qquad\qquad\qquad\qquad\qquad\qquad$ □

1.4 The Abstract Field $\mathbb{C}$ Versus the $\mathbb{R}$-Algebra $\mathbb{C}$ of Complex Numbers

A *division algebra* over a field k is a pair (V, m) consisting of a finite-dimensional k-vector space V and a bilinear map $m \colon V \times V \to V$ such that for all $u, v \in V$ with $u \neq 0$ and $v \neq 0$ the image $m(u, v)$ satisfies $m(u, v) \neq 0$. A division algebra (V, m) is called *commutative* if the map m is symmetric, i.e. if $m(u, v) = m(v, u)$, $u, v \in V$, holds. A division algebra (V, m) is called *associative* if $m(m(u, v), w) = m(u, m(v, w))$, $u, v, w \in V$, holds.

Let k be a field. The pair (V, m) with $V = k$ and m being the multiplication of k is a k-division algebra, which moreover is associative and commutative.

The field of complex numbers $\mathbb{C}$ is commonly constructed as the real associative commutative division algebra (V, m), where V is the $\mathbb{R}$-vector space of expressions $a + bi$, $a, b \in \mathbb{R}$, with multiplication

$$m(a + bi, a' + b'i) = (aa' - bb') + (ab' + a'b)i.$$

It is convenient to identify a real number a with the complex number $a + 0i$. Also it is convenient to think of the symbol i as the complex number $0 + 1i$. Observe that $0 + bi = m(b + 0i, 0 + 1i)$. The product $m(u, v)$ of two complex numbers u, v will be written as uv.

This construction induces on $\mathbb{C}$ even more structure than just the structure of a field $(\mathbb{C}, +, \cdot)$, namely the topology $T_{\mathbb{C}}$ that is given by the absolute value $|a + bi| = \sqrt{a^2 + b^2}$.

Remember that the order relation $\leq$ on $\mathbb{R}$ is encoded in the multiplication: one has $a \leq b$ if and only if the difference $b - a$ is a square number. Hence, also the topology of $\mathbb{R}$ is encoded in its field structure. The topology $T_{\mathbb{R}}$ is given by the absolute value

$$t \in \mathbb{R} \mapsto |t| := \text{Maximum}\{\pm t\} \in \mathbb{R}.$$

This allows us to think of $\mathbb{R}$ as a field with the topology $T_{\mathbb{R}}$ that is given by the absolute value $|\cdot|$. The absolute value satisfies the multiplicative property $|ts| = |t||s|$, $t, s \in \mathbb{R}$, and the triangle inequality $|t + s| \leq |t| + |s|$, $t, s \in \mathbb{R}$.

Since the order relation on $\mathbb{R}$ is encoded in its multiplication, every field automorphism ϕ of $\mathbb{R}$ will be strictly monotonously increasing and continuous. Moreover, $\phi(t) = t$ holds for every rational number t. Hence, ϕ is the identity since the rationals are dense in $\mathbb{R}$.

Every real vector space V has a canonical topology, namely the coarsest topology T_V on V such that all linear maps $L \colon (V, T_V) \to (\mathbb{R}, T_{\mathbb{R}})$ are continuous. The canonical topologies on real vector spaces have the property that every linear map $A \colon V \to W$ between real vector spaces V and W is continuous for the canonical topologies T_V and T_W on V and W.

In the case of finite-dimensional vector spaces V, the topology T_V is given by a norm. Let $L_1, \ldots, L_n$ be a basis L of the dual space V^*. The topology T_V is given by the norm

$$u \in V \mapsto \|u\|_{V,L} = \text{Max}_{1 \le i \le n} |L_i(u)| \in \mathbb{R}.$$

The norm $\| \cdot \|_{V,L}$ depends upon the chosen basis, but not its topology T_V.

The above construction of the field of complex numbers $\mathbb{C}$ as the real division algebra $\mathbb{C} = (V, m)$ provides a topology $T_{\mathbb{C}} = T_V$. Also an involution appears: the map $a + bi \in \mathbb{C} \mapsto a - bi \in \mathbb{C}$ is an automorphism of fields and is commonly called complex conjugation. Moreover the map

$$a + bi \in \mathbb{C} \mapsto |a + bi| := \sqrt{(a+bi)(a-bi)} = \sqrt{a^2 + b^2} \in \mathbb{R}_{\ge 0}$$

is an absolute value $|\cdot|$ on $\mathbb{C}$ that is multiplicative and satisfies the triangle inequality. The topology $T_{\mathbb{C}}$ is given by this absolute value. Moreover, multiplication together with complex conjugation define a Hermitian scalar product $(u, v)_{\mathbb{C}} = u\bar{v}$.

From now on we adopt the convention that the *field of complex numbers*, denoted by $\mathbb{C}$, refers to the construction as an $\mathbb{R}$ division algebra. So the field $\mathbb{C}$ is endowed with the canonical topology of its underlying real vector space. Also the field $\mathbb{C}$ comes with the involution given by complex conjugation. On the (very rare) occasions when we want to consider the field $\mathbb{C}$ without these extra structures, we refer to the *abstract field of complex numbers*.

At this point one may ask whether the topology $T_{\mathbb{C}}$ is encoded in the structure of the abstract field $(\mathbb{C}, +, \cdot)$ of complex numbers. If this were the case, then any automorphism of the field $\mathbb{C}$ would preserve the topology $T_{\mathbb{C}}$.

In fact, this is not true. Indeed, it is far from true. More precisely:

Theorem 1.44. *The group* $\text{Aut}(\mathbb{C}, T_{\mathbb{C}})$ *of* $T_{\mathbb{C}}$-*continuous automorphisms of the field* $\mathbb{C}$ *has two elements, namely the identity and complex conjugation.*

Proof. Let $\phi: \mathbb{C} \to \mathbb{C}$ be a $T_{\mathbb{C}}$-continuous automorphism of the field $\mathbb{C}$. Observe that $\phi(1 + 0i) = 1 + 0i$ and that $\phi(r + 0i) = r + 0i$ for rational numbers r, hence also for real numbers r by continuity. The polynomial $t^2 + 1 = 0$ of degree 2 has at most two roots in any commutative field and $0 \pm 1i$ are two roots in $\mathbb{C}$. The polynomial has real coefficients fixed by ϕ, so ϕ permutes the two roots $0 \pm 1i$. Hence $\phi(0 + 1i) = 0 \pm 1i$. Clearly $\phi(a + bi) = \phi(a) + \phi(b)\phi(i) = a \pm bi$. $\square$

Theorem 1.45. *The group* $\text{Aut}(\mathbb{C})$ *of automorphisms of the abstract field* $\mathbb{C}$ *of complex numbers has the cardinality of the power set* $P(\mathbb{R})$, *i.e. of the set of all subsets of the set of real numbers.*

Proof. A transcendence basis of $\mathbb{C}$ is a subset $\beta \subset \mathbb{C}$ minimal with respect to inclusion and such that the only algebraically closed subfield in $\mathbb{C}$ that contains the set β is the field $\mathbb{C}$.

Using the axiom of choice one can prove that a transcendence basis β exists. The cardinalities of $\mathbb{R}$, $\mathbb{C}$ and β are equal.

Every permutation of the set β induces a field automorphism of the abstract field $\mathbb{C}$. An automorphism of the field $\mathbb{C}$ is in particular a permutation of the set $\mathbb{C}$. The claim follows, since the set of permutations of the sets $\mathbb{R}$, or of $\mathbb{C}$, or of β and the power set $P(\mathbb{R})$ of subsets of $\mathbb{R}$ have equal cardinalities.   □

It is rare in mathematics to encounter a cardinality as big as the power set of the real numbers. In contrast, up to isomorphism of fields, the abstract field of complex numbers has a very short characterization by rather elementary properties. This remarkable characterization [114, 115] is due to Ernst Steinitz.[21]

Theorem 1.46 ($\mathbb{C}$-Characterization Theorem). *Every algebraically closed (commutative) field F of characteristic 0 and of cardinality equal to the cardinality of the power set of the natural numbers is isomorphic to the abstract field of complex numbers $\mathbb{C}$.*

Proof. Choose a transcendence basis ϕ for F and β for $\mathbb{C}$. The algebraic closure $\bar{P}$ of the prime field P in F is isomorphic to the algebraic closure $\bar{Q}$ in $\mathbb{C}$ of the prime field Q. The choice of an isomorphism $\bar{P} \to \bar{Q}$ and of a bijection $\phi \to \beta$ determines an isomorphism of fields $F \to \mathbb{C}$.   □

Imagine the frustration of a person who wants to buy a bike, only to be told by the shopkeeper that he only sells e-bikes. The same frustration appears when asking for a construction of the abstract field $\mathbb{C}$, only to be given a construction of the $\mathbb{R}$-algebra $\mathbb{C}$.

Of course a construction of the abstract field $\mathbb{C}$ needs material to start with. What is amazing is that such a construction is possible starting from all finite fields. The key tools for this construction are the set-theoretic devices of filter, ultrafilter and of ultraproducts. Filters and ultrafilters were invented by Henri Cartan and Jean Dieudonné,[22] and used by Nicolas Bourbaki in a spectacular proof of Tychonov's Theorem[23] [15, 33]. Seminal contributions using ultraproducts in logic are made by Łoś[24] in [69]. See the book [110] by Hans Schoutens for many other interesting applications of ultraproducts and ultrapowers.

A *filter* F on a set X is a subset $F \subset P(X)$ satisfying:

(i) $\emptyset \notin F$, $X \in F$,
(ii) if $A \in F$ and $A \subset B \subset X$ then $B \in F$,
(iii) if $A, B \in F$ then $A \cap B \in F$.

The inclusion relation on the power set $\mathcal{P}(X)$ induces an order relation on the set $F(X)$ of filters on X. An *ultrafilter* on X is a maximal element in $F(X)$. Ultrafilters exist by Zorn's lemma and every filter F on X is equal to the intersection of all ultrafilters on X that contain F. Ultrafilters U on X are filters with the property that for every subset A of X one of the possibilities $A \in U$ or $X \setminus A \in U$ holds.

A filter F on X is *elementary* if there exists a finite subset A in X with $A \in F$. It follows that for an elementary ultrafilter U on X a point $p \in X$ with $\{p\} \in U$ exists. On an infinite set X there do exist non-elementary ultrafilters. Indeed, let F

---

[21] Ernst Steinitz (1871–1928), German mathematician.
[22] Jean Dieudonné (1906–1992), French mathematician.
[23] Named after Russian mathematician Andrej Nikolaevič Tychonov (1903–1993).
[24] Jerzy Łoś (1920–1998), Polish mathematician.

be the Fréchet filter of all subsets $A \subset X$ with finite complement $X \setminus A$. By Zorn's lemma the Fréchet filter F is the intersection of ultrafilters, none of which can be elementary.

Let $P = \{2, 3, 5, \ldots\}$ be the set of prime numbers. Let U be a non-elementary ultrafilter on P. For $p \in P$ let $\bar{F}_p$ be the algebraic closure of the finite field F_p with p elements. The field $\bar{F}_p$ can be seen in a non-canonical way as the union of all finite fields F_q of characteristic p. Let E be the disjoint union of all fields $\bar{F}_p$. Let $\mathbb{P}$ be the ring of all maps $s: P \to E$ such that for all $p \in P$ the value $s(p)$ belongs to $\bar{F}_p$. Addition and multiplication in $\mathbb{P}$ is pointwise: $(s + s')(p) = s(p) + s'(p)$, $(ss')(p) = s(p)s'(p)$. Let $m_U \subset \mathbb{P}$ be the ideal of all maps $s \in \mathbb{P}$ such that its zero set $Z_s = \{p \in P \mid s(p) = 0 \in \bar{F}_p\}$ belongs to U. The ideal m_U is maximal, so the quotient ring $\mathbb{P}/m_U$ is a field, which, by the $\mathbb{C}$-Characterization Theorem 1.46, is a model of the abstract field $\mathbb{C}$.

Would the cyclist be really happy with this construction? Probably not, the group of automorphisms of the field $\mathbb{P}/m_U$ remains very well hidden! The "visible" automorphisms of $\mathbb{P}/m_U$ seem to commute since the groups $\mathrm{Aut}(\bar{F}_p)$ are isomorphic to the profinite completion $\hat{\mathbb{Z}}$ of the commutative group $\mathbb{Z}$.

The above characterization of the abstract field $\mathbb{C}$ allows many appearances of fields isomorphic to the field of complex numbers. Let $(X_\alpha)_{\alpha \in A}$ be a family of commuting symbols. Assume that the indexing set A has the cardinality of the real numbers. Let F be the algebraic closure of the fraction field of the ring $\mathbb{Q}[X_\alpha]$ of polynomials with unknowns (X_α) and with rational coefficients. This field F is isomorphic to the field of complex numbers. The same holds for the algebraic closure of the fraction field of the ring $\mathbb{R}[X, Y, Z]$. Also the algebraic closure $\bar{\mathbb{Q}}_p$ of the p-adic completion $\mathbb{Q}_p$ is isomorphic to the abstract field $\mathbb{C}$.

Exercise 1.47. Show the existence of a family $(F_r)_{r \in \mathbb{R}}$ of subfields of $\mathbb{C}$ such that for all $a < b$ the strict inclusion $F_a \subsetneq F_b$ holds and moreover for all a the fields F_a and $\mathbb{C}$ are isomorphic.

Exercise 1.48. Show that no proper subfield F of $\mathbb{R}$ is isomorphic to $\mathbb{R}$.

Remember that by convention the multiplication of a field is commutative. This convention does not hold for *skew fields*.

Theorem 1.49. *No real division algebra* (V, m) *such that the triple* $(V, +, m)$ *is a field and* $\mathrm{Dim}(V) > 2$ *exists.*

Proof. First we remark that no $\mathbb{R}$-division algebra (V, m) of odd dimension > 1 exists. Indeed, let $u, v \in V$ be linearly independent in V with $\mathrm{Dim}(V)$ odd and ≥ 3. Consider the real line $L = \{tu + (1 - t)v \mid t \in \mathbb{R}\}$. Observe $0 \notin L$. The determinant $\Delta(t)$ of the linear map

$$w \in V \mapsto M_{tu+(1-t)v}(w) = m(tu + (1 - t)v, w) \in V$$

is a real polynomial of odd degree, hence has a root $\lambda \in \mathbb{R}$. It follows that $\lambda u + (1 - \lambda)v \neq 0$ has no m-inverse, a contradiction.

Let (V, m) be a real division algebra of even dimension $n \geq 2$, such that $(V, +, m)$ is a field. Let $\Sigma \subset V$ be those elements $u \in V$ with $\mathrm{Det}(M_u) = 1$. The set Σ is closed and bounded in V. Moreover, Σ intersects each half-ray in one point, so Σ is homeomorphic to the sphere of dimension $n - 1$. Also the unity 1_V belongs to Σ. Also $-1_V \in \Sigma$.

Let $q: \Sigma \to \Sigma$ be the map $q(u) = u^2$. Since m is commutative, we compute the differential of q at $u \in \Sigma$ by

$$(Dq)_u(h) = 2uh \text{ for } h \in T_u\Sigma = \{h \in V \mid \text{Trace}M_{u^{-1}h} = 0\}.$$

The differential Dq is of maximal rank, so the mapping q is open by the inverse mapping theorem. Hence $q(\Sigma)$ is open in Σ. The image $q(\Sigma)$ is also closed since Σ is compact. It follows that $q(\Sigma) = \Sigma$ since the sphere Σ of dimension ≥ 1 is connected. Let $i \in \Sigma$ be such that $q(i) = i^2 = -1_V$.

The sub-vector space $W = \{a + bi \mid a, b \in \mathbb{R}\}$ in V spanned by 1_V and i is a copy of the field of complex numbers. Now via W, the space V is also a complex vector space of complex dimension $n/2$. Assuming $n/2 > 1$, let u, v be two complex linearly independent elements in V. For $t \in W = \mathbb{C}$ let $\Delta(t)$ be the determinant of the complex linear mapping $M_{tu+(1-t)v}$. As before, since the polynomial $\Delta(t)$ has a complex root λ, we run into a contradiction if $n/2 > 1$. $\square$

Exercise 1.50. Again assume that (V, m) is a real division algebra such that $(V, +, m)$ is a field. Let Σ be as in the above proof and let $i \in \Sigma$ be such that $\text{Trace}M_i = 0$. Show that $\omega_u(h) = \text{Trace}M_{u^{-1}ih}$ defines a closed 1-form on Σ. The 1-form vanishes nowhere and is exact if the sphere Σ is of dimension > 1. This yields a contradiction since the function f with $df = \omega$ has a maximum on Σ.

Hint: The 1-differential form α on $V \setminus \{0\}$ with values in V and defined by $\alpha_u(h) = m(u^{-1}, h)$ is near the unity element $1 \in V$ the differential of the map $F(w) = \sum_{k=1}^{\infty} (w - 1)^k / k$. Hence, α being exact near 1 is also closed near 1. The same holds for $\omega = \text{Trace}(i\alpha)$, which is a 1-differential form with real values. Deduce that the restriction of ω to Σ is closed. If the sphere is of dimension > 1, show that this restriction is exact.

The smallest subfield in $\mathbb{C}$ is its prime field $\mathbb{Q}$ of rational numbers. A model of the algebraic closure of the field $\mathbb{Q}$ is provided by the smallest algebraic closed subfield $\bar{\mathbb{Q}}$ in $\mathbb{C}$. Number fields are subfields of $\bar{\mathbb{Q}}$ that as $\mathbb{Q}$-vector spaces are of finite dimension.

1.5 Coordinates and Locally Smooth Rigidity Theorems

Smooth local Rigidity Theorems provide local normal forms for smooth fields that satisfy some conditions and are defined on an open subset $U \subset V$ of a vector space V. In order to express these local normal forms, well-adapted differentiable coordinate systems need to be constructed.

The following definition is important. Let $p \in U$ be a point in an open subset $U \subset V$ of a vector space V of finite dimension n. A *locally differentiable system of coordinates* at p is a pair (U', x) where U' is an open subset of U that contains the point p and $x = (x_1, x_2, \ldots, x_n)$ on U' is a system of differentiable functions such that the differentiable map $x: U' \to \mathbb{R}^n$, $q \in U' \mapsto (x_1(q), x_2(q), \ldots, x_n(q)) \in \mathbb{R}^n$ maps U' bijectively with open image $x(U')$ in $\mathbb{R}^n$. Moreover, the inverse map $x^{-1}: x(U') \to U'$ is differentiable.

The following theorem provides locally differentiable coordinate systems and shows the importance of the concept of the differential of maps.

Theorem 1.51 (Construction of Local Coordinates). *Let V be a vector space of dimension $n \in \mathbb{N}$. Let $U \subset V$ be an open neighborhood of $p \in V$ and let $F: U \to \mathbb{R}^n$ be a continuously differentiable map such that the differential $(DF)_p: V \to \mathbb{R}^n$ is an isomorphism of vector spaces. Denote by $x_i: U \to \mathbb{R}$ the composition of F with the ith linear coordinate function on $\mathbb{R}^n$, $i = 1, 2, \ldots, n$. Then there exists an open neighborhood $U' \subset U$ of p such that the restrictions of the functions x_i to U' build a locally continuously differentiable system of coordinates at p. Denoting again by x_i these restrictions, $x = (x_1, x_2, \ldots, x_n)$ is this system. If the map F is assumed to be smooth, the resulting local system of coordinates is smooth too.*

Proof. Let $(v_1, v_2, \ldots, v_n)$ be the vectors in V with $(DF)_p(v_i) = e_i$, $i = 1, 2, \ldots, n$, where e_i is the ith canonical basis vector in $\mathbb{R}^n$. We choose to work with the maximum norms $\| \cdot \|_V$ and $\| \cdot \|_{\mathbb{R}^n}$ in the based vector spaces V and $\mathbb{R}^n$. Observe that the differential $(DF)_p$ preserves norms. Let U' be the open ball $B(p, r)$ in V with $r > 0$ and $B(p, r) \subset U$ such that for all $q \in B(p, r)$ the inequality

$$\|(DF)_p - (DF)_q\|_{\text{OperatorNorm}} < 1/2$$

holds.

For $q \in U'$ and $v \in V$ the inequality

$$\|((DF)_p^{-1} \circ (DF)_q - \text{Id}_V)(v)\|_V \leq 1/2 \|v\|_V$$

follows.

The restriction of F to U' is injective. Indeed, for $q, q' \in U'$, $q \neq q'$,

$$\|F(q') - F(q)\|_{\mathbb{R}^n} = \left\| \int_0^1 (DF)_{(1-t)q+tq'}(q' - q) dt \right\|$$

$$\geq \|(DF)_p(q' - q)\|_{\mathbb{R}^n} - \int_0^1 \|((DF)_{(1-t)q+tq'} - (DF)_p)(q' - q)\|_{\mathbb{R}^n} dt$$

$$\geq 1/2 \|q' - q\|_V > 0,$$

showing $F(q) \neq F(q')$.

For all $q \in U'$ the differential $(DF)_q: V \to \mathbb{R}^n$ is injective, and hence an isomorphism of vector spaces. Indeed, for $v \in V$, $v \neq 0$,

$$\|(DF)_q(v)\|_{\mathbb{R}^n} = \|(DF)_p(v) - ((DF)_p - (DF)_q)(v)\|_{\mathbb{R}^n}$$

$$\geq \|v\|_V - 1/2 \|v\|_V > 0.$$

The image $F(U')$ is open in $\mathbb{R}^n$. First observe that the image $F(U')$ contains the ball $B_{\mathbb{R}^n}(F(p), r/2)$. Indeed, for a given $a \in B_{\mathbb{R}^n}(F(p), r/2)$ recursively define points $p_k \in B_V(p, r)$ by $p_0 = p$ and

$$p_{k+1} = p_k - (DF)_p^{-1}(F(p_k) - a).$$

Observe that $p_1 \in B_V(p, r/2)$ and

$$p_{k+1} - p_k = p_{k+1} - p_k - (DF)_p^{-1}(F(p_{k+1} - F(p_k))),$$
$$\|p_{k+1} - p_k\|_V \leq 1/2\|p_k - p_{k-1}\|_V,$$

which shows that the sequence (p_k) is a Cauchy sequence with limit $q = \lim_{k\to\infty} p_k \in B_V(p, r)$.

Observe that

$$(DF)_p^{-1}(F(q) - a) = \lim_{k\to\infty}(DF)_p^{-1}(F(p_k) - a) = \lim_{k\to\infty} p_k - p_{k+1} = 0,$$

therefore $F(q) = a$, showing $B_{\mathbb{R}^n}(F(p), r/2) \subset F(U')$ as claimed. So, $F(U')$ is a neighborhood of $F(p)$.

The analogous argumentation applies to show that for all $q \in U'$ the image $F(U')$ is a neighborhood of $F(q)$ since $(DF)_q$ is an isomorphism, as seen above. So, $F(U')$, being a neighborhood of all its points, is open in $\mathbb{R}^n$.

Let the map $G : F(U') \to U'$ be the inverse of the map F. For $a, b \in F(U')$ observe that

$$\|b - a\|_{\mathbb{R}^n} = \|F(G(b)) - F(G(a))\|_{\mathbb{R}^n} \geq 1/2\|G(b) - G(a)\|_V,$$

proving that G is Lipschitz continuous.

For $a, a + h \in F(U')$ put $a = F(q), a + h = F(q + v)$. The maps F, G being Lipschitz bijections inverse to each other, with uniform Lipschitz constants 2, it follows that $1/2 \leq \|h\|_{\mathbb{R}^n}/\|v\|_V \leq 2$. From the three-term expansion

$$F(q + v) = F(q) + (DF)_q(v) + R_q^F(v)$$

it follows that $h = (DF)_q(v) + R_q^F(v)$. One obtains a three-term expansion with the typical properties for G at the point a as follows

$$G(a + h) = G(F(q + v)) = q + v = G(a) + (DF)_q^{-1}(h) - (DF)_q^{-1}(R_q^F(v)).$$

In particular, the differential of G at a is given by

$$(DG)_a = (DF)_q^{-1} = (DF)_{G(a)}^{-1},$$

which proves that the map G is continuously differentiable if the map F is continuously differentiable. Also G is smooth if F is so. Composing F with the canonical linear coordinates of $\mathbb{R}^n$ yields a differentiable system of coordinates $x = (x_1, x_2, \ldots, x_n)$ on U'. □

Let $U \subset V$ be an open subset of a vector space V of dimension n. Let $p \in U$ be a point. Basic examples of rigidity of fields are:

Theorem 1.52 (Basic Smooth Rigidities). *Let $U \subset V$ be an open subset of a finite-dimensional vector space V.*

a) *Let $f: U \to \mathbb{R}$ be a smooth function with $(Df)_p \neq 0$, $p \in U$. Then there exist at p locally smooth coordinates (U', x) with $f(q) = x_1(q)$, $q \in U'$.*

b) *Let $f: U \to \mathbb{R}^k$, $1 \leq k \leq n$ be smooth with $\text{Rank}((Df)_p) = k$, $p \in U$. Then there exist at p locally smooth coordinates (U', x) with*

$$f(q) = (x_1(q), x_2(q), \ldots, x_k(q)), \ q \in U'.$$

c) *Let $\omega: U \to V^*$ be a closed 1-differential form with $\omega_p \neq 0$, $p \in U$. Then there exist at p locally smooth coordinates (U', x) with $\omega = dx_1$ on U'.*

d) *Let $X_1, X_2, \ldots, X_n: U \to V$ be smooth vector fields such that at $p \in U$ the fields are linearly independent and the Lie brackets $[X_i, X_j]_{\text{Lie}}$, $1 \leq i < j \leq n$, vanish on U. Then there exist at p locally smooth coordinates (U', x) with $X_i = \frac{\partial}{\partial x_i}$, $i = 1, 2, \ldots, n$ on U'.*

e) *Let $X: U \to V$ be a smooth vector field with $X_p \neq 0$, $p \in U$. Then there exist at p local smooth coordinates (U', x) with $X_q = \frac{\partial}{\partial x_1}$, $q \in U'$.*

Proof. a) Choose a basis $v_1, v_2, \ldots, v_n$ for V such that $(Df)_p(v_1) = 1$. Let $(y_1, y_2, \ldots, y_n)$ be the linear function on V with $y_i(v_j) = \delta_i^j$. Let $F: U \to \mathbb{R}^n$ be defined by $F(q) := (f(q), y_2(q), \ldots, y_n(q)) \in \mathbb{R}^n$. Now apply Theorem 1.51 (Construction of Local Coordinates) to F and get local coordinates (U', x) at p. Clearly $f = x_1$ on U'.

b) Choose a basis $v_1, v_2, \ldots, v_n$ for V such that $(Df)_p(v_i) = e_i$, $i = 1, 2, \ldots, k$. Define $F: U \to \mathbb{R}^n$ by $F(q) = (f(q), y_{k+1}, \ldots, y_n) \in \mathbb{R}^k \times \mathbb{R}^{n-k} = \mathbb{R}^n$. Apply Theorem 1.51 (Construction of Local Coordinates) to F and get local coordinates (U', x) at p with $f = (x_1, x_2, \ldots, x_k)$ on U'.

c) Choose $U' \subset U$ to be an open star-shaped neighborhood of p. Let $f: U' \to \mathbb{R}$ be such that $df = \omega$. Apply a) and get local coordinates (U'', x) with $\omega = dx_1$ on U''.

d) Choose a star-shaped open neighborhood $U' \subset U$ of p such that at $q \in U'$ the vector fields are linearly independent. Define differential 1-forms ω_i on U' by $\omega_i(X_j) = \delta_i^j$. The forms are closed. Let $f_i: U' \to \mathbb{R}$ be such that $df_i = \omega_i$. Apply Theorem 1.51 (Construction of Local Coordinates) to $F = (f_1, f_2, \ldots, f_n)$ and get local coordinates (U'', x). Observe that $x_i = f_i$. Hence, $X_i = \frac{\partial}{\partial x_i}$, $i = 1, 2, \ldots, n$ on U''.

e) Choose a basis $v_1, v_2, \ldots, v_n$ for V with $v_1 = X_p$. Let $B_\epsilon(\mathbb{R}^n)$ be the "box" $\{h \in \mathbb{R}^n \mid |x_i(h)| < \epsilon, i = 1, 2, \ldots, n\}$. Define the map $\beta: B_\epsilon(\mathbb{R}^n) \to V$ by $h \mapsto p + \sum_{i=2}^n x_i(h)v_i$. Let $\epsilon_0 > 0$ be such that for all $h \in B_{\epsilon_0}(\mathbb{R}^n)$ the solution curve $t \in]-\epsilon_0, \epsilon_0[\mapsto \phi_X(\beta(h), t) \in U$ of the differential equation corresponding to the vector field X with initial condition $\beta(h)$ exists. Define the map $G: B_{\epsilon_0}(\mathbb{R}^n) \to V$ by $h \mapsto \phi_X(\beta(h), x_1(h))$. The map G is continuously differentiable and $(DG)_p(e_i) = v_i$. So, the main part of the proof of Theorem 1.51 (Construction of Local Coordinates) applies, hence G maps a smaller box $B_\epsilon(\mathbb{R}^n)$ injectively onto an open neighborhood U' of p and $F = G^{-1}$ is continuously differentiable too. The composition of F with the canonical coordinate functions of $\mathbb{R}^n$ provides a local coordinate system (U', x) such that $X = \frac{\partial}{\partial x_1}$ on U' holds. $\qquad \square$

Point e) of the previous theorem affirms that the solution curves of the differential equation corresponding to a smooth vector field X near a point p with $X_p \neq 0$ can be viewed as segments. More precisely, using the local coordinates (U', x) the solution curves are the segments with x_1 running in $]-\epsilon, \epsilon[$ and the other coordinates fixed. In particular, a solution curve leaves the flow box after at most every 2ϵ-time interval. The coordinates (U', x) are called *local flow box coordinates* for the vector field X.

proof The local rigidity of smooth non-vanishing vector fields leads to a global result for vector fields on the plane.

Theorem 1.53 (Poincaré–Bendixson Theorem). [25] *Let $X : \mathbb{R}^2 \to \mathbb{R}^2$ be a smooth vector field and let $p \in \mathbb{R}^2$ be a point. Assume that the forward solution $t \in [0, +\infty[\mapsto \phi_X(p, t)$ exists and is a bounded planar curve. Let K be the closure of $\{\phi_X(p, t) \mid t \in [0, +\infty[\}$ in $\mathbb{R}^2$. Assume that $X_q \neq 0$, $q \in K$, holds. Then there exists a periodic solution $\gamma : \mathbb{R} \to \mathbb{R}^2$ with image in K of the differential equation corresponding to X.*

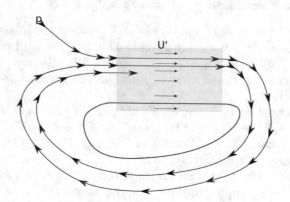

Fig. 1.6 The red limiting periodic trajectory.

Proof. The closure K is compact. Cover K with finitely many flow boxes. The solution $t \in [0, +\infty[\mapsto \phi_X(p, t)$ leaves each flow box after a time interval of positive length that is uniformly bounded from below. Hence one flow box (U', x) is run through infinitely many times by the trajectory $t \in [0, +\infty[\mapsto \phi_X(p, t)$. If the trace of that trajectory on U' has finitely many connected components, the trajectory $t \in [0, +\infty[\mapsto \phi_X(p, t)$ is periodic. If there are infinitely many connected components, those components accumulate in $U' \cap K$ to the trace on U' of a periodic trajectory . Indeed, we number those components $c_n, n \in \mathbb{N}$, according to the entrance times $t_n < t_{n+1}$ in the flow box U'. Let $a_n = \phi_X(p, t_n)$ be the entrance point in U' of c_n. The sequence (a_n) converges in local flow box coordinates (x, y), since the coordinate $y(a_n)$ is strictly monotone by the Jordan Curve Theorem and $x(a_n) = 0$. The limit point $a' = \lim_{n \to \infty} a_n$ is the entrance point in U' of a periodic solution. $\square$

---

[25] Named after French mathematician Jules Henri Poincaré (1854–1912) and Swedish mathematician Ivar Otto Bendixson (1861–1935).

Exercise 1.54. Let X be a smooth vector field on $\mathbb{R}^2$ for which there exists a bounded forward solution $t \in [0, +\infty[\mapsto \phi_X(p, t)$. Show that X vanishes at a point $q \in \mathbb{R}^2$.

The local flow box coordinates for a vector field X are useful when proving the so-called *Cartan Formula*, which itself is a coordinate-free formula.

Theorem 1.55 (Cartan Formula). *Let X_t, $t \in]-T, T[$, be a time-dependent vector field on an open subset U of a finite-dimensional vector space V. Assume that the flow $\phi_t^X : U \to U$ is defined for $t \in]-T, T[$, $T > 0$. Let α be a k-differential form on U. Then*

$$\frac{d}{dt}(\phi_t^X)^*\alpha = i_{X_t} d\alpha + d i_{X_t}\alpha$$

holds.

Proof. For given $t \in]-T, T[$ the LHS and RHS only depend upon the vector field $Y = X_t$. So it suffices to prove $\frac{d}{dt}(\phi_t^Y)^*\alpha = i_Y d\alpha + d i_Y \alpha$.

Let $U_0 \subset U$ be the interior of the complement in U of the support of Y, the support being the closure of $U_1 := \{q \in U \mid Y_q \neq 0\}$. The open set $U_0 \cup U_1$ is dense in U. The formula holds on U_0 since ϕ^Y restricts to the identity on U_0.

For $q \in U_1$ choose local flow box coordinates

$$x = x_1, x_2, \ldots, x_n : U' \to \mathbb{R}^n$$

centered at q with $Y = \frac{\partial}{\partial x_1}$. A k-form α is a linear combination of basic forms $a dx_{i_1} \wedge \cdots \wedge dx_{i_k}$. For a basic form $\beta = a dx_{i_1} \wedge \cdots \wedge dx_{i_k}$ we have

$$\frac{d}{ds}(\phi_{t+s}^Y)^*\beta_{|s=0} = (i_Y da)\beta$$

and one checks

$$(i_Y da)\beta = i_Y d\beta + d i_Y \beta$$

for both cases $i_1 > 1$ and $i_1 = 1$.

We conclude by linearity, since the formula holds on a dense subset. □

A symplectic form on an open subset $U \subset \mathbb{R}^{2n}$ is a closed differential 2-form ω that moreover is *regular* as a two form at every $p \in U$, i.e. for $u \in \mathbb{R}^{2n}$, $u \neq 0$, there exists $v \in \mathbb{R}^{2n}$ with $\omega_p(u, v) \neq 0$. A contact form on an open subset $U \subset \mathbb{R}^{2n+1}$ is a 1-differential form α on U such that $\alpha \wedge (d\alpha)^n$ is a volume form on U. The following theorem states local rigidity results due to Gaston Darboux[26] for volume, symplectic and contact forms [27].

Theorem 1.56 (Local Darboux Rigidities). *Let ω be a volume or a symplectic form, and let α be a contact form on an n-, 2n- or $(2n + 1)$-manifold respectively. Then at each point there exists a coordinate chart $(x_1, \cdots, x_n)$ or $(x_1, \cdots, x_n, y_1, \cdots, y_n)$ or $(x_1, \cdots, x_n, y_1, \cdots, y_n, z)$ such that the volume form is expressed by $\omega = dx_1 \wedge \cdots \wedge dx_n$, the symplectic form by $\omega = dx_1 \wedge dy_1 + \cdots + dx_n \wedge dy_n$ and the contact form by $\alpha = dz - y_1 dx_1 - \cdots - y_n dx_n$.* □

---

[26] Gaston Darboux (1842–1917), French mathematician, Professor in Paris.

The presentation of the above rigidities by Darboux is very interesting and differs from today's presentation of local symplectic rigidity. In fact Darboux obtained a local rigidity theorem for Pfaff forms: let $a_1, \cdots, a_n, X_1, \cdots, X_n$ be smooth functions in a neighbourhood of a point $p \in \mathbb{R}^{2n}$ such that for the Pfaff form $\alpha = a_1 dX_1 + \cdots + a_n dX_n$ the expression $da_1 \wedge dX_1 + \cdots da_n \wedge dX_n$ is regular. Then, the functions $a_1, \cdots, a_n, X_1, \cdots, X_n$ are near p a local smooth coordinate system.

1.6 Differentiation in Banach Spaces

A *Banach space* is a vector space E together with a norm $\| \cdot \|_E : V \to \mathbb{R}$ such that E is *complete*, i.e. such that every Cauchy sequence in E converges to a limit belonging to E. Clearly, a finite-dimensional normed vector space $(V, \| \cdot \|_V)$ is a Banach space. A typical example of an infinite-dimensional Banach space is the space $C^0([0,1], \mathbb{R})$ of all continuous functions $f : [0,1] \to \mathbb{R}$ equipped with the supremum norm $\|f\|_{C^0} = \sup_{t \in [0,1]} |f(t)|$. The choice of norm is important, for instance $C^0([0,1], \mathbb{R})$ equipped with the L^1-norm $\|f\|_{L^1} = \int_0^1 |f(t)| dt$ is not a Banach space. The same remark holds for the L^2-norm $\|f\|_{L^2} = (\int_0^1 |f(t)|^2 dt)^{1/2}$. Both the L^1 and L^2-norm compare only on one side with the C^0-norm. In fact they compare from below as follows: $\|f\|_{L^2} \leq \|f\|_{L^1} \leq \|f\|_{C^0}$.

An important result, due to Baire–Banach–Schauder that we will study later (see Theorem 12.14) concerns two norms $\| \cdot \|_1$ and $\| \cdot \|_2$ on a vector space E such that the first compares from below or from above to the second. If E equipped with the norm $\| \cdot \|_1$ and E equipped with the norm $\| \cdot \|_2$ are both Banach spaces, then the norms compare, i.e. there exists a constant $C > 1$ such that $1/C\|f\|_1 \leq \|f\|_2 \leq C\|f\|_1$, $f \in E$, holds. In particular the topologies on E provided by the norms $(E, \| \cdot \|_1)$ and $(E, \| \cdot \|_2)$ coincide. This result is elementary for Banach spaces of finite dimension, and not at all elementary in general.

Let $(E, \| \cdot \|_E)$ and $(F, \| \cdot \|_F)$ be normed spaces and let $A : E \to F$ be a linear map. In the case of finite-dimensional spaces a linear map is continuous. This is no longer true in general. The following theorem characterizes the continuous maps among the linear ones.

Theorem 1.57. *Let $A : E \to F$ be a linear map between normed spaces $(E, \| \cdot \|_E)$ and $(F, \| \cdot \|_F)$. The following are equivalent:*

  (i) *A is continuous,*
 (ii) *A is continuous at $0 \in E$,*
(iii) *there exists a constant $C \geq 0$ such that the inequality*

$$\|A(u)\|_F \leq C\|u\|_E, \; u \in E,$$

holds.   □

The proof follows immediately from the definitions.

Let $(E, \| \cdot \|_E)$ and $(F, \| \cdot \|_F)$ be normed spaces. Let $U \subset E$ be open in E and let $f : U \to F$ be a map. Let $p \in U$ be a point. The map is *differentiable* at p if there exists a three-term expansion with the typical properties

$$f(p + h) = f(p) + L(h) + R(h), \ h \in E, \ a + h \in U$$

such that moreover the linear map $L : E \to F$ is continuous.

Theorem 1.51 (Construction of Local Coordinates) uses as input a continuously differentiable map $F : U \to \mathbb{R}^n$ defined on an open subset of a vector space V of finite dimension n and moreover at a point $p \in U$ the differential $(DF)_p : V \to \mathbb{R}^n$ was assumed to be an isomorphism of vector spaces. A first step in the proof was the construction of Cauchy sequences in V and using the fact that in a finite-dimensional vector space every Cauchy sequence converges in V. This implied the existence of an open neighborhood U' of p in U such that the image $F(U')$ is open in $\mathbb{R}^n$ together with the construction of a continuously differentiable map $G : F(U') \to U'$ such that G and the restriction of F to U' are reciprocal bijections. This part of the proof works in the more general setting of Banach spaces and proves:

Theorem 1.58 (Inverse Mapping Theorem). *Let $U \subset E$ be open in a Banach space E. Let $p \in U$ be a point. Let $f : U \to F$ be a differentiable map with values in a normed space F, such that the differential $(Df)_p : E \to F$ is bijective, continuous with continuous inverse $((Df)_p)^{-1} : F \to E$. Then there exists an open neighborhood $U' \subset U$ of p such that f maps bijectively U' onto an open neighborhood $f(U')$ of $f(p)$ in F. Moreover the inverse map $g : f(U') \to U'$ is differentiable.* $\square$

The differential at a point of a function with vectors as arguments and as values was introduced in the more general context of Banach and Fréchet spaces by Maurice Fréchet,[27] see the books of Henri Cartan [22, 23] and Jean Dieudonné [32].

1.7 Sard's Theorem

Let $f :]a, b[\to \mathbb{R}$ be a differentiable function. Moreover assume that the derivative $f' :]a, b[\to \mathbb{R}$ is continuous. Remember $f'(t) := (Df)_t(1)$. At a point $t \in]a, b[$ with $f'(t) \neq 0$ we can say something about the local behavior of f at t, namely in a neighborhood $]t - \epsilon, t + \epsilon[\cap]a, b[$ the function f is strictly monotone. At a point $t \in]a, b[$ with $f'(t) = 0$ we cannot say much about the local behavior of f, except if moreover f'' is continuous and $f''(t) \neq 0$ holds. In this case f has at t a local maximum or minimum and is near t strictly convex or concave.

A point t with $f'(t) = 0$ is called a critical point of f. Critical since a statement, without making further assumptions, about the local behavior at t of f is not possible. A point t with $f'(t) = 0$ and $f''(t) \neq 0$ is called a good critical point.

A function can have many critical points. Take for example a constant function.

---

[27] Maurice Fréchet (1878–1973), French mathematician, Professor in Poitiers, Strasbourg and Paris.

A critical value of f is the value $f(t)$ at a critical point t of f. Can a function have many critical values? For instance, a constant function has many critical points but only one critical value.

In order to state a first result of Arthur Sard[28] [109] we need the following definition. A subset X in $\mathbb{R}$ is *Jordan small*[29] if for every $\epsilon > 0$ there exist finitely many intervals $[a_i, b_i]$, $1 \leq i \leq N$, that cover X and have total length $\sum_{i=1}^{N}(b_i - a_i) < \epsilon$.

Theorem 1.59 (Sard in dimension 1). *Let $f: [a, b] \to \mathbb{R}$ be continuously differentiable. Then the set of critical values of f is Jordan small.*

Proof. Given $\epsilon > 0$, let $m(f', \frac{\epsilon}{b-a}) > 0$ be a continuity modulus for f' and $\frac{\epsilon}{b-a}$, i.e. for $t, s \in [a, b]$ with $|t - s| < m(f', \frac{\epsilon}{b-a})$ the inequality $|f'(t) - f'(s)| < \frac{\epsilon}{b-a}$ holds. For $N \in \mathbb{N}$ with $\frac{b-a}{N} < m(f', \frac{\epsilon}{b-a})$ we subdivide the interval $[a, b]$ into N intervals of length $\frac{b-a}{N}$. Let C_i, $1 \leq i \leq K$, be the list of intervals in this subdivision that contain a critical point of f. We have $K \leq N$. We also have

$$\text{Length}(f(C_i)) \leq \int_{C_i} |f'(t)|dt \leq \frac{b-a}{N}\frac{\epsilon}{b-a} = \frac{\epsilon}{N}.$$

The finitely many intervals $f(C_i)$, $1 \leq i \leq K$, cover the set of critical values of f and have total length less than ϵ. $\qquad\square$

Theorem 1.60. *Let $f: [a, b] \to \mathbb{R}$ be twice continuously differentiable. Let $L: \mathbb{R} \to \mathbb{R}$, $L(t) = \lambda t$ be linear with λ not a critical value of f'. The function $g := f - L: [a, b] \to \mathbb{R}$ has finitely many critical points, all being good.*

Proof. Let t be a critical point of g, i.e. $g'(t) = f'(t) - \lambda = 0$. Since λ is not a critical value of f', we conclude $f''(t) \neq 0$, hence $g''(t) = f''(t) - L''(t) = f''(t) \neq 0$ and the critical point t of g is good. At an accumulation point s of critical points of g we would have $g'(s) = 0$ by the continuity of g' and also $g''(s) = 0$ by Rolle's Theorem together with the continuity of g''. This is impossible since s would be a non-good critical point of g. An infinity of critical points of g is excluded, since otherwise by the compactness of $[a, b]$ an accumulation of critical points of g would follow. $\qquad\square$

As a consequence we see that for a continuously twice differentiable function $f: [a, b] \to \mathbb{R}$ many linear functions $L: \mathbb{R} \to \mathbb{R}$ exist such that $f - L$ has only good critical points. In particular we can choose $L(t) = \lambda t$ with λ arbitrarily small. Moreover we can choose L such that $f - L$ has different values at different critical points or boundary points a, b.

It follows that the counting function

$$s \in \mathbb{R} \mapsto \#\{t \in [a, b] \mid f(t) - L(t) = s\} \in \mathbb{N}$$

changes by ± 2 or ± 1 for $s_{\pm} := s \pm \epsilon$ if s is a critical value or if s is a boundary value.

---

[28] Arthur Sard (1909–1980), American mathematician.
[29] Named after French mathematician Marie Ennemond Camille Jordan (1838–1922).

Let $\mathrm{Graph}(f) \subset \mathbb{R} \times \mathbb{R}$ be the graph of the function f in the Cartesian plane with X and Y axes. We can change the X-axis to a new X'-axis with $X' := \mathrm{Graph}(L)$ and observe that the graph $\mathrm{Graph}(f)$ of f with respect to the axes X, Y is also the graph of $f - L$ with respect to the axes X', Y.

Changing the system of axes from X, Y to X', Y amounts to changing the coordinate functions x, y to the functions $x, y - L(x)$. We have learned from the above that a function $y = f(x)$ with continuous second derivative cannot be made responsible for having a "bad graph", it is rather due to the choice of coordinates.

Here we think of a graph as being "good" if it has only finitely many critical points, all maxima and minima with strictly concave or convex local behavior. This is the kind of graph that is studied at high school. We did not know at high school that every graph of a C^2-function becomes "good" by turning the X-axis a little. This applies, for instance, to the function

$$y = f(x) := (1 - \cos(x^2)) \sin\left(\frac{1}{x}\right).$$

Does the above remain true for mappings defined on an open subset U of a real vector space V of finite dimension and with values in a real vector space W of the same dimension? The answer is very short: yes. In order to formulate it we need to adapt some definitions.

Let $g : U \to W$ be a continuously differentiable mapping. If the differential $(Dg)_p : V \to W$ is an isomorphism, we understand the local behavior of g at p by the Inverse Mapping Theorem (Theorem 1.58). More generally, if the differential $(Dg)_p : V \to W$ is injective or surjective the local behavior at p of the mapping g is equal to the local behavior of the linear map $(Dg)_p : V \to W$. So, a point p is called a critical point if the differential $(Dg)_p$ is not of maximal rank, i.e. not an isomorphism, not injective or not surjective. Let $f : U \to \mathbb{R}$ be a continuously differentiable function. If $(Df)_p \neq 0$ at a point $p \in U$ we understand the local behavior of f at p by the implicit function theorem. So p is a critical point of f if $(Df)_p = 0$. Remember, each time a main tool for the local study of functions/mappings fails, we have a critical situation.

Let W be equipped with a basis $e_1, e_2, \ldots, e_n$. The basic cube in W is the subset $B := \{t_1 e_1 + t_2 e_2 + \cdots + t_n e_n \mid 0 \leq t_i \leq 1\}$. An affine cube C in W is the image $C := \phi(B)$ under an affine map $w \in W \mapsto A(w) + w_0 \in W, w_0 \in W, A \in \mathrm{Hom}(W, W)$. We declare the volume of $C := \phi(B)$ to be the absolute value of the determinant of A. A subset X of W is *Jordan small* if for every $\epsilon > 0$ there exists a finite family $C_1, C_2, \ldots, C_N$ of affine cubes in W that cover X and which have total volume less than ϵ. The high-dimensional (both range and target) Sard Theorem [109] is:

Theorem 1.61 (Arthur Sard). *Let V and W be real vector spaces of finite dimensions n and m and let U be open in V. Let $g : U \to W$ be k-times continuously differentiable, $k \geq \max\{n - m + 1, 1\}$. Let $C \subset U$ be an affine cube. Then the set of critical values of the restriction $g_{|C}$ to C is Jordan small in W.*

For the proof in the case $n \neq m$, see John Milnor's book [75].

Proof. We consider only the case $n = m$. We choose in V a basis such that C is a translation of the basic cube. Also we work in V with the maximum norm $\|\cdot\|_V$ relative to this basis. In W we will work with a norm that derives from a scalar product. For $p \in U$ we denote by $\|(dg)_p\|_{\text{Operator}}$ the norm of $(dg)_p : V \to W$. Let b be an upper bound for $\|(dg)_p\|_{\text{Operator}}, p \in C$. Given $\epsilon > 0$, let $m(df, \epsilon) > 0$ be a continuity modulus for the restriction of dg to C, i.e. for $p, q \in C$ with $\|p - q\|_V < m(dg, \epsilon)$ the inequality $\|(dg)_p - (dg)_q\|_{\text{Operator}} < \epsilon$ holds. For $N \in \mathbb{N}$ with $\frac{1}{N} < m(dg, \epsilon)$ we subdivide the cube C into $N^{\text{Dim}(V)}$ in sub-cubes by cutting C along hyperplanes parallel to the faces of C in such a way that the sub-cubes are congruent pairwise by translations. Let $C_i, 1 \le i \le K$, be the list of sub-cubes that contain a critical point p_i of g. We have $K \le N^{\text{Dim}(V)}$. We wish to cover $g(C_i)$ by finitely many affine cubes of W and control the total volume. We use for $p_i + h \in C_i$ the three-term expansion for g at the point p_i:

$$g(p_i + h) = g(p_i) + (Dg)_{p_i}(h) + \int_0^1 ((Dg)_{p_i+th} - (Dg)_{p_i})(h)dt.$$

Since $\|h\|_V \le \frac{1}{N} < m(dg, \epsilon)$, we have

$$\left\| \int_0^1 ((Dg)_{p_i+th} - (Dg)_{p_i})(h)dt \right\|_W \le \epsilon/N.$$

We see that the image $g(C_i)$ is contained in the ϵ/N-tubular neighborhood of $g(p_i) + (Dg)_{p_i}(C_i)$. The cube C_i has $2\text{Dim}(V)$ faces $F_{ij}, 1 \le j \le 2\text{Dim}(V)$. Since $(Dg)_{p_i}$ is not injective the image $(Dg)_{p_i}(C_i)$ is covered by the $g(p_i)$-translates of the images $(Dg)_{p_i}(F_{ij})$, which are cubes of dimension $\text{Dim}(W) - 1$ in W. The spanning vectors of such a cube $(Dg)_{p_i}(F_{ij})$ have $\| \cdot \|_W$ norm less than $\frac{b}{N}$. Hence, since we are working in W with norm and volume deriving from a scalar product, its ϵ/N-tubular neighborhood is contained in an affine cube of volume less than

$$\left(\frac{b}{N} + \frac{2\epsilon}{N} \right)^{\text{Dim}(W)-1} \frac{2\epsilon}{N}$$

by the Hadamard inequality. We conclude that $g(C_i)$ is covered by $2\text{Dim}(V)$ affine cubes of total volume

$$2\text{Dim}(V) \left(\frac{b}{N} + \frac{2\epsilon}{N} \right)^{\text{Dim}(W)-1} \frac{2\epsilon}{N}.$$

The image $g(\cup C_i)$ contains the critical values of the restriction of g to C and is covered in W by fewer than $2\text{Dim}(V)N^{\text{Dim}(V)}$ affine cubes of total volume less than

$$2N^{\text{Dim}(V)}\text{Dim}(V) \left(\frac{b}{N} + \frac{2\epsilon}{N} \right)^{\text{Dim}(W)-1} \frac{2\epsilon}{N}.$$

It suffices to work with $0 < \epsilon \le 1$. Our total volume estimate then simplifies to

$$4\mathrm{Dim}(V)(b+2)^{\mathrm{Dim}(W)-1}\epsilon,$$

which can be made arbitrarily small.   □

Theorem 1.62. *We use the notations U, V, C as above. Let $f: U \to \mathbb{R}$ be twice continuously differentiable. Let $L: V \to \mathbb{R}$ be linear with $L \in V^*$ not being a critical value of $\mathrm{d}f: C \to V^*$. Then the function $g := f - L: U \to \mathbb{R}$ has finitely many critical points $p \in C$ and moreover for each such p the map $(\mathrm{D}(\mathrm{d}f))_p: V \to V^*$ is an isomorphism.*

Proof. Let $p \in C$ be a critical point of g, i.e. $(\mathrm{D}g) = (\mathrm{D}f)_p - L = 0$. Since L is not a critical value of $\mathrm{d}f$, we conclude that $(\mathrm{D}(\mathrm{d}f))_p: V \to V^*$ is an isomorphism, hence $(\mathrm{D}(\mathrm{d}g))_p = (\mathrm{D}(\mathrm{d}f))_p - (\mathrm{D}(\mathrm{d}L))_p = (\mathrm{D}(\mathrm{d}f))_p$ is also an isomorphism. The remaining part will be proved in the next section.   □

Exercise 1.63. Let f be the differentiable function on $\mathbb{R}$ defined by the formula $x^4 \sin(\frac{1}{x})$. For which $a \in \mathbb{R}$ does $x^4 \sin(\frac{1}{x}) + ax$ have only good critical points?

1.8 The Morse Lemma and Morse Functions

Let $f:]-1, +1[\to \mathbb{R}$ be a twice continuously differentiable function with $f'(0) = 0$, $f''(0) \neq 0$. Assume $f''(t) > 0$ on $]-\epsilon, +\epsilon[$. Define the function $y:]-\epsilon, +\epsilon[\to \mathbb{R}$ by $y(t) := \sqrt{f(t) - f(0)}$, $t \geq 0$ and $y(t) := -\sqrt{f(t) - f(0)}$, $t < 0$. The function y will be twice differentiable with $f_{|]-\epsilon,+\epsilon[} = f(0) + y^2$. If $f''(t) < 0$, put $y(t) := \sqrt{f(0) - f(t)}$, $t \geq 0$ and $y(t) := -\sqrt{f(0) - f(t)}$, $t < 0$. We then have $f_{|]-\epsilon,+\epsilon[} = f(0) - y^2$. In the first case f has a local minimum at 0, and in the second, a local maximum. The following is a generalization due to H.C.M. Morse[30] for functions on open sets in vector spaces [80].

Theorem 1.64 (Morse Lemma). *Let $f: U \to \mathbb{R}$ be a twice continuously differentiable function defined on an open subset U in a real vector space V of finite dimension n. Assume f is critical at $p \in U$ and that $(\mathrm{D}(\mathrm{d}f))_p: V \to V^*$ is an isomorphism. Then there exists an open subset $U' \subset U$ with $p \in U'$, and twice differentiable functions $y_1, y_2, \ldots, y_n: U' \to \mathbb{R}$ with $y_i(p) = 0$, which are on U' a system of differentiable coordinates, such that*

$$f_{|U'} = f(p) \pm y_1^2 \pm y_2^2 \pm \cdots \pm y_n^2.$$

A twice continuously differentiable function $f: U \to \mathbb{R}, U \subset V$, such that at every critical point p the second differential $(\mathrm{D}(\mathrm{d}f))_p: V \to V^*$ is an isomorphism is called a *Morse function*. From the Morse Lemma it follows that critical points of Morse functions do not accumulate. This observation finishes the proof of Theorem 1.62 at the end of the previous section.

The Morse Lemma follows by induction from its stronger version.

---

[30] Harold Calvin Marston Morse (1892–1977), American mathematician.

Theorem 1.65 (Strong Morse Lemma). *Let $f : U \to \mathbb{R}$ be a twice continuously differentiable function defined on an open subset U of a real vector space V of finite dimension n. Assume f is critical at $p \in U$ and that $(D(df))_p \neq 0$. Then there exist an open subset $U' \subset U$ with $p \in U'$, and twice differentiable functions $y_1, x_2, \ldots, x_n : U' \to \mathbb{R}$ with $y_i(p) = 0$, which are on U' a system of differentiable coordinates, and a function $g : \mathbb{R}^{n-1} \to \mathbb{R}$*

$$f_{|U'} = f(p) \pm y_1^2 + g(x_2, x_3, \ldots, x_n).$$

Proof (of the Morse Lemma). If $(D(df))_p$ is an isomorphism, $(D(df))_p \neq 0$ and the Strong Morse Lemma applies. The Hessian of f at p now splits as a direct sum of the Hessian of $\pm y_1^2$ at p and the Hessian of g at $0 \in \mathbb{R}^{n-1}$. So, if $n > 1$, the Strong Morse Lemma applies again for the critical point $0 \in \mathbb{R}^{n-1}$ of g. Repeating $n - 1$ times we get the Morse Lemma.   □

Proof (of the Strong Morse Lemma). The Hessian $(D(df))_p$ is symmetric and by assumption not 0. Let $e_1, e_2, \ldots, e_n$ be a basis of V such that $((D(df))_p(e_1))(e_1) \neq 0$. Let $z_1, x_2, \ldots, x_n$ be linear coordinates with $z_1(p) = 0$, $dz_1(e_1) = 1$, $dz_1(e_i) = 0$, $x_i(p) = 0$, $i = 2, 3, \ldots$, and $dx_i(e_j) = \delta_j^i$, $i, j = 2, 3, \ldots$. The function f restricted to the curve $C_0 := \{q \in V \mid x_2(q) = x_3(q) = \cdots = x_n(q) = 0\}$ is near p strictly convex or concave with its extremum at p. Think of the curve C_0 locally near p as the z_1-axis and moreover parametrized by $s \in \,]a, b[\,\mapsto \gamma(t) \in C_0$ with $\gamma(0) = p$ and $\frac{d}{ds} z_1(\gamma(t)) \equiv 1$. The composed function $f \circ \gamma$ will be convex or concave near $0 \in \,]a, b[$. We may assume

$$\frac{\partial^2 f}{\partial z_1^2}(p) = ((D(df))_p(e_1))(e_1) > 0$$

and $f_{|C_0}$ will be strictly convex with a minimum at p.

Let $T := \{t \in V \mid z_1(t) = 0\}$ be the $x_2, x_3, \ldots, x_n$ coordinate hyperplane through p. For $t \in T$ and t near p let C_t be the curve $C_t := \{q \in V \mid x_i(q) = x_i(t)\}$. The restriction $f_{|C_t}$ is strictly convex near a point $q_t \in C_t$. Let $k(z_1, x_2, \ldots, x_n) = f$ be the local expression of f in the coordinates $z_1, x_2, \ldots, x_n$. Let $m(x_2(t), x_3(t), \ldots, x_n(t))$ be the z_1 coordinate value of q_t. The number $m(x_2(t), x_3(t), \ldots, x_n(t))$ is implicitly defined by the equation $\frac{\partial k}{\partial z_1}(m(x_2(t), \ldots, x_n(t)), x_2(t), x_3(t), \ldots, x_n(t)) = 0$. The implicit function theorem applies since $\frac{\partial^2 f}{\partial z_1^2}(p) \neq 0$. We make a first coordinate change, only changing the first coordinate function: $z_1 := z_1 - m(x_2, x_3, \ldots, x_n)$, $x_2 := x_2, \ldots, x_n := x_n$. In the new coordinate z_1 the minimum of f restricted to C_t is achieved at the point q_t with $z_1(q_t) = 0$.

We make a second coordinate change by changing again only the new coordinate function z_1. For $q \in C_t$ we put $y_1(q) := \mathrm{Sign}(z_1(q))\sqrt{f(q) - f(q_t)}$. We have for $q \in V$ locally at p

$$f(q) = y_1(q)^2 + k(m(x_2(q), x_3(q), \ldots, x_n(q)), x_2(q), \ldots, x_n(q)).$$

Finally, putting $g(x_2, \ldots, x_n) := k(m(x_2, x_3, \ldots, x_n), x_2, \ldots, x_n)$ we get

$$f = f(p) + y_1^2 + g(x_2, \dots, x_n),$$

as required.   □

Applying the Strong Morse Lemma r-times, the co-rank $n - r$ Morse Lemma follows. The Morse Lemma is in fact the special case $r = n$.

Theorem 1.66 (Co-rank Morse Lemma). *Let $f: U \to \mathbb{R}$ be a twice continuously differentiable function defined on an open subset U of a real vector space V of finite dimension n. We assume f is critical at $p \in U$ and that the Hessian $(D(df))_p: V \to V^*$ has rank $r \geq 1$. Then there exist an open subset $U' \subset U$ with $p \in U'$, twice differentiable functions $y_1, y_2, \dots, y_r, x_{n-r}, \dots, x_n: U' \to \mathbb{R}$ with $y_i(p) = 0$, which are on U' a system of differentiable coordinates, and a function $g: \mathbb{R}^{n-1} \to \mathbb{R}$ such that*

$$f_{|U'} = f(p) \pm y_1^2 \pm \cdots \pm y_r^2 + g(x_{n-r}, x_{n-r+1}, \dots, x_n).$$ □

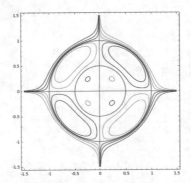

Fig. 1.7 The $-0.05, -0.02, -0.0065, 0, +0.0065, +0.02, +0.05$-levels of $f = -xy(x^2 + y^2 - 1)(x^2 + y^2 - 1/4)$.

Exercise 1.67. Study the 17 critical points of $f = xy(x^2 + y^2 - 1)(x^2 + y^2 - 1/4)$ on $\mathbb{R}^2$ (Fig. 1.7).

Exercise 1.68. Two Morse functions $a, b: M \to \mathbb{R}$ are called family equivalent if there exists a smooth function $c: M \times [0, 1] \to \mathbb{R}, (p, t) \mapsto c_t(p) \in \mathbb{R}$ such that the functions $c_t: M \to \mathbb{R}, t \in [0, 1]$, are Morse functions with $a = c_0, b = c_1$. A maximal set of family equivalent functions is called a Morse family. Count the families of Morse functions on $\mathbb{R}$ with 5 critical points. A Morse function is called general if the number of critical points equals the number of critical values. Count the number of families of general Morse functions with 5 critical points. Count also those families with $n \in \mathbb{N}$ critical points and values.

Exercise 1.69. A Morse function is called exceptional if it has two critical values. How many families of exceptional Morse functions on $\mathbb{R}$ exist? Do such families have polynomial members? Are those polynomials famous?

Exercise 1.70. How many families of exceptional Morse functions on the one and two-dimensional spheres S^1 and S^2 exist?

Chapter 2
The Geometry of Manifolds

2.1 Differentiable Manifolds

We wish to paste together open subsets of vector spaces in order to get more general spaces on which fields and their operations are defined. We proceed in four or five steps.

1. Let A be a set. Let $(U_\alpha)_{\alpha \in A}$ be a family, indexed by A, of open subsets of a real vector space V of finite dimension. Let X be the disjoint union $X := \dot\cup_{\alpha \in A} U_\alpha$ with its topology.
2. Let $\sim$ be an equivalence relation on X. Let Y be the quotient topological space $Y := X/\sim$.

On X we wish to carry out all the constructions concerning fields described in Section 1.1. In the next steps 3 and 4 we impose properties on the equivalence relation $\sim$ that will enable us to execute all of these constructions on Y.

3. For each pair $(\alpha, \beta) \in A \times A$ there exist (possibly empty) open subsets $U_{\alpha\beta} \subset U_\beta$ and $U_{\beta\alpha} \subset U_\alpha$ together with a differentiable map $\phi_{\alpha\beta} : U_{\alpha\beta} \to U_{\beta\alpha}$ such that the composition $\phi_{\beta\alpha} \circ \phi_{\alpha\beta}$ is the identity map of $U_{\alpha\beta}$. For a pair $(\alpha, \beta) \in A \times A$ with $\alpha = \beta$ the equality $U_{\alpha\alpha} = U_\alpha$ holds and the differential map $\phi_{\alpha\alpha}$ is the identity of $U_{\alpha\alpha}$. The equivalence relation $\sim$ is generated by $p \in U_{\alpha\beta} \sim \phi_{\alpha\beta}(p) \in U_{\beta\alpha}$.
4. For each triple $(\alpha, \beta, \gamma) \in A \times A \times A$ the restrictions to $U_{\alpha\gamma} \cap U_{\beta\gamma}$ of the composition $\phi_{\alpha\beta} \circ \phi_{\beta\gamma}$ and of $\phi_{\alpha\gamma}$ are equal maps into $U_{\gamma\alpha} \cap U_{\beta\alpha}$.

Let $\pi : X \to Y$ be the quotient map. It follows from properties 3 and 4 that the map $\pi : X \to Y$ is open, and that the restrictions π_α of π to each U_α are injective. So, the family of images $\pi(U_\alpha) \subset Y$ is an open cover of Y. The family of pairs $(\pi(U_\alpha), \phi_\alpha)$, where $\phi_\alpha : \pi(U_\alpha) \to U_\alpha$ is the right inverse of the restriction of π to U_α, form an atlas with differentiable transition maps for the space Y.

If we wish the space Y to have the Hausdorff separation property we moreover impose on the equivalence relation $\sim$:

5. For $p \in U_\alpha$ and $q \in U_\beta$ with $\pi(p) \neq \pi(q)$ there exist open subsets $U'_\alpha \subset U_\alpha$ and $U'_\beta \subset U_\beta$ with $p \in U'_\alpha$, $q \in U'_\beta$ and $\pi(U'_\alpha) \cap \pi(U'_\beta) = \emptyset$.

© The Author(s), under exclusive license to Springer Nature Switzerland AG 2021
N. A'Campo, *Topological, Differential and Conformal Geometry of Surfaces*, Universitext,
https://doi.org/10.1007/978-3-030-89032-2_2

In this text a differentiable manifold M is the data of families $(U_\alpha)_{\alpha \in A}$ and $(\phi_{\alpha\beta} \colon U_{\alpha\beta} \to U_{\beta\alpha})_{(\alpha,\beta) \in A \times A}$ as above with the properties from steps 3, 4 and 5.

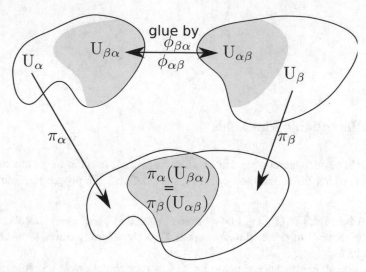

Fig. 2.1 Gluing makes a manifold: $\pi_{\alpha|U_{\beta\alpha}} = \pi_{\beta|U_{\alpha\beta}} \circ \phi_{\beta\alpha}$.

Let M be a manifold. The dimension of M is by definition the dimension of the vector space V in which the open sets $(U_\alpha)_{\alpha \in A}$ of the first family lie. The topological space Y of step 2 is the underlying topological space of M. We say that a manifold M has a topological property if its underlying topological space has the property in question. Examples of topological properties include being connected, simply connected, compact, homeomorphic to some topological space T, homotopic to some homotopy type H, etc.

The family $(\pi_\alpha(U_\alpha))_{\alpha \in A}$ together with the family $(\phi_\alpha)_{\alpha \in A}$ of right inverses of the maps π_α is called the family of *charts* of M or is an *atlas* for M.

Imagine that the underlying topological space is the surface of some planet Y. Different institutions, academies, armies organize expeditions and explore Y locally, Institution $\alpha \in A$ explores a region $\pi(U_\alpha)$ of Y. Important for building the appropriate theoretical language is that the expeditions also take place in the imagination of humans. The result of the efforts of Institution α is a paper map $U_\alpha \subset \mathbb{R}^2$ of the region $\pi(U_\alpha)$. The other Institutions, other expeditions, explore other regions $\pi(U_\beta), \pi(U_\gamma), \dots$ of planet Y. Finally, the explored regions $\pi(U_\alpha), \pi(U_\beta), \pi(U_\gamma), \dots$ cover the planet Y and a lot of money has been spent. All paper maps are stored by an Academy in their library. A critical voice suddenly proposes to compare the results of the different expeditions. Regions $\pi(U_\alpha)$ and $\pi(U_\beta)$ of the planet sometimes overlap, so in the library of an important Academy one can draw subsets $U_{\beta\alpha} := (\pi_\alpha)^{-1}(\pi(U_\alpha) \cap \pi(U_\beta)) \subset U_\alpha$ and $U_{\alpha\beta} := (\pi_\beta)^{-1}(\pi(U_\alpha) \cap \pi(U_\beta)) \subset U_\beta$, and while deploring the waste of research resources, discover maps $(\phi_{\alpha\beta}: U_{\alpha\beta} \to U_{\beta\alpha})_{(\alpha,\beta)\in A\times A}$ that compare the paper maps drawn by the Institutions. Many years later a headline appears in a newspaper:

Planet Y is a differentiable manifold.

Today humanity tries to describe not only the universe, but also our brain, music and the space of all DNA configurations as a manifold.

Fig. 2.2 Our moon, seen in November 2017, from ICTS Bangalore, and in December 2019, from BHU Varanasi.

Let M be a differentiable manifold given by families $(U_\alpha)_{\alpha\in A}$ and by maps $(\phi_{\alpha\beta}: U_{\alpha\beta} \to U_{\beta\alpha})_{(\alpha,\beta)\in A\times A}$. Let U_σ be a chart and assume that the underlying topological space Y is already covered by the family $\pi(U_\alpha), \alpha \neq \sigma$. Strictly speaking, we now have two manifolds, M and also M' given by the families $(U_\alpha)_{\alpha\in A'}$ and $(\phi_{\alpha\beta}: U_{\alpha\beta} \to U_{\beta\alpha})_{(\alpha,\beta)\in A'\times A'}$ with $A' := A \setminus \{\sigma\}$. We declare that the manifolds M and M' are equivalent. More generally we declare that two manifolds N and M are equivalent if one can go from N to M by adding charts to the atlas, or deleting them, keeping the underlying topological space fixed. Logically speaking, a manifold is an equivalence class of manifolds given by families $(U_\alpha)_{\alpha\in A}$ and $(\phi_{\alpha\beta}: U_{\alpha\beta} \to U_{\beta\alpha})_{(\alpha,\beta)\in A\times A}$ that satisfy the above properties 1, 2, 3, 4 and maybe the property 5.

A manifold M is called *orientable* or is *oriented* if M admits a presentation or is presented with families $(U_\alpha)_{\alpha\in A}$ and $(\phi_{\alpha\beta}: U_{\alpha\beta} \to U_{\beta\alpha})_{(\alpha,\beta)\in A\times A}$ such that for all $\alpha, \beta, p \in U_{\alpha\beta}$, the determinant of the differentials $(D\phi_{\alpha\beta})_p: V \to V$ is > 0.

The real projective plane $\mathbb{P}^2(\mathbb{R})$ is an example of a non-orientable manifold. The real projective space $\mathbb{P}^3(\mathbb{R})$ is an orientable manifold.

Example 2.1. Let $Y \subset \mathbb{R}^N$ be a subset of N-dimensional Euclidean space. We assume the existence of a differentiable proper map $F: \mathbb{R}^N \to \mathbb{R}^k$ having $0 \in \mathbb{R}^k$ as regular value with $Y = F^{-1}(0)$. Then using the implicit function theorem, one can find an atlas with differentiable transition maps for the space Y.

Example 2.2. Let $Y \subset \mathbb{R}^N$ be a subset of N-dimensional Euclidean space. We assume the existence of a family $(S_\alpha)_{\alpha\in A}$ of open subsets of $\mathbb{R}^N$ covering Y and a family of regular differentiable maps $(F_\alpha: S_\alpha \to \mathbb{R}^k)_{\alpha\in A}$, proper over 0, with $Y \cap S_\alpha = F_\alpha^{-1}(0)$. Then again using the implicit function theorem, one can find an atlas with differentiable transition maps for the space Y.

Example 2.3. Let M be a manifold given by families

$$(U_\alpha \subset V)_{\alpha \in A} \text{ and } (\phi_{\alpha\beta}: U_{\alpha\beta} \to U_{\beta\alpha})_{(\alpha,\beta) \in A \times A}.$$

We define the manifold TM by families

$$(U_\alpha \times V \subset V \times V)_{\alpha \in A} \text{ and } (T\phi_{\alpha\beta}: U_{\alpha\beta} \times V \to U_{\beta\alpha} \times V)_{(\alpha,\beta) \in A \times A},$$

where $T\phi_{\alpha\beta}$ is the map

$$(p, u) \in U_{\alpha\beta} \times V \mapsto (\phi_{\alpha\beta}(p), (D\phi_{\alpha\beta})_p(u)) \in U_{\beta\alpha} \times V.$$

The manifold TM is called the tangent space of M. It comes with a differentiable map $\pi: TM \to M$ which is induced by the projection maps $\pi_\alpha: U_\alpha \times V \to U_\alpha$. For each point $p \in M$ the fiber $\pi^{-1}(p)$ is a vector space isomorphic to V. The isomorphism of $\pi^{-1}(p)$ to V is not canonical, but depends on the chosen chart $p \in U_\alpha$.

Example 2.4. We define the manifold T^*M by families $(U_\alpha \times V^* \subset V \times V^*)_{\alpha \in A}$ and $(T^*\phi_{\alpha\beta}: U_{\alpha\beta} \times V^* \to U_{\beta\alpha} \times V^*)_{(\alpha,\beta) \in A \times A}$, where $T^*\phi_{\alpha\beta}$ is the map

$$(p, L) \in U_{\alpha\beta} \times V^* \mapsto (\phi_{\alpha\beta}(p), L \circ (D\phi_{\beta\alpha})_p^{-1}) \in U_{\beta\alpha} \times V^*$$

The manifold T^*M is called the co-tangent space of M. Observe that the linear part of the transition map at $p \in U_{\alpha\beta}$ is the transpose of the inverse of $(D\phi_{\alpha\beta})_p$.

The manifolds in Examples 2.1 and 2.2 are sub-manifolds of the Euclidean space $\mathbb{R}^N$. Every manifold with a countable atlas of dimension n is diffeomorphic to a sub-manifold of $\mathbb{R}^{2n+1}$.

Please consider the following exercises:

Exercise 2.5. Describe the real projective plane $\mathbb{P}^2(\mathbb{R})$ as a sub-manifold of $\mathbb{R}^5$.

Exercise 2.6. Let $F: \mathbb{C}^2 \to \mathbb{C}$ be the complex function $F := z_1 z_2$. Show that the manifolds $L := F^{-1}(1)$, $\mathbb{C} \setminus \{0\}$ and $S^1 \times \mathbb{R}$ are diffeomorphic.

Exercise 2.7. The value 1 is a regular value of the determinant function on square matrices with real or complex entries. Conclude that the spaces of $n \times n$ matrices $\mathrm{Sl}(n, \mathbb{R})$ and $\mathrm{Sl}(n, \mathbb{C})$ of determinant 1 are manifolds.

The group operation of multiplication $(A, B) \mapsto AB$ and the inverse map $A \mapsto A^{-1}$ are smooth.

Exercise 2.8. Show that the manifolds $S^1 \times \mathbb{R}^2$, $\mathrm{SL}(2, \mathbb{R})$ and $\mathrm{PSL}(2, \mathbb{R})$ are diffeomorphic.

Exercise 2.9. Show that the manifolds $S^3 \times \mathbb{R}^3$ and $\mathrm{SL}(2, \mathbb{C})$ are diffeomorphic.

Exercise 2.10. Show that the manifolds $\mathrm{SL}(2, \mathbb{C})$ and $\mathrm{PSL}(2, \mathbb{C})$ are not diffeomorphic.

Exercise 2.11. Let $ac, vel \subset \mathbb{R}^2$ be the planar subsets (Fig. 2.3)

$$ac := \{p \in \mathbb{R}^2 \mid x(p) \geq 0 \wedge y(p) \geq 0\},$$

$$vel := \{p \in \mathbb{R}^2 \mid x(p) \geq 0 \vee y(p) > 0\}.$$

The logical symbol "$\wedge$" indicates logical conjunction and the symbol "$\vee$" indicates logical inclusive disjunction. Does there exist a homeomorphism $\phi: \mathbb{R}^2 \to \mathbb{R}^2$ with $\phi(ac) = vel$? Does there exist a diffeomorphism $\phi: \mathbb{R}^2 \to \mathbb{R}^2$ with $\phi(ac) = vel$?

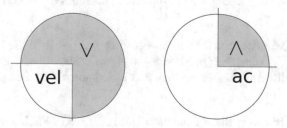

Fig. 2.3 Intersection of the planar regions ac and vel with the unit disk.

Let the manifold M be given by families

$$(U_\alpha \subset V)_{\alpha \in A} \text{ and } (\phi_{\alpha\beta}: U_{\alpha\beta} \to U_{\beta\alpha})_{(\alpha,\beta) \in A \times A}.$$

We assume M to be compact and wish to realize M explicitly as a sub-manifold of a finite-dimensional real vector space W. First we equip V with a scalar product and work with the induced norm $\| \cdot \|_V$ on V. For every point $p \in M$ choose in a chart U_α an open ball $B(p, r_p) \subset U_\alpha$. The open balls $B(p, r_p/2)$ cover M, so let finitely many among them with centers $p_1, p_2, \ldots, p_k$ already cover M. We define mappings $\phi_i: M \to V \times \mathbb{R}$ as follows. For $q = p_i + h \in B(p_i, r_{p_i}/2) \subset U_{\alpha_i} \subset V$ put $\phi_i(q) = (q, 0)$, for $q = p_i + h \in B(p_i, r_{p_i}) \setminus B(p_i, r_{p_i}/2)$ put $\phi_i(q) = (q - \chi(\|h\|_V)/r_{p_i})h, \chi(\|h\|_V)/r_{p_i}))$. Finally for $q \notin B(p_i, r_{p_i})$ put $\phi_i(q) = (p_i, 1)$. Here χ is a C^∞-function, with $\chi(t) = 0, t < 1/2, \chi(t) = 1, t > 1$, χ positive and strictly monotone on $[1/2, 1]$. The image $\phi_i(M)$ is a topological sphere of dimension $\text{Dim}(M) = \text{Dim}(V)$ in $V \times \mathbb{R}$.

Let $\phi: M \to (V \times \mathbb{R})^k$ be the product mapping

$$p \in M \mapsto \phi(p) = (\phi_1(p), \phi_2(p), \ldots, \phi_k(p)) \in (V \times \mathbb{R})^k.$$

The mapping ϕ is injective and its differential at every point is of rank $\text{Dim}(M)$. So it is an embedding of M into a finite-dimensional vector space $W = (V \times \mathbb{R})^k$.

Theorem 2.12. *Let M be a manifold given by a countable family $(U_\alpha \subset V)_{\alpha \in A}$. Then there exists a Morse function $f: M \to \mathbb{R}$.*

The following proof uses an embedding of the manifold into a vector space of finite dimension, so still needs to be completed for non-compact manifolds.

Proof. First embed M by $\phi: M \to W$ into a vector space W as above. For each cube $C \subset U_\alpha$, the set of all $L \in W^*$ such that the restriction to $\phi(C)$ is a Morse function is of full Lebesgue[1] measure in W^*. Since M is covered by countably many cubes, the set of $L \in W^*$ such that the restriction to $\phi(M)$ is a Morse function is of full Lebesgue measure in W^*.   □

Exercise 2.13. Let $p = (p_0, p_1, \ldots, p_n)$ be a configuration of $(n+1)$ labeled points in $\mathbb{R}^2$, $n > 1$. We assume p_0 and p_n are at fixed positions with $x(p_0) = y(p_0) = 0 = y(p_n)$ and $x(p_n) = n - \frac{1}{2}$. Further we assume that the points p_i, $1 \le i \le n - 1$, can move in the plane, only constrained by $(x(p_i) - x(p_{i-1})^2 + (y(p_i) - y(p_{i-1}))^2 = 1$, $1 \le i \le n$. Let P the space of all possible positions of the points p_i. Show that P is a manifold.

Exercise 2.14. (*Exercise 2.13 continued*) Let $f: P \to \mathbb{R}$ be the function defined by $f(p) := \sum_{i=0}^{n}(x(p_i) - x(p_{i+1}))(y(p_i) + y(p_{i+1}))$. Show that f is a Morse function with two critical points. Prove that P is diffeomorphic to the sphere of dimension $n - 2$.

Exercise 2.15. Show that the function Trace: $\mathrm{Sl}(2, \mathbb{R}) \to \mathbb{R}$ is a Morse function with two critical points. Study the Morse index at the critical points.

Exercise 2.16. Show that the function Trace: $\mathrm{Sl}(2, \mathbb{C}) \to \mathbb{C}$ is a complex Morse function with two critical points. Study the Morse index at the critical points of $A \mapsto \mathrm{Re}(\mathrm{Trace}(A))$.

2.2 Fields on Manifolds

Let M be a manifold given by a family $(U_\alpha \subset V)_{\alpha \in A}$ of open subsets of a fixed finite-dimensional vector space V and a family of gluing maps $(\phi_{\alpha\beta}: U_{\alpha\beta} \to U_{\beta\alpha})_{(\alpha,\beta) \in A \times A}$. A family $(X_\alpha: U_\alpha \to V)_{\alpha \in A}$ of vector fields on each U_α defines a vector field on M if for all $(\alpha, \beta) \in A \times A$ the restriction of X_α to $U_{\beta\alpha}$ is transformed by the differential of $\phi_{\beta\alpha}$ to the restriction of X_β to $U_{\alpha\beta}$. If so, the family (X_α) defines a section of $\pi: TM \to M$. We call a map $X: M \to TM$ a *section* of the tangent bundle $\pi: TM \to M$, or also for short a section of TM, if $\pi \circ X$ is the identity map of M.

Sections of TM are called vector fields on M. Sections ω of T^*M are called 1-differential forms on M.

We have, as before, the insertion of a vector field in a 1-differential form ω. We have to check that the value of the function $i_X \omega$ at a point $p \in M$ is independent of the chosen chart U_α containing p. We also check that the differential of a function f on M defines a 1-differential form df on M. The same is true for the Lie bracket of vector fields on M.

Our first definition of $d\omega$ uses the linear structure on U_α, which is not preserved by the maps $\phi_{\beta\alpha}$. Hence it is not immediately clear that the exterior differential is

---

[1] Named after French mathematician Henri Lebesgue (1875–1941).

well defined on a manifold. The alternative definition from Exercise 1.29 helps out and the exterior differential is also well defined on manifolds. Indeed, we can work with the three-term definition:

$$d\omega(X,Y) := X\omega(Y) - Y\omega(X) - \omega([X,Y]_{\text{Lie}}).$$

A new difficulty appears: why does $(d\omega(X,Y))_p$ only depend on the values X_p, Y_p of the vector fields X, Y at the point p?

To answer this question, let X' be another vector field with $X_p = X'_p$. Then the difference vector field $X - X'$ in the chart U_α equals the linear combination $Z = X - X' = a_1 E_1 + a_2 E_2 + \cdots + a_n E_n$ of coordinate vector fields E_i, where the coefficients are functions $a_1, a_2, \ldots, a_n$ having at p a common zero. Hence

$$(d\omega(X,Y))_p - (d\omega(X',Y))_p = (d\omega(X - X',Y))_p = \sum_i (d\omega(a_i E_i, Y))_p.$$

The three terms of the above definition are $\mathbb{R}$-linear in the arguments X and Y. Only one term satisfies a stronger linearity property: namely, for an expression $T(X,Y)$, the property $T(fX,Y) = fT(X,Y)$, where f is not a number, but a smooth real-valued function. Clearly, the first term satisfies this stronger linearity property. The two others don't, but their sum does! Indeed, by using $[fX,Y] = f[X,Y] - (Yf)X$ and $\omega(gX) = g\omega(X)$,

$$\begin{aligned} d\omega(fX,Y) &= fX\omega(Y) - Y\omega(fX) - \omega([fX,Y]) \\ &= fX\omega(Y) - Y(f\omega(X)) - \omega(f[X,Y] - (Yf)X) \\ &= fX\omega(Y) - fY\omega(X) - (Yf)\omega(X) - f\omega([X,Y]) + (Yf)\omega(X) \\ &= fd\omega(X,Y). \end{aligned}$$

In particular at the point p it follows that

$$(d\omega(a_i E_i, Y))_p = a_i(p)(d\omega(E_i, Y))_p = 0.$$

So, $(d\omega(X,Y))_p - (d\omega(X',Y))_p = 0$. In conclusion, $d\omega$ is indeed a 2-differential form.

We give more examples of fields on manifolds. Let M be a manifold given by families $(U_\alpha \subset V)_{\alpha \in A}$ and $(\phi_{\alpha\beta}: U_{\alpha\beta} \to U_{\beta\alpha})_{(\alpha,\beta) \in A \times A}$.

A k-differential form η will be a field given by a family

$$\eta_\alpha: U_\alpha \to \Lambda^k(V)$$

such that the restriction of η_α to $U_{\alpha\beta}$ transforms by $\Lambda^k(\phi_{\alpha\beta})$ to the restriction of η_β to $U_{\beta\alpha}$. Here $\Lambda^k(V)$ is the vector space of all bilinear alternating maps with k arguments from V^k to $\mathbb{R}$. The space of k-differential forms on M is denoted by $\Omega^k(M)$. The *exterior differential* is the map

$$d: \Omega^k(M) \to \Omega^{k+1}(M)$$

defined by

$$d\omega(X_0, X_1, \ldots, X_k) := \sum_{0 \le i \le k} (-1)^i X_i \omega(X_0, \ldots, \hat{X}_i, \ldots, X_k)$$

$$+ \sum_{0 \le i < j \le k} \omega([X_i, X_j], X_0, \ldots, \hat{X}_i, \ldots, \hat{X}_j, \ldots, X_k).$$

Exercise 2.17. Please check $d\omega \in \Omega^{k+1}(M)$. Find a more geometric definition in charts using the linear structure and integration of k-forms over faces of a hypercube.

An almost complex structure J on a manifold M with charts $U_\alpha \subset V$ is a family of fields $J_\alpha : U_\alpha \to J(V)$ such that the restriction of J_α to $U_{\alpha\beta}$ transforms by $J \mapsto \phi_{\alpha\beta *}^* J$ to the restriction of J_β to $U_{\beta\alpha}$. What is $J(V)$? What is $\phi_{\alpha\beta *}^*$? $J(V)$ is the space of all linear maps $J : V \to V$ with $J^2 = -\mathrm{Id}_V$. The transformation $J \mapsto \phi_{\alpha\beta *}^* J$ has to be the only natural one we can think of: for $p \in U_\alpha$ the value $J_{\alpha,p}$ of J_α at p transforms by $(D\phi_{\alpha\beta})_p \circ J_{\alpha,p} \circ (D\phi_{\alpha\beta})_q^{-1}$ to the value $J_{\beta,q}$ at $q = \phi_{\alpha\beta}(p)$ of J_β. We denote by $J(TM)$ the space $J(TM) = \cup_{p \in M} J(T_p M)$. The space $J(TM)$ comes with a natural projection $\pi : J(TM) \to M$. Over a chart U_α observe that $\pi^{-1}(U_\alpha) = U_\alpha \times J(V)$. The gluings organize π as a locally trivial bundle map with typical fiber $J(V)$. So, J-fields on M are sections of the bundle $\pi : J(TM) \to M$.

New operations between fields appear. For instance, an almost complex structure J makes out of a 1-differential form ω a new 1-differential form ω^J by defining $(\omega^J)_p := \omega_p \circ J_p$. For a function f we can formulate in the presence of a J-field a new condition, namely $dd^J f = 0$, where $d^J f := (df)^J = (df) \circ J$. This leads to the very interesting and useful class of J-harmonic functions satisfying $dd^J f = 0$. A closely related class of functions with complex values are the J-holomorphic functions satisfying: $(Df)_p(Jh) = i(Df)_p(h), i := \sqrt{-1}$.

Of special importance is the case of surfaces S with J-field J. The operator $f \mapsto \Delta_J = d(df \circ J)$ is called the J-Laplace operator[2] and maps functions to 2-differential forms (see Section 6.5).

A *Riemannian metric* on M is a family of fields $SP_\alpha : U_\alpha \to SP(V)$ which is naturally compatible with the family of transitions between charts $(\phi_{\alpha\beta} : U_{\alpha\beta} \to U_{\beta\alpha})_{(\alpha,\beta) \in A \times A}$. Here $SP(V)$ is the space of scalar products on V. (Exercise: Describe the natural transition between charts for a Riemannian metric.)

Please consider the following exercises.

Exercise 2.18. $J(V)$ is empty for an odd-dimensional real vector space V. If $\mathrm{Dim}(V) = 2n$ then $J(V)$ is a manifold of dimension $(2n)^2 - 2n^2$.

Exercise 2.19. Let X, Y, A, B be four 2-dimensional sub-vector spaces of a 4-dimensional real vector space V. We assume that each pair of distinct subspaces intersect only in the origin. Construct out of the quadruple X, Y, A, B an automorphism of X. Show that two systems X, Y, A, B and X', Y', A', B' are linearly equivalent if and only if the corresponding automorphisms of X and X' are conjugate.

---

[2] Named after French mathematician, physician and astronomer Pierre-Simon Laplace (1749–1827).

Exercise 2.20. Characterize those quadruples X, Y, A, B of two-dimensional sub-vector spaces in general position in a real 4-dimensional vector space V that are fixed by some $J \in J(V)$. Is this set X of quadruples a sub-manifold of $\text{Grass}_2(V)^4$? Of which dimension? Equivalently, characterize those quadruples X, Y, A, B for which there exists an $\mathbb{R}$-linear map $\phi : V \to \mathbb{C}^2$ such that the images $\phi(X), \phi(Y), \phi(A), \phi(B)$ are $\mathbb{C}$-sub-vector spaces of $\mathbb{C}^2$. Consider the same questions, but now for quintuples X, Y, A, B, C.

Exercise 2.21. Let J be the constant field on $\mathbb{R}^2$ with $J_p = \begin{pmatrix} 0 & 1 \\ -1 & 0 \end{pmatrix}$. Use standard coordinate functions x, y and write the operator $\mathrm{dd}^J f$ explicitly.

Exercise 2.22. Put a differentiable system of charts on the space $\text{Grass}_k(n)$ of all k-dimensional subspaces of $\mathbb{R}^n$.

2.3 Frobenius' Integrability Condition

Let V, B be vector spaces of finite dimension. Suppose that at every point $p = (q, b) \in V \times B$ is given a vector space $G_p \subset T_p(V \times B) \cong T_q V \times T_b B \cong V \times B$ satisfying

(i) $\text{Dim}(G_p) = \text{Dim}(B)$,
(ii) $G_p \cap V \times \{0\} = \{0\}$.

Observe that for every $p \in V \times B$ the direct sum decomposition $T_p(V \times B) = V \oplus G_p$ holds.

A field $G : p \in V \times B \mapsto G_p \subset V \times B$ as above is called a *distribution supplementary to the factor V*.

Let $\pi_p^G : T_p(V \times B) \to T_b V \cong V$ be the linear projection with kernel G_p. Think of π^G as a 1-differential form on $V \times B$ with values in V.

The following global and local questions, which are versions of the so-called *Pfaff Problem*, need answers. Answers were given by Jacobi, Clebsch, Deahna and Frobenius, see [108].

Global question: Given $p = (q, b_0) \in V \times B$, does there exist a smooth map $f : B \to V$ such that $f(b_0) = q$ and $\text{Graph}((Df)_b) = G_{(f(b), b)}$, $b \in B$?

Local question: Given $p = (q, b_0) \in V \times B$, does there exist an open neighborhood U of b in B and a smooth map $f : U \to V$ such that $f(b_0) = q$ and $\text{Graph}((Df)_b) = G_{(f(b), b)}$, $b \in B$?

Global question: Given $p = (q, b_0) \in V \times B$, does there exist a smooth map $F : V \times B \to V$ such that $F(p) = q$ and $\ker(dF) = \ker(\pi^G) = G$?

Local question: Given $p = (q, b_0) \in V \times B$, does there exist an open neighborhood U of p in $V \times B$ and a map $F : U \to V$ such that $F(p) = q$ and $\ker(dF) = \ker(\pi^G) = G$ hold?

For the first global and local questions, the desired map f appears in the condition $\text{Graph}((Df)_b) = G_{(b, f(b))}$, $b \in B$, on both sides of the equality. This question is the most difficult one.

In the last two questions the desired map F appears only on one side. The local sufficient and necessary condition for the existence of $F \colon U \to V$ on a neighborhood of p is the condition $(d\pi^G)_q(u, v) = 0$ for $u, v \in \ker(\pi_q^G)$. The condition does not ask for π^G to be a closed 1-form, but asks that the restriction of $d\pi^G$ to the kernel of π^G vanishes.

The following theorem presents an interesting reformulation of this condition in terms of vector fields with values in the distribution G.

Theorem 2.23 (Jacobi–Clebsch[3]–Deahna[4]–Frobenius Integrability Theorem).
For finite-dimensional vector spaces V, B, let G be a smooth distribution supplementary to the factor V on $V \times B$. Assume the following condition: Lie brackets $[X, Y]_{\mathrm{Lie}}$ of vector fields X, Y on an open subset $U \subset V \times B$, both with values in G, take values in G. Then for every $p_0 = (q_0, b_0) \in U$ there exist an open neighborhood U' of p_0 in U and a smooth map $F \colon U' \to V$ with $\ker(dF) = \ker(\pi^G) = G$ and $F(p_0) = q_0$.

This theorem was published by Deahna in 1840 [28], and was referred to by Frobenius. If the hypothesis of above theorem holds, the distribution G is called *locally integrable* on U. A map $F \colon U' \to V$ is called a *local integral* of G.

Proof. Let $v_0, v_1, \ldots, v_m$ be a basis for V and $e_1, e_2, \ldots, e_n$ a basis for B. Let $z_0, z_1, \ldots, z_m, x_1, x_2, \ldots, x_n$ be the corresponding linear coordinate functions on $V \times B$.

Denote by $X_i = \frac{\partial}{\partial x_i}$ and $Z_k = \frac{\partial}{\partial z_k}$ the corresponding coordinate vector fields on $V \times B$. Denote by $a_{i,k} \colon V \times B \to \mathbb{R}$ the well-defined smooth functions such that the vector field $X_i^G = X_i + \sum_k a_{i,k} Z_k$ takes values in G.

For a piece-wise smooth curve $\gamma \colon [0, 1] \to B$, a G-lift along V is a smooth curve $\gamma^G \colon [0, 1] \to V \times B, t \mapsto \gamma^G(t) = (\lambda^G(t), \gamma(t)) \in V \times B$ with $\lambda^G(t) \in V$ and speed $\dot{\gamma}^G(t) = (\dot{\lambda}^G(t), \dot{\gamma}(t)) \in G$. For every $q \in V$, there exist a unique G-lift $\gamma_q^G = (\lambda_q^G, \gamma)$ along V with $\lambda_q^G(0) = q$.

The following key property relating G-lifts and Lie brackets of the G-lifted coordinate vector fields $[X_i^G, X_j^G]_{\mathrm{Lie}}$ should be compared with the definition of the exterior derivative of a differential 1-form given in Section 1.1.

For $p = (q, b) \in V \times B, s \in \mathbb{R}, s > 0$, define $\gamma_{s,i,j,b} \colon [0, 4] \to B$ to be the closed path that bounds the s-scaled square $P(b, b + se_i, b + se_i + se_j, b + se_j)$. Let $\gamma_{s,i,j,q,b}^G$ be the G-lift of $\gamma_{s,i,j,b}$ with $\gamma_{s,i,j,q,b}^G(0) = (q, b) \in V \times B$. Then $\gamma_{s,i,j,q,b}^G(4) = (q(s), b) \in V \times B$. The following holds:

$$([X_i^G, X_j^G]_{\mathrm{Lie}})_{(q,b)} = \lim_{s \to 0} \frac{q(s) - q}{s^2}.$$

The coordinate value $z_k(q(s) - q)$ is computed as a sum of 4 path integrals along $\gamma_{s,i,j,q,b}^G$ of functions $a_{i,k}$. We spell this out:

---

[3] Rudolf Friedrich Alfred Clebsch (1833–1872), German mathematician.
[4] Heinrich Wilhelm Feodor Deahna (1815–1844), German mathematician.

$$z_k(q(s) - q) = \int_0^1 s a_{i,k}(\gamma^G_{s,i,j,q,b}(t)) dt + \int_1^2 s a_{j,k}(\gamma^G_{s,i,j,q,b}(t)) dt$$

$$+ \int_2^3 s a_{i,k}(\gamma^G_{s,i,j,q,b}(t)) dt + \int_3^4 s a_{j,k}(\gamma^G_{s,i,j,q,b}(t)) dt$$

$$= s \int_0^1 a_{i,k}(\gamma^G_{s,i,j,q,b}(t)) - a_{i,k}(\gamma^G_{s,i,j,q,b}(3 - t)) dt$$

$$+ s \int_1^2 a_{j,k}(\gamma^G_{s,i,j,q,b}(t)) - a_{j,k}(\gamma^G_{s,i,j,q,b}(5 - t)) dt.$$

In particular, if at $p = (q, b)$ the Lie bracket $([X_i^G, X_j^G]_{\text{Lie}})_{(q,b)}$ vanishes, then there exists a constant A such that for $0 < s \leq 1$, $k = 1, 2, \ldots, m$, the estimate

$$|z_k(q(s) - q)| \leq A s^3$$

holds. The constant is controlled by second-order derivatives of the functions $a_{i,k}$. The constant A can be chosen uniformly over compact subsets.

We work with the norm $\|b\|_\Sigma = \sum_i |x_i(b)|$ on B. A zigzag path in B is a path $\gamma \colon [0, L] \to B$ that is continuous, piece-wise differentiable, with speed vectors belonging to $\{\pm e_1, \pm e_2, \cdots, \pm e_n\}$ and of length $\|\gamma(0) - \gamma(L)\|_\Sigma = L$.

For given $b \in B$ let $\gamma_b \colon [0, \|b - b_0\|_\Sigma] \to B$ be the unique continuous path from b_0 to b with norm-speed 1, piece-wise smooth, with constant speed vector $\pm e_1$ on the first piece, $\pm e_2$ on the second piece, $\ldots$, $\pm e_n$ on the last piece. Let $\gamma_b^{G,q}$ be its G-lift along V with $\gamma_b^{G,q}(0) = (q, b_0)$.

Define maps $\lambda^q \colon B \to V \times B$ by

$$b \in B \mapsto \lambda(b) = \gamma_b^{G,q}(\|b\|_\Sigma)) \in V.$$

The definition of γ_b and of λ^q depend on the order of the coordinates $x_1, x_2, \ldots, x_n$. Let $\lambda^{\phi,q}$ be the map obtained after permuting the coordinates by a permutation ϕ. So, $\lambda^{\phi,q}(b)$ is obtained by G-lifting along V the unique continuous path γ_b^ϕ from b_0 to b with norm-speed 1, piece-wise smooth, with constant speed vector $\pm e_{\phi(1)}$ on the first piece, $\pm e_{\phi(2)}$ on the second piece, $\ldots$, $\pm e_{\phi(n)}$ on the last piece to the path $\gamma_b^{G,\phi,q}$ with $\gamma_b^{G,\phi,q}(0) = (q, b_0)$.

For $\epsilon > 0$ define $U' = U_\epsilon = \{(q, b) \in V \times B \mid |z_k(q - q_0)| < \epsilon, \|b\|_\Sigma < \epsilon\}$. Fix $\epsilon > 0$ such that for all $(q, b) \in U_\epsilon$ all G-lifts γ^G with $\gamma^G(0) = q$ of zigzag paths γ_b from b_0 to b of length $\|b - b_0\|_\Sigma$ stay in U. If $U = V \times B$ then the choice $U' = V \times B$ is admissible.

Claim 1: The map $\lambda \colon B \to V$ is differentiable.

Indeed, one can construct f by solving the differential equation of the vector fields X_i^G. Solutions depend smoothly on the initial data, hence the claim.

Claim 2: For fixed $(q, b_0) \in U'$, the restrictions to $U' \cap B$ of the maps $\lambda^{\phi,q}$ coincide.

Indeed, G-lifts in U' of a closed path that bound a coordinate rectangle R of size ρ are closed too. Such a rectangle can be decomposed into N^2 coordinate rectangles, all equal up to translation and of size $\leq \frac{\rho}{N}$. So, since all Lie brackets vanish, there exists a constant A such that the above boundary path integral around a rectangle of the subdivision fails to close up to $\frac{A\rho}{N^3}$. It follows that the boundary path integral over the boundary of R fails to close up to a translation of size $N^2 \frac{A\rho}{N^3} = \frac{A\rho}{N}$. This estimate holds for all N. In conclusion, the G-lift of the boundary path is a closed path too.

Every permutation ϕ is a product of transpositions, so $\lambda^{\phi,q}$ does not depend upon ϕ.

For $(q, b_0) \in U'$ define Λ_q to be the graph of λ^q. For $p = (q, b) \in U'$ define $F(p) \in V \cap U'$ implicitly such that $p \in \Lambda^{F(p)}$.

Claim 3: The map F is smooth of maximal rank.

The value $F(p)$, $p = (q, b) \in U'$, can be computed by G-lifting the path γ_b to γ_b^G with $\gamma_b^G(\|b\|_\Sigma) = p$. Then $F(p) = \gamma_b^G(0) = p$.

Claim 4: $\mathrm{Ker}(DF)_p = G_p$, $p \in U'$.

The level sets of $\{p \in U' \mid F(p) = q\}$ are the graphs Λ^q of $\lambda^{\phi,q}$, for any permutation ϕ. At $p \in U'$, $(D\lambda^{\phi,q})_p((X_{\phi(n)}^G)_p) = 0$. Hence,

$$\mathrm{Ker}((DF)_p) = T_p\Lambda^{F(p)} = \mathrm{Span}\{(X_{\phi(n)}^G)_p\} = G_p. \qquad \square$$

Let M be a differentiable manifold of dimension n. A k-distribution G on M is a map $p \in M \mapsto G_p \subset T_pM$ where G_p is a linear subspace of dimension k in the tangent space T_pM. A k-distribution G is smooth if locally there exist k smooth vector fields such that pointwise G is the span of the values of the vector fields.

Exercise 2.24. Let G be a smooth k-distribution on a manifold M of dimension $n > k$. Show for every point $p \in M$ the existence of an open neighborhood U of p and a system $\alpha = (\alpha_1, \ldots, \alpha_{n-k})$ of smooth 1-differential forms on U such that at $q \in U$ the subspace G_q is the intersection of the kernels of the forms $(\alpha_i)_q$, $i = 1, \ldots, n - k$.

The system of 1-forms α of the previous exercise can be considered as a 1-form on $U \subset M$ with values in $\mathbb{R}^{n-k}$ by defining $\alpha_p : T_pM \to \mathbb{R}^{n-k}$ to be a linear map satisfying $\alpha_p(u) = ((\alpha_{1,p}(u), \ldots, \alpha_{n-k,p}(u)) \in \mathbb{R}^{n-k}$. The kernel of α_p is the intersection of the kernels of $\alpha_{i,p}$.

Exercise 2.25. Let G be a smooth k-distribution on an n-manifold M. Show the equivalence of the two conditions of integrability:

(i) Lie brackets of vector fields on M with values in G are vector fields with values in G.

(ii) For $p \in M$, let the distribution G be defined on an open neighborhood U of p as the kernel of the 1-form $\Omega^1(U, \mathbb{R}^{n-k})$. Then $d\alpha \in \Omega^2(U, \mathbb{R}^{n-k})$ vanishes on $G_{|U} = \mathrm{Ker}(\alpha)$.

Any smooth k-distribution G on an n-manifold is locally transversal to an $n - k$-dimensional submanifold. More precisely, for $p \in M$ there exists a chart $\phi_\alpha : U_\alpha \to \mathbb{R}^n$ with $p \in U_\alpha$ and the distribution $\phi_{\alpha,*}G$ being transverse to $\mathbb{R}^{n-k} \times \{0\} \subset \mathbb{R}^n$.

As a consequence, Theorem 2.23 holds for smooth k-distributions on n-manifolds.

2.4 Foliations on Manifolds

Let E be a topological space. A natural equivalence relation $\sim_{\text{path}}$ on E is generated by its topology as follows. Declare $p, q \in E$ to be equivalent for the relation $\sim_{\text{path}}$ if there exists a continuous path $\gamma\colon [0, 1] \to E$ with values $\gamma(0) = p$ and $\gamma(1) = q$. (Exercise: Show that this declaration does indeed define an equivalence relation.)

If additionally to its topology E carries another structure, then if needed one can impose extra conditions on the paths that generate an equivalence relation. As an example, let E be a differentiable n-manifold M equipped with a smooth k-distribution G. Declare $p, q \in M$ to be equivalent for $\sim_G$ if there exists a piece-wise smooth path $\gamma\colon [0, 1] \to M$ with speed vectors $\dot\gamma(t)$ in G and $\gamma(0) = p$, $\gamma(1) = q$. (Exercise: Show that this declaration also defines an equivalence relation.)

A piece-wise smooth path γ with speed vectors in G is called *G-horizontal*. The equivalence relation $\sim_G$ is called the *horizontal equivalence relation* of the distribution G.

The following are the main examples. Let G_1 be the 2-distribution on $\mathbb{R}^3$ given by the kernels of the 1-form dz and G_2 the 2-distribution on $\mathbb{R}^3$ given by the kernels of $dz + x\,dy$. The equivalence relations $\sim_{G_1}$ and $\sim_{G_2}$ are very different. The distribution G_1 is integrable, the distribution G_2 is not integrable.

Indeed, the G_1-horizontal paths are the paths such that the composition $t \mapsto z(\gamma(t))$ is constant. Hence, the equivalence classes of $\sim_{G_1}$ are the level sets of the function z, i.e. the planes parallel to the X, Y-coordinate plane.

In contrast, the entire $\mathbb{R}^3$ is the only equivalence class for the relation $\sim_{G_2}$! Indeed, $p = (a, b, c) \in \mathbb{R}^3$ is $\sim_{G_2}$-equivalent to $p' = (1, b, c)$ by the linear segment $[p, p']$, the point p' is equivalent to $p'' = (1, b-c, 0)$ by $[p', p'']$ and finally p'' is equivalent to $c = (1, 0, 0)$ by $[p'', c]$.

Exercise 2.26. Consider the space $E_{G,g} = \mathbb{R}^3$ equipped with the 2-distribution $G = G_2$ and the standard Riemannian metric $g(u, v) = x(u)x(v) + y(u)y(v) + z(u)z(v)$. Define a distance function $D_{G,g}$ on E by the infimum

$$D_{G,g}(p, q) = \text{Inf}_\gamma \text{Length}_g(\gamma)$$

taken over all G-horizontal paths γ from p to q. Draw the $D_{G,g}$-ball of radius 4 and center $(0, 0, 0)$. Study the topology of the metric $D_{G,g}$.

Exercise 2.27. Allow distance functions to take the value $+\infty$. Study the metric on $E_{G_1,g}$. Study the topology of the metric $D_{G_1,g}$.

A *k-foliation* F on an n-manifold M is the partition of M by the equivalence classes of the horizontal relation $\sim_G$ of an integrable k-distribution G on M. The equivalence classes of $\sim_G$ are $(n - k)$-dimensional submanifolds of M and are the *leafs* of the foliation F. Through each point $p \in M$ runs a leaf Λ_p.

Be aware, the converse is not true. Consider in $\mathbb{R}^2$ the following submanifolds R_θ and L_h. For $\theta \in \,]0, \pi[$ define $R_\theta = \{(e^t \sin(\theta), e^t \cos(\theta)) \mid t \in \mathbb{R}\}$ and for $h \in \mathbb{R}$, $h \leq 0$, define $L_h = \{p \in \mathbb{R}^2 \mid y(p) = h\}$. Then $\mathbb{R}^2$ is the disjoint union of the submanifolds R_θ and L_h. This partition is not a foliation. The manifold $\mathbb{R}^3$ is the disjoint union of the spheres $S_r = \{p \in \mathbb{R}^3 \mid x(p)^2 + y(p)^2 + z(p)^2 = r\}, r > 1$, and the discs $D_c = \{p \in \mathbb{R}^2 \mid z(p) = c, x(p)^2 + y(p)^2 < 1 - c^2\}$, $-1 < c < 1$. This partition of $\mathbb{R}^3$ by smooth 2-manifolds is not a 2-foliation.

For a k-distribution G on an n-manifold M the property of being integrable is local. The Integrability Theorem (Theorem 2.23) affirms for $p \in M$ the existence of an open neighborhood U of p and a map $F \colon U \to \mathbb{R}^{n-k}$ such that $G_{|U}$ coincides with the kernels of the differential of F. A more careful choice of the neighborhood U and map F allows us to assume that the image $F(U)$ is an open ball in $\mathbb{R}^{n-k}$ and that the preimages $F^{-1}(v)$, $v \in F(U)$, are diffeomorphic to open balls in $\mathbb{R}^k$. Such a choice (U, F) is called a *pile box* for the foliation. A preimage is called a *plate* of the pile box (U, F). If one wishes to remember the local map F we use the term F-plate. For $p \in U$ one has the inclusion of the F-plate in a leaf: $F^{-1}(F(p)) \subset \Lambda_p$. Two points a, b in the same F-plate are clearly in the same leaf. Be aware: the converse is not true, as is shown in the following example. On $M = \mathbb{R}^2 \setminus \{0\}$ with Euclidean norm define the 1-distribution G to be the kernels of the 1-form $\alpha = (y - f)\mathrm{d}x - (x + f)\mathrm{d}y$ for $f(p) = 0$, $\|p\| \le 1$, and $f(p) = \mathrm{e}^{\frac{-1}{\|p\|-1}}$, $\|p\| > 1$. The leafs of the foliation induced by $\sim_G$ are circles C_r with center $0 \in \mathbb{R}^2$ and radius $0 < r \le 1$ and lines L that spiral to the circle C_1. No choice of a pile box (U, F) is possible for $p \in C_1$ such that two points $a, b \in U$ belonging to the same leaf also belong to the same F-plate. The intersection of a spiralling leaf L and U is an infinite union of plates. For all points $p \notin C_1$ such a choice is possible.

Exercise 2.28. On the 3-torus T^3 construct a 2-foliation such that every leaf is dense in T^3. Every leaf intersects every pile box along infinitely many plates.

The multiplicative group $T = \{z \in \mathbb{C} \mid |z| = 1\}$ acts on the sphere $S^3 = \{(u, v) \in \mathbb{C}^2 \mid u\bar{u} + v\bar{v} = 1\}$ by $(\lambda, (u, v)) \mapsto (\lambda^2 u, \lambda^3 v)$. The orbit partition of S^3 is a 1-foliation for which all points $p = (u, v)$ admit a pile box (U, F) such that the intersection, if non-empty, of a leaf Λ with U consists of one, two or three plates.

The local structure of a k-foliation on an n-manifold is rigid and is described by a typical pile box. The more global structure of a foliation is reflected in how pile boxes and leafs intersect each other.

2.5 The Topology of Connected, Compact Surfaces

In this section a surface is defined as a differentiable manifold of dimension two. In particular, we do not study surfaces with boundary. See the treatise of André Gramain [41] for a more general topological study of surfaces. Let S be a compact connected surface. How do we get control on the topology of S? One possibility is the method of slicing: let $f \colon S \to \mathbb{R}$ be a function. Study the level sets $L_t := f^{-1}(t)$, $t \in \mathbb{R}$, and consider how they change by varying $t \in \mathbb{R}$. This method will work particularly well if we already understand the function f locally. Examples of such functions are the Morse functions.

So, let $f \colon S \to \mathbb{R}$ be a Morse function. We assume that different critical points of f have different critical values and call such a Morse function *elementary*. Since S is assumed to be connected and compact, the set of values of f is an interval $[m, M]$.

The set of critical values is finite, which we number in a monotonely increasing way: $m = c_1 < c_2 < \cdots < c_k = M$. Let $p_1, p_2, \ldots, p_k$ be the critical points with $f(p_i) = c_1$. For $t \in \mathbb{R}$, $t < c_1$, the level set $L_t = \emptyset$. For $t = c_1$ the level set L_t is the one point set $\{p_1\}$ where f takes its minimal value. According to the Morse Lemma (Theorem 1.64) locally near p_1 on a neighborhood U_1 of p_1 there exist differentiable coordinates x_1, y_1 with $x_1(p_1) = y_1(p_1) = 0$ and $f_{|U_1} = m + x_1^2 + y_1^2$. It follows that all levels $L_t, t \in \,]m, c_2[$, are diffeomorphic to a circle. What happens at $t = c_2$? Again by the Morse Lemma, locally near the unique critical point p_2 with $f(p_2) = c_2$, on a neighborhood U_2 there exist differentiable coordinates x_2, y_2, with $x_2(p_2) = y_2(p_2) = 0$ and $f_{|U_1} = c_2 \pm x_2^2 \pm y_2^2$. Up to permuting the names of the coordinates x_2, y_2, we have three possibilities.

1. $f_{|U_1} = c_2 - x_2^2 - y_2^2$. In this case, $c_2 = M$ and the surface is diffeomorphic to the sphere S^2.
2. $f_{|U_1} = c_2 - x_2^2 + y_2^2$. In this case the level set L_{c_2} looks like the numeral "8". This is a closed planar curve with one double point where locally two branches cross with transversal tangents. The circle L_t, $t < c_2$, pinches as t converges to c_2. The level sets L_t, $c_2 < t < c_3$, are diffeomorphic to the disjoint union of two circles.
3. $f_{|U_1} = c_2 + x_2^2 + y_2^2$. In this case the L_{c_2} looks like the disjoint union of a circle and a 1-point set. At p_2 the function f has a local minimum. The level sets L_t, $c_2 < t < c_3$, are diffeomorphic to the disjoint union of two circles.

In cases 2. and 3. the slicing continues: at c_3 one of the circles will be pinched like in case 2. at c_2 or will be stopped by a local maximum like in case 1. at c_2 or the two circles will merge to one circle if the critical point p_3 with $f(p_3) = c_3$ is connected in $S_{c_3} := \{p \in S \mid f(p) \le c_3\}$ by a path to p_2 and also by another path to p_1.

Fig. 2.4 Morse transitions up-down.

Using an elementary Morse function $f: S \to \mathbb{R}$ with $k \ge 2$ critical points and values we decompose the surface S into three types of standard pieces. Let $r_1, r_2, \ldots, r_{k-1}$ be regular values with $c_i < r_i < c_{i+1}$, $i = 1, 2, \ldots, k - 1$. The connected components of the complement of the union of the level sets L_{r_i} are open surfaces in S and their closures are closed surfaces "with boundary". We call a closure of a connected component a piece. Such a pieces can be a cylinder, a disk or a pair of pants. We can reduce the number of pieces in the following way. If a disk and a cylinder meet along a common boundary circle, we replace the disk and the cylinder by their union, which is again a disk. So we have one piece less. If a disk and a pair of pants have a common boundary circle, we replace the disk and the pair of pants by their union, which is a new cylinder. So again one piece less. If two cylinders have a common boundary we replace them by their union, which is again a cylinder. So again one piece less. If a cylinder and a pair of pants have one common boundary component, we replace by the union of the cylinder and the pair of pants, which is again a pair of pants. Again one piece less.

This process will stop. We have four possibilities:

1. We have one disk, which touches its boundary from both sides. In this case the surface is the real projective plane $\mathbb{P}^2(\mathbb{R})$ obtained by identifying the points on the boundary circle of the disk by $\theta \sim -\theta$.
2. We have two discs with common boundary. In this case the surface S is obtained by gluing the boundary point θ of the first disk to the boundary point θ of the second disk. The surface S is diffeomorphic to the two-sphere S^2.
3. We have one cylinder, the two boundaries coincide. In this case the surface S is diffeomorphic to the torus $T := S^1 \times S^1$ or to the Klein K bottle $K := [-1, +1] \times S^1/\sim$, where $\sim$ is generated by $(-1, \theta) \sim (+1, \bar{\theta})$.
4. We have an even number $e = 2(g-1)$, $g \geq 2$, of pairs of pants with $3e = 6(g-1)$ boundary components, glued pairwise such that the resulting surface is connected. Up to diffeomorphism we have two possibilities for the resulting surface S: the orientable surface $S_g = S_g^+$ of genus g and the non-orientable surface S_g^-.

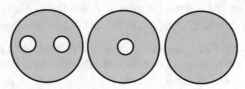

Fig. 2.5 Pairs of pants, cylinder, disk.

With the use of a Morse function one can cut a surface into discs, cylinders and pairs of pants. This can be seen as the beginning of a triangulation of the surface: first introduce two vertices on each of the boundary circles that appear in the decomposition. Each circle gets decomposed into two edges. Introduce in the cylinders 4 more edges and in the pairs of pants 9 more edges without introducing extra vertices. Introduce in the interior of each disk a vertex and two edges. The surface is now triangulated with vertices, edges and triangles. Euler[5] introduced the number

$$e(S) = -\#\text{vertices} + \#\text{edges} - \#\text{triangles}.$$

The number $e(S)$ is independent of the chosen triangulation, so $e(S)$ is a topological invariant of the surface S. It follows that the genus $g(S) = \frac{1}{2}e(S) + 1$ is also a topological invariant. Nowadays one works with the opposite sign convention and defines the Euler characteristic $\chi(S)$ of a surface to be the number $-e(S)$.

The Euler characteristic $\chi(S)$ is very powerful for the problem of classifying surfaces. It is known for which values of e a connected, oriented, compact surface S with Euler characteristic $\chi(S) = e$ exists. Moreover two such surfaces with equal Euler characteristic are diffeomorphic. One says that the Euler characteristic with

---

[5] Leonhard Euler (1707–1783), born in Basel, was a Swiss, Russian, Prussian mathematician, physicist, astronomer, logician and engineer.

values in $\{2, 0, -2, -4, \ldots\}$ is a complete invariant of classification up to homeomorphism or diffeomorphism for connected, oriented, compact surfaces (again without boundary).

For a connected, oriented compact surface S the natural number $g(S) = \frac{1}{2}(2 - \chi(S))$ is called the *genus* of the surface.

Theorem 2.29. *For every even integer $e \leq 2$ or natural number $g = 0, 1, 2, \ldots$ there exists a connected, oriented, compact surface S with Euler characteristic e or with genus g. Two connected, oriented, compact surfaces with equal Euler characteristic or genus are diffeomorphic.* $\qquad\square$

The Greek word γένος was introduced 23 centuries ago by Aristotle. As concept, it emphasizes a property that different objects may have in common. See the recent book *What is the Genus?* by Patrick Popescu-Pampu [97]. For the further study of topology enjoy [31].

Exercise 2.30. Construct a Morse function with 3 critical values on the torus T^2, the two-sphere S^2 and on the surface S_g of genus $g \geq 2$.

Exercise 2.31. Let P be a regular polygonal region in the plane. We assume that P has 10 sides. Let c be a pairing of the sides of P. We glue the sides together by isometries according the pairing c. How many topological types of surfaces do we get?

Exercise 2.32. Assume that there exists an embedding of the real projective space $\mathbb{P}^2(\mathbb{R})$ into $\mathbb{R}^3$. Let F be the image of such an embedding. Construct a compact 1-dimensional manifold C in $\mathbb{R}^3$ that intersects F transversally in one point. For $v \in \mathbb{R}^3$, $v \neq 0$, let $(C_t)_{t \in \mathbb{R}}$ be the family $C_t = C + tv$ of translates of C. Show that there exists a vector v, $v \neq 0$, such that all intersections $C_t \cap F$, $t \in \mathbb{R}$, are transversal or with tangencies of order 1. Conclude that $\#C_T \cap F$, T large, is an odd integer equal to 0. So, the assumption has to be rejected. The real projective plane admits no embedding in $\mathbb{R}^3$.

The real projective plane $\mathbb{P}^2(\mathbb{R})$ is a surface according to our definition, but it is not a surface in the usual, everyday language sense since it is not the *surface of a body*.

Exercise 2.33. Find an embedding of $\mathbb{P}^2(\mathbb{R})$ in the 4-dimensional sphere.

Exercise 2.34. Let M be the set of subsets in S^1 having one or two elements. Define a topology on M such that M is homeomorphic to a differentiable surface with boundary.

2.6 Thoughts

The Euler characteristic is a counting with signs. A Morse function f on a compact surface has a finite number of critical points of three types: At a maximum p the value of f is lowered by moving p a little, at a minimum p the value is augmented by moving p a little, and at a saddle point p moving p a little can lower or augment the value of f. The Euler count $s - m - M$, s being the number of saddle points and m, M the number of minima/maxima, gives the opposite of the (modern version of the) Euler characteristic of the surface.

In societies one tries to measure trends: for instance "optimistic", "pessimistic", "undecided" by $o - p$ and forgets u. Situations which invite a response of "yes", "no", or "doubt" tend to be measured by $y - n$, forgetting the d, which is a counting that systematically forgets those who want to think or think more.

Unfortunately, many countries on today's Earth consider individuals as critical points. This is true especially in the context of power and truth. In a society with only one source of truth, say by a religion or by a dominant, or only, political party, the count is even more restricted. An individual with an opinion that does not question the official truth is counted, those that do question it, but keep silent are not counted, the others are stored in camps and prisons, if not eliminated.

Chapter 3
Hyperbolic Geometry

3.1 The Hyperbolic Plane $\mathbb{H} = \mathbb{H}_I$

A set can be defined by its elements. In my opinion a richer approach is to present
a definition which gives to those elements, also called points, a personality, which
share a culture and interact with other personalities from the same set or from other
sets.

Let us start out with the ring of polynomials in one indeterminate X and with real
coefficients. A common notation for this ring is $\mathbb{R}[X]$. It is endowed with the addition
and multiplication of polynomials. The triple $(\mathbb{R}[X], +, \cdot)$ satisfies the axioms of a
commutative ring with unit element. It is not a field, non-zero polynomials of degree
≥ 1 are not invertible. A natural question is how to compare the ring $\mathbb{R}[X]$ with
fields. More precisely, we study all surjective ring homomorphism from $\mathbb{R}[X]$ to a
field F. So here a personality is a surjective ring homomorphism $p\colon \mathbb{R}[X] \to F$,
where F is any field. We call two such ring homomorphisms $p\colon \mathbb{R}[X] \to F$ and
$p'\colon \mathbb{R}[X] \to F'$ equivalent if there exists a field isomorphism $f\colon F \to F'$ with
$p' = f \circ p$. Let us also assume that for some reason surjective ring homomorphisms
up to equivalence from $\mathbb{R}[X]$ to fields are interesting for our civilization. Observe
that the similar map 上海-Rank: {University} $\to$ {Number} seems to be of interest
nowadays.

Let us start the study of $\mathbb{R}[X]$ following [2]. The first step will be to declare A
to be the set of all surjective ring homomorphisms up to equivalence from $\mathbb{R}[X]$
to some field. It turns out that such a ring homomorphism $p\colon \mathbb{R}[X] \to F$ up to
equivalence is characterized by the kernel of p, which is the ideal

$$I_p := \{P(X) \in \mathbb{R}[X] \mid p(P(X)) = 0\}$$

in the ring $\mathbb{R}[X]$. Not every ideal in $\mathbb{R}[X]$ is the kernel of a surjective ring homomorphism $p\colon \mathbb{R}[X] \to F$ to a field. Those kernels are precisely the ideal equal to the
whole ring $I = \mathbb{R}[X]$ and the maximal proper ideals, i.e. ideals I that are maximal
with respect to inclusion and satisfy $I \neq \mathbb{R}[X]$. Further it turns out that every ideal I
in $\mathbb{R}[X]$ is generated as an ideal by one polynomial $P(X)$. The ideal $(P(X)) \subset \mathbb{R}[X]$
generated by $P(X)$ is the set of all its multiples

© The Author(s), under exclusive license to Springer Nature Switzerland AG 2021
N. A'Campo, *Topological, Differential and Conformal Geometry of Surfaces*, Universitext,
https://doi.org/10.1007/978-3-030-89032-2_3

$$(P(X)) := \{Q(X)P(X) \mid Q(X) \in \mathbb{R}[X]\}.$$

If $1 \in I = \mathbb{R}[X]$ the ring homomorphism takes its values in the field F_1 with only one element. Usually a field is assumed to have at least two elements $0, 1, 0 \neq 1$. The study of the field F_1 was initiated by Jacques Tits in 1956 [120]. If $1 \notin I = (P(X))$, the degree of $P(X)$ exceeds 0. If the degree of $P(X)$ equals $d > 0$, we may replace $P(X) = a_d X^d + a_{d-1} x^{d-1} + \cdots + a_0$ by the monic polynomial $\dfrac{P(X)}{a_d}$ without changing the ideal generated by $P(X)$. The monic generator of an ideal I in $\mathbb{R}[X]$ is uniquely defined. The ideal $I = (P(X))$ will be maximal and proper if the polynomial $P(X)$ is of degree > 0 and $P(X)$ is not the product of two polynomials of lower degree > 0. Now we see that for $P(X)$ monic and generating a maximal ideal not containing $1 \in \mathbb{R}[X]$ there are only two possibilities.

1. The polynomial $P(X)$ is of degree 1: $P(X) = X - a$, $a \in \mathbb{R}$.
2. The polynomial $P(X)$ is of degree 2 and without real zeros: $P(X) = X^2 + bX + c$, $b, c \in \mathbb{R}$, $b^2 - 4c < 0$.

We redefine the set A as the set of maximal ideals, proper or not proper, in the ring $\mathbb{R}[X]$. We can also define the set A as the set of monic polynomials $P(X)$ such that the ideal $(P(X))$ is maximal. We see, using the second redefinition, that the set A is the disjoint union of three sets:

$$A_0 := \{(1)\}, \; A_1 := \{(X - a) \mid a \in \mathbb{R}\},$$

$$A_2 := \{(X^2 + bX + c) \mid b, c \in \mathbb{R}, b^2 - 4c < 0\}.$$

More intrinsically, it turns out that for a surjective ring homomorphism $p: \mathbb{R}[X] \to F$ to a field F the field F can only be isomorphic to the field with only one element F_1, the field of real numbers $\mathbb{R}$ or the field of complex numbers $\mathbb{C}$. Points in A_1 correspond to surjections to the field of the real numbers, as points in A_2 correspond to surjections to a field isomorphic to the field of complex numbers. The unique point in A_0 is denoted by the symbol ∞.

We will begin with the study of personalities and cultures in A_1 and A_2. We have a lot of symmetries. Substituting the indeterminate X by the indeterminate $X - t$, $t \in \mathbb{R}$, defines a map $P(X) \in \mathbb{R}[X] \mapsto P(X - t) \in \mathbb{R}[X]$. We see that the additive group $\mathbb{R}$ of the real numbers acts transitively on A_1 by

$$(t, (X - a) \in \mathbb{R} \times A_1 \mapsto (X - a - t) \in A_1.$$

The multiplicative group $\mathbb{R}^*$ of real numbers ($\neq 0$) acts also by substituting for the indeterminate X the indeterminate λX as

$$(\lambda, (P(X)) = (\lambda, (X - a)) \mapsto (Q(X)) = \left(X - \frac{a}{\lambda}\right).$$

The first action suggests that the notion of "distance" $d((X - a), (X - b)) := |a - b|$ is invariant under the natural symmetries of A_1. The second action shows that this notion of distance is not invariant, hence not natural.

Is there some other *structure=culture* in A_1? The answer is "yes"! We can define the symmetry invariant 3-point function of Michel Chasles[1]

$$CH: A_1 \times A_1 \times A_1 \setminus \text{first diagonal} \to \mathbb{R}.$$

Let $(P(X)), (Q(X)), (R(X))$ be three different points in A_1. We assume that the three ideals are given by monic generators $P(X) = X - a$, $Q(X) = X - b$, $R(X) = X - c$. We define

$$CH((P(X)), (Q(X)), (R(X)) := \frac{c - a}{b - a}.$$

The three-point function CH is natural, i.e. the function CH is invariant under the above actions of the additive and multiplicative groups of real numbers on $A_1 \times A_1 \times A_1 \setminus$ first diagonal.

Exercise 3.1. Let A_1 be equipped with the coarsest topology such that the map CH is continuous. Is A_1 homeomorphic to $\mathbb{R}$? What other natural structures does A_1 have? The structure of a differentiable manifold? Hint: A_1 is called the real affine line $\mathbb{A}^1(\mathbb{R})$.

Exercise 3.2. Determine the group of all bijections $\phi: A_1 \to A_1$ that preserve the structure given by CH.

The union $A_0 \cup A_1 = \mathbb{R} \cup \{\infty\}$ is a model of the real projective line $\mathbb{P}^1(\mathbb{R})$. Indeed, let V be a real vector space of dimension 2, let L be a line in V not passing through the origin. The line L is a model of the real affine line and can be identified with $A_1 = L$. Let l_∞ be the one-dimensional sub-vector space in V that does not intersect L. The map $\lambda: \mathbb{P}(V) \to A_0 \cup A_1$ given by $l \in \mathbb{P}(V) \mapsto \lambda(l) \in A_0 \cup A_1$, $\{\lambda(l)\} = l \cap L$ if $l \neq l_\infty$ and $\lambda(l_\infty) = \infty$ realizes the claimed identification. The group $\mathrm{PGL}(2, \mathbb{R})$ is the group of projective automorphisms of $\mathbb{P}^1(\mathbb{R})$. This group acts simply transitively on triples (p, q, r) of distinct points in $\mathbb{P}^1(\mathbb{R})$, hence has no invariant triple point like CH on the real affine line $\mathbb{A}^1(\mathbb{R})$.

What about A_2? Again substitutions of $X - t$ for X, $t \in \mathbb{R}$, give symmetries. At High School we learned that $(P(X)) \in A_2$, $P(X) = X^2 + bX + c$ transforms to $(Q(X))$ with the substitution $t = \frac{-b}{2}$ to $Q(X) = X^2 + (\frac{b^2}{4} + c)$. The substitution of λX for X with $\lambda = \sqrt{\frac{b^2}{4} + c}$ will transform the ideal $(Q(X))$ to the ideal $(X^2 + 1)$. This shows that the space A_2 is homogeneous. Moreover, the point $(X^2 + 1) \in A_2$ is the unique fixed point of the composition σ of the substitution $\frac{-1}{X}$ for X followed by multiplication by X^2. The composition σ is an involution that transforms $(X^2 + bX + c) \in A_2$ to $(cX^2 - bX + 1) = (X^2 - \frac{b}{\sqrt{c}}X + \frac{1}{c}) \in A_2$.

Do we have a more natural structure? Natural means invariant under the above symmetries. The following shows that there exists a natural two-point function:

$$CR: A_2 \times A_2 \to \mathbb{R}.$$

For $((P(X), (Q(X)) \in A_2 \times A_2$ given by monic polynomials $P(X) \neq Q(X)$ let $I_{P(X), Q(X)}$ be the set $\{t \in \mathbb{R} \mid tP(s) + (1 - t)Q(s) > 0, s \in \mathbb{R}\}$. The set $I_{P(X), Q(X)}$

---

[1] Michel Chasles (1793–1880), French mathematician.

is an interval $I_{P(X),Q(X)} =]t^-, t^+[$. Possibly we have $t^- = -\infty$, $t^+ = +\infty$. We define

$$CR((P(X)),(Q(X))) := \operatorname{cr}(t^-, 0, 1, t^+) := \frac{(1-t^-)(t^+ - 0)}{(0-t^-)(t^+ - 1)},$$

where the following expression $\operatorname{cr}(a, b, c, d)$ is used that involves four distinct numbers $a, b, c, d \in \mathbb{R} \cup \{\infty\}$ and applies the rules $\frac{\infty}{\infty} = 1$ and $\pm\infty + t = \infty$, $t \in \mathbb{R}$,

$$\operatorname{cr}(a, b, c, d) := \frac{(c-a)(d-b)}{(b-a)(d-c)}.$$

This expression is well known in geometry and was already used by Menelaus of Alexandria and Pappus of Alexandria, Greek mathematicians working more than 1900 years ago. It can also be written as a ratio of ratios: $\frac{c-a}{b-a}/\frac{d-c}{d-b}$, and is called the $\Delta\iota\pi\lambda\delta\varsigma$ $\lambda\delta\gamma o\varsigma$, *cross-ratio*, *birapport*, 交比, *Doppelverhältnis* or 非和比 of a, b, c, d. The cross-ratio plays a central role in geometry, vision, image recognition, finance, sociology and many other activities. In fact, the function cr defines a 4-point function on the real projective line $\mathbb{R} \cup \{\infty\} = \mathbb{P}^1(\mathbb{R})$. In their book on Menelaus' Spherics [100, pages 340, 357, 360], Roshdi Rashed and Athanase Papadopoulos study properties of invariance of the cross-ratio in relation to the Theorem "Figure Secteur".

The structure on A_2 that comes from the map CR is very rich. We first change the name of the set A_2 to $\mathbb{H}_I$. Remember $\mathbb{H}$ stands for hyperbolic, and I for ideal. In short we set $\mathbb{H} = \mathbb{H}_I$. The set $\mathbb{H}$ with all the structure that we will discover soon is a model of the *hyperbolic plane*.

Theorem 3.3 (Hyperbolic Distance in $\mathbb{H} = \mathbb{H}_I$). *The two-point-function* $d_{\mathbb{H}} \colon \mathbb{H} \times \mathbb{H} \to \mathbb{R}$

$$d_{\mathbb{H}}((P(X),(Q(X))) := \frac{1}{2}\log(CR((P(X),(Q(X))))$$

is a distance function on $\mathbb{H}$.

For distinct points $p = (P(X))$, $q = (Q(X))$ in $\mathbb{H}$, $(P(X)) \neq (Q(X))$ given by monic polynomials $P(X)$ and $Q(X)$, we define the *hyperbolic line* $L_{pq} = L_{P(X),Q(X)} \subset \mathbb{H}$ through $p = (P(X))$, $q = (Q(X))$ as the subset

$$L_{pq} = L_{P(X)Q(X)} := \{(t(P(X) + (1-t)Q(X)) \in \mathbb{H} \mid t \in I_{P(X),Q(X)}\}.$$

The *hyperbolic segment* $[(P(X)),(Q(X))]_{\mathbb{H}}$ is the subset

$$[(P(X)),(Q(X))]_{\mathbb{H}} := \{(t(P(X) + (1-t)Q(X)) \in \mathbb{H} \mid t \in [0,1]\}.$$

For hyperbolic lines $L_{pq} = L_{P(X)Q(X)}$ with $P(X) = X^2 + bX + c$, $Q(X) = X^2 + b'X + c'$ and $b \neq b'$ there exist $i_{pq}, i_{qp} \in \mathbb{R}$ such that $L_{pq} = \{(t(X - i_{pq})^2 + (1-t)(X - i_{qp})^2) \mid t \in]0,1[\}$. Indeed let $i_{pq}, i_{qp} \in \mathbb{R}$ be the numbers such that $(X - i_{pq})^2 = t_-(p,q)P(X) + (1 - t_-(p,q))Q(X)$ and $(X - i_{qp})^2 = t_+(p,q)P(X) + (1-t_+(p,q))Q(X)$. Putting $L_{i_{pq}i_{qp}} = \{(t(X-i_{pq})^2 + (1-t)(X - i_{qp})^2) \mid t \in]0,1[\}$, we have $L_{i_{pq}i_{qp}} = L_{pq}$.

If instead of $b \neq b'$ one has $b = b'$ and $c < c'$, put $i_{pq} = \frac{b}{2}$ and $i_{qp} = \infty$ with

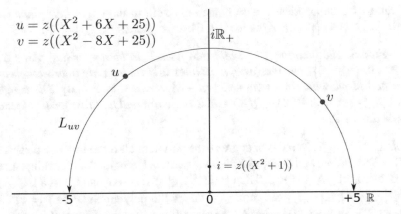

Fig. 3.1 The hyperbolic line L_{uv} in the upper half plane $\mathbb{C}_+$ through $u = z((X^2 - 6X + 25))$ and $v = z((X^2 - 8X + 25))$.

$$L_{i_{pq}\infty} = L_{i_{pq}i_{qp}} = \{(t(X - i_{pq})^2 + (1 - t)(0X - 1)^2) \mid t \in \,]0, 1[\}.$$

Again $L_{i_{pq}i_{qp}} = L_{pq}$ holds. The points $i_{pq}, i_{qp} \in A_1 \cup \{\infty\}$ are the *ideal extremities* of the line L_{pq}.

Observe that for any pair of distinct points p, q in $A_2 \cup A_1 \cup \{\infty\}$ there exists a unique line L_{pq} in $A_2 = \mathbb{H}_I$ passing through or having as ideal extremity the points p and q.

The following is surprising. We re-discover a new geometry, Hyperbolic Geometry,[2] that is very young compared to Euclidean Geometry.

Theorem 3.4 ($\mathbb{H}_I$ as model of the Hyperbolic Plane). *The geometry on $\mathbb{H}_I$ defined by the hyperbolic lines and by the metric $d_{\mathbb{H}}$ satisfies the axioms I, II, III, IV of Euclid and the axiom V': given a line $L \subset \mathbb{H}$ and a point $p \in \mathbb{H}$, $p \notin L$, there exist more than one line in $\mathbb{H}$ through p that does not intersect the line L.*

In the Euclidean plane, a figure consisting of a triple of lines A, B, C with $A \neq B$, $A \cap B \neq \emptyset$, $A \cap C = B \cap C = \emptyset$ does not exist, but in the hyperbolic plane $\mathbb{H}$ this can occur.

Proof. That the axioms I, II, III, IV hold in $\mathbb{H}$ will follow from the forthcoming study of triangles and trigonometry in $\mathbb{H}$. The symmetries of $\mathbb{H}_I$, see below, provide the needed motions.

For a triple of pairwise distinct points $p, q, r \in \mathbb{H}$ with $r \notin L_{pq}$ the two distinct lines $L_{ri_{pq}}, L_{ri_{qp}}$ pass through r, and are disjoint from L_{pq}. Hence V' holds. $\qquad\square$

On $\mathbb{H} = \mathbb{H}_I$ we define a coordinate function $z \colon \mathbb{H} \to \mathbb{C}$ by

$$(P(X)) = (X^2 + bX + c) \mapsto z((P(X))) = \frac{-b + i\sqrt{-b^2 + 4c}}{2}.$$

[2] Discovered by Russian mathematician Николай Иванович Лобачевский, Nikolai Ivanovich Lobachevsky (1792–1856) and Hungarian mathematician János Bolyai (1802–1860).

The value of z on an ideal $I \in \mathbb{H}$ is the only zero in the *upper half plane* $\mathbb{C}_+ :=$ $\{a + bi \mid a, b \in \mathbb{R}, \ b > 0\}$ of the ideal I. Observe: $\mathrm{Re}(z((P(X))) = \frac{-b}{2}$.

Theorem 3.5. *The function z is a bijection from $\mathbb{H}$ to the upper half plane $\mathbb{C}_+ :=$ $\{u \in \mathbb{C} \mid \mathrm{Im}(u) > 0\}$. The images of most lines in $\mathbb{H}$ under z are intersections of the upper half plane with circles $\{u \in \mathbb{C} \mid |u - a| = r, \ a \in \mathbb{R}, \ r > 0\}$. The images of lines $L_{P(X)Q(X)}$ with $P(X) - Q(X) \in \mathbb{R}$ under z are real half rays perpendicular to the real line in $\mathbb{C}_+$.*

Proof. If $P(X) - Q(X) \neq 0$ is of degree 0 the polynomials differ by a real constant, so $P(X) = X^2 + bX + c$ and $Q(X) = X^2 + bX + c + d$. Without loss of generality, assume $d < 0$, therefore $-c < d < 0$. Then $t^+(P(X), Q(X)) = +\infty$ and $t^-(P(X), Q(X)) = \frac{b^2 - 4c - 4d}{4d}$. For $t \in]t^-, t^+[$ the root $z_t \in \mathbb{C}_+$ of the polynomial $P_t(X) := tP(X) + (1 - t)Q(X) = (X + b/2)^2 - b^2/4 + c + d - td$ runs through the half line in $\mathbb{C}_+$ with $\mathrm{Re}(z_t) = -b/2$. Indeed the term $-b^2/4 + c + d - td$ is positive for $t > t^-$ and vanishes at $t = t^-$.

If $P(X) - Q(X) \neq 0$ is of degree 1 then $C = \frac{-P(0) + Q(0)}{P'(0) - Q'(0)} \in \mathbb{R}$ satisfies $P(C) = Q(C) > 0$. For $t \in [t^-, t^+]$ we verify the equality $|z(P_t(X)) - C|^2 = P(C)$ by observing:

$$P_t(X + C) = X^2 + m_t X + P(C),$$
$$z(P_t(X + C)) = z(P_t(X)) - C,$$
$$4P(C) - m_t^2 \geq 0,$$

$$|z(P_t(X)) - C|^2 = |z(P_t(X + C))|^2 = \left| \frac{-m_t + i\sqrt{4P(C) - m_t^2}}{2} \right|^2 = P(C).$$

It follows that the z-image of the line $L_{P(X)Q(X)}$ is the half circle in $\mathbb{C}_+$ with center $C \in \mathbb{R}$ and radius $\sqrt{P(C)}$. $\qquad\square$

Let $\Pi : \mathbb{C}_+ \to \mathbb{H}$ be the inverse bijection:

$$u \in \mathbb{C}_+ \mapsto \Pi(u) = ((X - u)(X - \bar{u})) \in \mathbb{H}.$$

We wish to understand natural symmetries of $\mathbb{H}$. For instance the substitution $X - t$ for X, $t \in \mathbb{R}$, corresponds via z to the translation $u \mapsto u + t$ on $\mathbb{C}_+$. The substitution $\frac{X}{\lambda}$ for X, $\lambda > 0$, corresponds via z to the stretching $u \mapsto \lambda u$ on $\mathbb{C}_+$.

The following is a hidden symmetry, hidden if one only thinks in terms of polynomials. It is the symmetry $(P(X)) \mapsto (X^2 P(-1/X))$ of $\mathbb{H}$ which fixes the ideal $(X^2 + 1) \in \mathbb{H}$. This symmetry corresponds via the coordinate z to the holomorphic symmetry $u \mapsto \frac{-1}{u}$ of $\mathbb{C}_+$, which is an involution and fixes $i \in \mathbb{C}_+$.

The coefficient "b" of $P(X) = X^2 + bX + c$ equals the derivative $P'(0)$ evaluated at 0. The map $P(X) \mapsto P(X) - 2P'(0)X$ is a symmetry of the 2-point function CR on $\mathbb{H}$, which fixes the line $\{(X^2 + c) \mid c > 0\} \subset \mathbb{H}$ pointwise. This symmetry corresponds via the coordinate z to the anti-holomorphic involution $u \mapsto -\bar{u}$. This symmetry does not preserve any orientation of $\mathbb{H}$.

Theorem 3.6. *The symmetries of substituting $X - t$ for X, $t \in \mathbb{R}$, $\dfrac{X}{\lambda}$ for X, $\lambda \in \mathbb{R}^*$, and the symmetries $(P(X)) \mapsto (X^2 P(-1/X))$ and $(P(X) \mapsto ((X) - 2P'(0)X))$ described above, generate the group of symmetries of the map $CR \colon \mathbb{H} \times \mathbb{H} \to \mathbb{R}$.*

We call symmetries of the map $CR \colon \mathbb{H} \times \mathbb{H} \to \mathbb{R}$ symmetries of the geometric space $\mathbb{H}$, which are precisely the isometries of the metric $d_{\mathbb{H}}$.

Theorem 3.7. *The group of orientation-preserving automorphisms of $\mathbb{H}$ is isomorphic to the group $\mathrm{PSL}(2, \mathbb{R})$. Its action on $\mathbb{H}$ corresponds via the coordinate $z \colon \mathbb{H} \to \mathbb{C}_+$ to the modular action $u \mapsto \dfrac{au + b}{cu + d}$ on $\mathbb{C}_+$.*

A point $p \in A_2 = \mathbb{H}_I$ defines an involution θ_p of $A_0 \cup A_1 = \mathbb{P}^1(\mathbb{R})$ in the following way: for $l \in A_0 \cup A_1$ define $\theta_p(l) \in A_0 \cup A_1$ by requiring that $l \neq \theta_p(l)$ and that the hyperbolic line in A_2 with end points $l, \theta_p(l)$ passes through p. This construction, as explained to me by Jacques Tits, leads to the following model $\mathbb{H}_\theta$ of the hyperbolic plane: The hyperbolic plane is the space of all involutions $\theta \colon \mathbb{P}^1(\mathbb{R}) \to \mathbb{P}^1(\mathbb{R})$ of the real projective line $\mathbb{P}^1(\mathbb{R})$ that preserve the cross-ratio cr, i.e. the 4-point function, and that act fixed-point-free. Clearly it follows that the groups of automorphisms of the hyperbolic planes $\mathbb{H} = \mathbb{H}_I = \mathbb{H}_\theta$ are all isomorphic to the group of automorphisms $\mathrm{PGL}(2, \mathbb{R})$ of the real projective line $\mathbb{P}^1(\mathbb{R})$.

The structure of $\mathbb{H}$ is really very rich, for instance:

Theorem 3.8. *Let Hess_p be the Hessian at p of the function $q \in \mathbb{H} \mapsto d_{\mathbb{H}}^2(p, q)$. The field $p \mapsto \dfrac{1}{2}\mathrm{Hess}_p$ defines a Riemannian metric $\rho = \rho_{\mathbb{H}}$ on $\mathbb{H}$. The Riemannian metric ρ is isotropically homogeneous.*

Proof. The coordinate $z \colon \mathbb{H} \to \mathbb{C}_+$ is a bijection. The transported Riemannian metric $z_* \rho$ on the upper half plane $\mathbb{C}_+$ evaluates at $q \in \mathbb{C}_+$ and the tangent vectors $u, v \in T_q \mathbb{C}_+ = \mathbb{C}$ to $\rho_{\mathbb{C}_+}(u, v) = \frac{\mathrm{Re}(u\bar{v})}{\mathrm{Im}^2(q)}$. The modular action is isotropically homogeneous by isometries, hence so is the action of $\mathrm{PSL}(2, \mathbb{R})$ on $\mathbb{H}$. The group $\mathrm{PSL}(2, \mathbb{R})$ realizes by the modular action the group of conformal automorphisms of $\mathbb{C}_+$, so is isomorphic via the isometry z to the group of orientation-preserving isometries of $\mathbb{H}$. $\qquad \square$

The coordinate $z \colon \mathbb{H} \to \mathbb{C}_+$ transports the Riemannian metric $\rho_{\mathbb{H}}$ to the metric $\rho_{\mathbb{C}_+}$, which is conformal to the Euclidean metric of the Gaussian plane $\mathbb{C}$. The length $\|h\|$ of a tangent vector $h \in T_p \mathbb{H}$ is given by $\|h\| = \frac{|(dz)_p(h)|}{\mathrm{Im}(z(p))}$. By integration one can get a closed formula for the distance between points $p, q \in \mathbb{H}$. The proof of the following theorem is not by integration.

Theorem 3.9.

$$d_{\mathbb{H}}(p, q) = \mathrm{arcCosh}\left(1 + \frac{|z(p) - z(q)|^2}{2\mathrm{Im}(z(p))\mathrm{Im}(z(q))}\right).$$

Proof. The following proof uses the group of isometries of $(\mathbb{H}, d_{\mathbb{H}})$. Both sides are invariant under the above four basic symmetries of $\mathbb{H} = \mathbb{H}_I$. Moreover, we claim that the formula holds for $p = (X^2 + 1)$ and $q_t = (t(X-1)^2 + (1-t)(X+1)^2)$, $t \in [\frac{1}{2}, 1[$. Notice that $t \in]0, 1[\mapsto q_t \in \mathbb{H}_I$ is a line, moreover $p = q_{\frac{1}{2}}$, $q_0, q_1 \notin \mathbb{H}$ and

$$z(p) = i, \; z(q) = (-1 + 2t) + i\sqrt{1 - (1 - 2t)^2}.$$

So by equivariance and isotropical transitivity the formula will be proved. For the claim observe that

$$\text{LHS} = d_{\mathbb{H}}(p, q_t) = \frac{1}{2} \log \left(cr \left(0, \frac{1}{2}, t, 1 \right) \right) = \frac{1}{2} \log \left(\frac{t}{1-t} \right)$$

and

$$\text{RHS} = \text{arcCosh} \left(1 + \frac{1 - \sqrt{1 - (1 - 2t)^2}}{\sqrt{1 - (1 - 2t)^2}} \right) = \text{arcCosh} \left(\frac{1}{\sqrt{1 - (1 - 2t)^2}} \right).$$

Further observe that

$$\cosh^2(\text{LHS}) = \frac{1}{4t(1-t)}, \; \cosh^2(\text{RHS}) = \frac{1}{1 - (1 - 2t)^2} = \frac{1}{4t(1-t)}.$$

Hence $\cosh^2(\text{LHS}) = \cosh^2(\text{RHS})$, which finishes the proof since cosh is a strictly monotone positive function on the positive reals. □

Theorem 3.10. *Let $J^{\pm}$ be the continuous fields on $\mathbb{H}$ of turning by $\pm\pi/2$ corresponding to the Riemannian metric ρ of the previous theorem. These fields correspond via the coordinate $z \colon \mathbb{H} \to \mathbb{C}$ to the constant J-fields $m_{\pm i} \colon h \mapsto \pm ih$ on $\mathbb{C}_+$.* □

So, the m_i-field that we have in $\mathbb{C}_+$ corresponds via the coordinate z to a natural J-field on $\mathbb{H}$. We say that the coordinate z is holomorphic. The same holds for a variant of the coordinate z. The coordinate

$$w \colon \mathbb{H} \to \mathbb{D}, \; w = \frac{z-i}{z+i}$$

maps $\mathbb{H}$ to the open unit disk $\mathbb{D} = \{v \in \mathbb{C} \mid |v| < 0\}$, which in this role is called the Poincaré disk. Most lines L in $\mathbb{H}$ are mapped by the coordinate w to the intersection of the circles

$$C_c = \{v \in \mathbb{C} \mid |v - c| = r, \; r^2 = (|c| - 1)(|c| + 1), \; |c| > 1\}$$

with $\mathbb{D}$. Here, the exceptions are the lines in $\mathbb{H}$ through $(X^2 + 1)$ which are mapped by w to straight diameters of the disk $\mathbb{D}$. The circles C_c have their center c outside the closure of $\mathbb{D}$ and meet orthogonally the boundary of the closure of $\mathbb{D}$. Circles with "centers" at ∞ correspond to the exceptions. From the coordinate w on $\mathbb{D}$ we can retrieve the coordinate z on $\mathbb{H}$ by

$$z = -i \frac{w+1}{w-1}.$$

The spaces $A_0 \cup A_1 = \mathbb{P}^1(\mathbb{R})$ and $A_2 = \mathbb{H}_I = \mathbb{H}_\theta$ carry natural topologies. The following properties define a topology T on the union $\bar{\mathbb{H}} = A_0 \cup A_1 \cup A_2$. The first property requires that $A_0 \cup A_1$ is a closed subspace of $(\bar{\mathbb{H}}, T)$ and that the induced topology is the natural topology of $\mathbb{P}^1(\mathbb{R})$. The second property requires that A_2 is open in $(\bar{\mathbb{H}}, T)$ and that the induced topology is the topology of $\mathbb{H}_I$. The third property requires that the only sequences in $(p_n)_{n \in \mathbb{N}} \in \mathbb{H}_I$ with limit $l \in A_0 \cup A_2$ are those sequences that do not have accumulation points in $\mathbb{H}_I$. Moreover, it is required that there exists a sequence of lines $(L_n)_{n \in \mathbb{N}}$ with $p_n \in L_n$ and with endpoints $a_n, b_n \in A_1 \subset \mathbb{P}^1(\mathbb{R})$ such that $\lim_{n \to \infty} a_n = l$, $\lim_{n \to \infty} b_n = l$.

The coordinate $w \colon A_2 = \mathbb{R}_I \to \mathbb{D}$ extends to a homeomorphism $\bar{w} \colon \bar{\mathbb{H}} \to \bar{\mathbb{D}}$. Here $\bar{\mathbb{D}}$ is the closure of the open unit disk $\mathbb{D}$ in $\mathbb{C}$. It follows that the space $\bar{\mathbb{H}} = (A_0 \cup A_1 \cup A_2, T)$ is a compactification of the space $\mathbb{H}$.

The closed disk $\bar{\mathbb{D}}$ is a compact, convex subspace with non-empty interior in $\mathbb{C}$. Its interior $\mathbb{D}$ inherits a metric by the following construction.

Let $K \subset \mathbb{R}^n$ be a compact convex set with non-empty interior K°. The Hilbert metric d_K on K° is the two-point-function defined for $p, q \in K^\circ$, $p \neq q$, by

$$d_K(p, q) = \frac{1}{2} \log(\mathrm{cr}(P, p, q, Q)),$$

where $\mathrm{cr}(P, p, q, Q)$ is the cross-ratio of the four points P, p, q, Q on the line $L(p, q)$ through p and q, where P, Q are the boundary points of the interval $L(p, q) \cap K$ with p in between P and q.

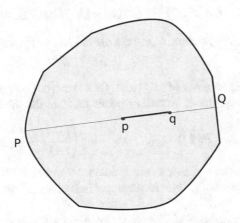

Fig. 3.2 The Hilbert metric $d_K(p, q) = \frac{1}{2} \log \left(\frac{(q-P)(Q-p)}{(p-P)(Q-q)} \right)$ on K°.

The space $\mathbb{D}$ with its Hilbert metric $d_{\mathbb{D}}$ is isometric to the hyperbolic plane $\mathbb{H}$ and hence provides a further model $\mathbb{K}$ of the hyperbolic plane which is due to Felix Klein. The map between the geometries of the Poincaré disk $\mathbb{D}$ and the Kleinian disk $\mathbb{K}$ is made visible in figure 246 on page 225 of the second edition [51] of *Anschauliche Geometrie*. Interestingly, the lines in this Kleinian model are just the straight chords. The price for this advantage is that the measure of angles differs from the Euclidean measure of angles, which is in contrast to the Poincaré model $\mathbb{D}$.

Exercise 3.11. Show that d_K defines a metric on K°.

Exercise 3.12. Re-prove the Perron–Frobenius Theorem, see Exercise 1.40, by showing that the self map $\bar{A}$ on the simplex K that is spanned by the standard basis of $\mathbb{R}^n$ is contracting with respect to the Hilbert metric on K°.

The Hilbert metric

$$d_{\mathrm{Hilbert}}^I(a, b) = \frac{1}{2}\left|\log\left(\frac{(b - t^-)(t^+ - a)}{(a - t^-)(t^+ - b)}\right)\right|, \quad a, b \in \,]t^-, t^+[$$

on the interior I° of an interval $I = [t^-, t^+]$ defines an isometric parametrization of I° by the real line $\mathbb{R}$ such that translations of the parameter become d_I-isometries:

Theorem 3.13. *The lines in $\mathbb{H}_I$ are precisely the geodesics for the metric ρ on $\mathbb{H}_I$. Moreover, the parametrization*

$$t \in \,]t^-, t^+[\,\mapsto (tP(X) + (1 - t)Q(X))$$

of the line $L_{P(X)Q(X)}$ is geodesic if one uses on the interval $]t^-, t^+[$ its Hilbert metric $d_{\mathrm{Hilbert}}^{t^- t^+}(a, b)$. $\qquad\square$

The distance in $\mathbb{H}_I$ between $(P(X))$ and $(Q(X))$ can be computed as follows. First compute the lower and upper boundary points t^-, t^+ of the interval $I_{(P(X)),(Q(X))}$. Then

$$d_{\mathbb{H}}((P(X)), (Q(X))) = \frac{1}{2}\log\left(\frac{(1 - t^-)t^+}{t^-(1 - t^+)}\right). \tag{3.1}$$

The coordinate functions z and w are holomorphic, which allows us to measure angles of sectors bounded by lines in the upper half plane $\mathbb{C}_+$ or in the Poincaré disk $\mathbb{D}$.

We compute a few examples. Consider $(1 + h)i$, $i \in \mathbb{C}_+$, $h \in \,] - 1, \infty[$. We have, recalling $\Pi(z) = ((X - z)(X - \bar{z}))$,

$$d_{\mathbb{C}_+}((1 + h)i, i) = d_{\mathbb{H}}(\Pi((1 + h)i), \Pi(i))$$
$$= d_{\mathbb{H}}((X^2 + (1 + h)^2), (X^2 + 1)).$$

We have, if $h > 0$, $t^- = \dfrac{-1}{(1 + h)^2 - 1}$, $t^+ = +\infty$. Hence

$$d_{\mathbb{C}_+}((1+h)i, i) = \frac{1}{2} \log \left(\frac{1 + \dfrac{1}{(1+h)^2 - 1}}{\dfrac{1}{(1+h)^2 - 1}} \right) = \log(1+h).$$

If $h \in {]-1, 0]}$ the values of $t^{\pm}$ become $t^+ = \dfrac{-1}{(1+h)^2 - 1}$, $t^- = -\infty$, and

$$d_{\mathbb{C}_+}((1+h)i, i) = \frac{1}{2} \log \left(\frac{\dfrac{-1}{(1+h)^2 - 1}}{\dfrac{-1}{(1+h)^2 - 1} - 1} \right) = -\log(1+h).$$

For $i + h, i \in \mathbb{C}_+, h \in \mathbb{R}$, we find

$$d_{\mathbb{C}_+}(i + h, i) = d_{\mathbb{H}}(\Pi(i + h, \Pi(i)))$$
$$= d_{\mathbb{H}}((X^2 - 2hX + h^2 + 1), (X^2 + 1)),$$

$$t^{\pm} = \frac{1}{2} \pm \sqrt{\frac{1}{4} + \frac{1}{h^2}},$$

hence,

$$d_{\mathbb{C}_+}(i + h, i) = \frac{1}{2} \log \left(\frac{\left(\dfrac{1}{2} + \sqrt{\dfrac{1}{4} + \dfrac{1}{h^2}} \right) \left(\dfrac{1}{2} + \sqrt{\dfrac{1}{4} + \dfrac{1}{h^2}} \right)}{\left(-\dfrac{1}{2} + \sqrt{\dfrac{1}{4} + \dfrac{1}{h^2}} \right) \left(-\dfrac{1}{2} + \sqrt{\dfrac{1}{4} + \dfrac{1}{h^2}} \right)} \right)$$

$$= \log \left(1 + \sqrt{h^2 + \frac{h^4}{4}} + h^2 \right).$$

For $0, h \in \mathbb{D}, h \in {]-1, 1[} \subset \mathbb{D}$, we compute

$$z(h) = i\frac{1+h}{1-h}, \quad z(0) = i \in \mathbb{C}_+$$

and conclude

$$d_{\mathbb{D}}(0, h) = d_{\mathbb{C}_+}(z(h), z(0)) = \log \left(\frac{1 + |h|}{1 - |h|} \right)$$

by the previous computation. Rotations of $\mathbb{D}$ with center 0 are automorphisms of $\mathbb{D}$ as a model of the hyperbolic plane. It follows that for $u \in \mathbb{D}$ the hyperbolic distance from u to 0 is given by

$$d_{\mathbb{D}}(0, u) = \log \left(\frac{1 + |u|}{1 - |u|} \right).$$

One way of computing the distance $d_{\mathbb{H}}((P(X)), (Q(X)))$ is to bring the point $(P(X))$ to $(X^2 + 1)$ by a symmetry ϕ of $\mathbb{H}$ and next compute $d_{\mathbb{H}}((X^2 + 1), \phi((Q(X))))$

using the coordinate z and formula (3.1). By applying a rotation ρ with center $(X^2 + 1)$ one can achieve that $(\phi(Q(X))$ transforms to $(X^2 + \lambda^2), \lambda \geq 1$, so finally $d_{\mathbb{H}}((P(X)), (Q(X))) = \log(\lambda)$.

For further study and a history of the hyperbolic plane, see [2, 14, 106, 107].

3.2 Intermezzo: Higher Cross-Ratios

The automorphism group G of a k-vector space V of finite dimension acts on various spaces that are attached to the vector space V. A *figure* or also a *configuration of labeled points* in V is a map $A: \{1, 2, \ldots, n\} \rightarrow V$. The number n is called the number of points of the configuration A. For instance a triangle in V is a map from the labeled three-element set $\{1, 2, 3\}$ to V. We denote by $P(n, V)$ the space of n-point configurations of points in V. The group G acts on each of the spaces $P(n, V)$.

Two configurations $A, B \in P(n, V)$ are called G-congruent if there exists a $\gamma \in G$ with $B_i = \gamma(A_i), i = 1, 2, \ldots, n$.

Up to G-congruence there exist only two configurations in $P(1, V)$, assuming that V is of dimension ≥ 1, namely a configuration A with $A_1 = 0$ or with $A_1 \neq 0$.

The list of possibilities for configurations up to G-congruence in $P(2, V)$ is longer, even infinite, if the field k has infinitely many elements. Assuming now that the dimension of V is ≥ 2, the first three possibilities are $A_1 = A_2 = 0$, $A_1 = 0$, $A_2 \neq 0$ and $A_1 \neq 0$, $A_2 = 0$. A fourth possibility is that A_1, A_2 are linearly independent. Moreover for every $\lambda \in K$, $\lambda \neq 0$, there is the possibility $A_1 \neq 0$, $A_2 = \lambda A_1$. The last possibility is the case that A_1 and A_2 both differ from 0 but are linearly dependent. So if the field has q elements, the number of possibilities up to G-congruence, i.e. the number of G-orbits in $P(2, V)$ equals $q + 3$.

Exercise 3.14. Assuming that the field k has q elements, and that the dimension of V is ≥ 3, count the number of G-orbits on $P(3, V)$. Count $\#P(n, V)/G$.

A figure or a configuration of labeled lines in V is a map $L: \{1, 2, \ldots, n\} \rightarrow \mathbb{P}V$. A configuration of labeled lines in V is a labeled configuration of points in the projective space $\mathbb{P}V$ of all sub-vector spaces of V of dimension 1. We denote by $L(n, V)$ the space of labeled n-line configurations in V. The group G acts on the spaces $L(n, V)$. Two configurations $L, L' \in L(n, V)$ are called congruent if there exists a $\gamma \in G$ such that $L'_i = \gamma(L_i), i = 1, 2, \ldots, n$.

Assuming that the dimension of V exceeds 1, up to G-congruence the spaces $L(n, V), n \leq 3$, have only finitely many elements. The first cases where up to G-congruence infinitely many labeled line configurations may exist start from configurations with 4 lines in a vector space of dimension 2.

The cross-ratio, which is the theme of this section, associates to 4 distinct labeled lines $A = (L_1, L_2, L_3, L_4)$ in a k-vector space V of dimension 2 a scalar $\mathrm{CR}(L_1, L_2, L_3, L_4) \in k \setminus \{0, 1\}$, such that two labeled configurations A, B are G-congruent if and only if $\mathrm{CR}(A) = \mathrm{CR}(B)$.

The definition of the number $\mathrm{CR}(A)$ for an $A \in L(4, V)$ is as follows. Consider $X = L_1$ and $Y = L_2$ as cartesian coordinate axes on V. Consider L_3 as the graph of a linear map $\phi\colon X \to Y$. Consider L_4 as the graph of a linear map $\psi\colon Y \to X$. The composition $\psi \circ \phi\colon X \to X$ is a linear endomorphism of a k-vector space X of dimension 1. The maps ϕ, ψ are invertible and $\phi \neq \psi^{-1}$ since $L_3 \neq L_4$, so the composition is invertible too and is given by the stretching $u \in X \mapsto \psi(\phi(u)) = \lambda u \in X$, $\lambda \neq 0, 1$. The stretching factor $\lambda \in k \setminus \{0, 1\}$ is the scalar $\mathrm{CR}(A)$.

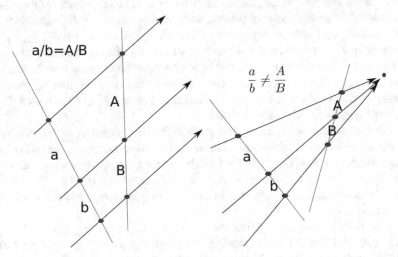

Fig. 3.3 Affine invariance of ratio, projective non-invariance of ratio.

Theorem 3.15. *The scalar* CR *detects in the above "if and only if" sense G-congruence of distinct labeled 4-line configurations in a two-dimensional k-vector space.* □

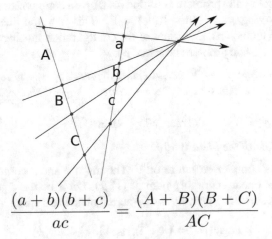

$$\frac{(a+b)(b+c)}{ac} = \frac{(A+B)(B+C)}{AC}$$

Fig. 3.4 Projective invariance of cross-ratio.

Exercise 3.16. Let a be a non-labeled configuration of 4 distinct lines in a two-dimensional vector space V. There are 24 different labelings $\pi \in S_4$ for a. Let $\lambda(a, \pi)$ be the cross-ratio of $A = (a, \pi)$. Study the map $\pi \in S_4 \to \lambda(a, \pi) \in k \setminus \{0, 1\}$.

The above cross-ratio construction can be used in many different situation. Here an example. Let $\mathrm{Gr}_{2,4}(\mathbb{R})$ be the Grassmannian of two-dimensional sub-vector spaces of a four-dimensional real vector space. Let $P(4, \mathrm{Gr}_{2,4}(\mathbb{R}))$ be the space of labeled configurations of 4 two-dimensional sub-vector spaces in a four-dimensional real vector space. For such a *generic* configuration of 4 labeled planes $A = (P_1, P_2, P_3, P_4)$ in a real vector space V of dimension 4 the cross-ratio $\mathrm{CR}_{2,4}(A)$ is defined analogously to the case of 4-line configurations and is an endomorphism of the two-dimensional vector space P_1. The term generic means, in the present context, that the planes P_i pairwise intersect only in the origin. The endomorphism $\mathrm{CR}_{2,4}(A)$ is invertible and moreover $\mathrm{CR}_{2,4}(A) - \mathrm{Id}_{P_1}$ is invertible too.

A labeled configuration C of four distinct 1-dimensional complex sub-vector spaces of a 2-dimensional complex vector space provides examples of generic labeled configurations $A \in P(4, \mathrm{Gr}_{2,4}(\mathbb{R}))$ by considering the involved vectors spaces as real vector spaces. So the configuration A is obtained by forgetting the structure of a complex vector space on the various spaces, and remembering only the structure of real vector spaces.

The cross-ratio $\mathrm{CR}_{2,4}(A)$ gives an answer to the following question. Which generic $A \in P(4, \mathrm{Gr}_{2,4}(\mathbb{R}))$ are obtained by forgetting a complex structure? More precisely, let $A = (P_1, P_2, P_3, P_4)$ a generic labeled configuration of planes in a real 4-dimensional vector space V. Under which conditions on A does there exist a complex structure $J : V \to V$ on V such that the planes satisfy $J(P_j) = P_j$, $j = 1, 2, 3, 4$?

Recall that a $\mathbb{R}$-linear map $J : V \to V$ endows the real vector space V with the structure of a complex vector space if J satisfies $J^2 = -\mathrm{Id}_V$ by declaring as multiplication

$$(a + bi)u = au + bJ(u), \ a, b \in \mathbb{R}, \ u \in V.$$

So very concretely the problem is to find for a given configuration A a linear map $J : V \to V$ satisfying $J^2 = -\mathrm{Id}_V, J(P_j) = P_j$. This is a non-linear but quadratic problem with 16 unknowns. The input of the configuration involves 32 parameters.

A similar more basic problem is solved by:

Theorem 3.17. *Let $\phi : V \to V$ be an endomorphism of a real vector space V of dimension two. There exists a complex structure $J : V \to V$ that commutes with ϕ if and only if*

$$\mathrm{Trace}(\phi)^2 < 4\mathrm{Det}(\phi)$$

or if ϕ is a multiple of the identity Id_V.

Proof. Let J be a complex structure on V. The maps ϕ and J commute if and only if ϕ is a complex linear endomorphism of (V, J). Let λ be the eigenvalue of ϕ on (V, J), so that $\lambda, \bar\lambda$ are the eigenvalues of ϕ acting on the real space V. The inequality

$$\mathrm{Trace}(\phi)^2 = (\lambda + \bar\lambda)^2 \leq 4\lambda\bar\lambda = 4\mathrm{Det}(\phi)$$

will follow if and only if a complex structure J commuting with ϕ exists. In the case of equality one observes that ϕ is moreover diagonalizable over $\mathbb{R}$. $\quad\square$

Remark. In the case of strict inequality only two complex structures $\pm J$ commuting with ϕ exist. For the other case, the space of solutions with its geometry turns out to be the union of two anti-oriented copies of the hyperbolic plane. See Section 5.2.

Theorem 3.18 (Four Complex Lines Theorem). *Let $A = (P_1, P_2, P_3, P_4)$ be a generic labeled configuration of planes in a real 4-dimensional vector space V. There exists a complex structure $J: V \to V$ with $J(P_j) = P_j$ if and only if*

$$\text{Trace}(CR_{2,4}(A))^2 < 4\text{Det}(CR_{2,4}(A))$$

or if $CR_{2,4}(A)$ is a multiple λId_{P_1}, $\lambda \neq 1$, of the identity Id_{P_1}.

Proof. Such a complex structure J exists if and only J preserves P_1 and $CR_{2,4}(A)$ commutes with the restriction of J to P_1. The result follows by the previous theorem. $\square$

A similar theorem concerning the opposite inequality holds in the symplectic setting. A bilinear anti-symmetric non-degenerate map $\omega: V \times V \to \mathbb{R}$ on a vector space V is called a linear symplectic form on V. Assume V is of dimension four. A two-dimensional linear subspace L of V is called a linear Lagrangian subspace if $\omega(u, v) = 0$ holds for all $u, v \in L$.

Theorem 3.19 (Four Lagrangians Theorem). *Let $A = (P_1, P_2, P_3, P_4)$ be a generic labeled configuration of planes in a real 4-dimensional vector space V. There exists a linear symplectic structure ω on V such that the planes P_i are ω-Lagrangian if and only if*

$$\text{Trace}(CR_{2,4}(A))^2 > 4\text{Det}(CR_{2,4}(A))$$

or if $CR_{2,4}(A)$ is a multiple λId_{P_1}, $\lambda \neq 1$, of the identity Id_{P_1}.

Proof. Assume that a linear symplectic structure ω exists for which the planes P_i become Lagrangian. Then ω allows us to identify P_4 with the dual space $\mathbb{P}_1^*$. Hence, P_2 and P_3 can be interpreted as maps $\alpha, \beta: P_1 \to \mathbb{P}_1^*$, and therefore as 1-differential forms on P_1. The forms α, β are closed since the graphs P_2, P_3 are Lagrangians. It follows that the maps α, β are symmetric, hence, define on P_1 non-degenerate quadratic forms. The Principal Axis Theorem applies, hence there exists a linear diagonalizable map $\Lambda: P_1 \to P_1$ with real eigenspaces and with $\alpha = \beta \circ \Lambda$. It follows that $CR_{2,4}(A): P_1 \to P_1$ is diagonalizable over the reals, hence the inequality. This argument can be reversed step by step. $\square$

The cross-ratio $CR_{n,2n}$ construction associates an endomorphism $CR_{n,2n}(A)$ of the first subspace to a labeled configuration $A \in P(4, \text{Gr}_{n,2n})$ of 4 n-dimensional subspaces of a $2n$-dimensional vector space.

Exercise 3.20. Given a configuration $A \in P(4, \text{Gr}_{n,2n})$ work out the condition on $CR_{n,2n}(A)$ that is equivalent to the existence of a linear complex structure or a linear symplectic structure that complexifies or symplectifies the configuration as in the above theorems.

The cross-ratio $\mathrm{CR}_{n,2n}$ can be useful in the following problem. Given a generic $A \in P(N, \mathrm{Gr}_{n,2n})$ determine the group G_A of linear maps $g \in \mathrm{GL}(2n, \mathbb{R})$ satisfying $g(A_k) = A_k$, $k = 1, 2, \ldots, N$. For $N = 1$ the group G_A is isomorphic to the subgroup of the matrix group $\mathrm{GL}(2n, \mathbb{R})$ consisting of the block matrices M with $M_{ij} = 0$ for $i > n$ and $j \leq n$. For $N = 2$ the group G_A is isomorphic to $\mathrm{GL}(n, \mathbb{R})^2$, and for $N = 3$ it is isomorphic to $\mathrm{GL}(n, \mathbb{R})$. For $N = 4$ G_A is isomorphic to the group of matrices in $\mathrm{GL}(n, \mathbb{R})$ that commute with $\mathrm{CR}_{n,2n}(A)$. For $N > 5$ the group is mostly reduced to the sub-group of multiples of the identity in $\mathrm{GL}(n, \mathbb{R})$.

Exercise 3.21. Given generic $A \in P(4, \mathrm{Gr}_{n,2n})$ study the group P_A of all linear transformation g with $g(\cup A_j) = \cup A_j$. Study P_A/G_A. The word *generic* has a meaning that can vary depending on context! "More generic" in the present context would be the case if moreover $P_A = G_A$ holds.

Exercise 3.22. Study all more generic $A \in P(4, \mathrm{Gr}_{1,2})$ and $A \in P(4, \mathrm{Gr}_{2,4})$.

The following question leads to a tri-ratio. In a 3-dimensional vector space V let C and C' be two given configurations of 3 pairwise distinct subspaces H_1, H_2, H_3 of dimension 2 and 3 subspaces L_1, L_2, L_3 of dimension 1 such that $L_i \subset H_i$ and $L_i \not\subset H_j$, $i, j \in \{1, 2, 3\}$, $i \neq j$. The problem is to construct a computable criterion that decides if there exists an automorphism of V sending the configuration C to C' respecting labels.

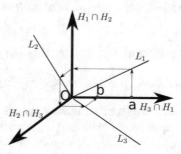

Fig. 3.5 $\tilde{E}_6$ tri-ratio invariant $\frac{b}{a}$.

Consider L_2 as graph of a linear map M_2 from $H_2 \cap H_1$ to $H_2 \cap H_3$. Accordingly, M_1 with graph L_1 maps $H_1 \cap H_3$ to $H_1 \cap H_2$ and M_3 with graph L_3 maps $H_3 \cap H_2$ to $H_3 \cap H_1$. The composition $M = M_1 \circ M_3 \circ M_2$ is an endomorphism of the 1-dimensional space $H_2 \cap H_1$. Define the tri-ratio $\mathrm{TR}(H_1, H_2, H_3, L_1, L_2, L_3)$ of the given configuration C of subspaces in V as the stretch factor of M. The configuration C corresponds to three labeled complete flags in generic position in the space V of dimension 3. The stretch factor $\mathrm{TR}(C)$ is an invertible element in the ground field. One has the following answer to the above question.

Theorem 3.23. *Two labeled generic configurations C and C' of flags in a 3-dimensional vector space are linearly equivalent if and only if* $\mathrm{TR}(C) = \mathrm{TR}(C')$ *holds.* $\square$

A flag $L \subset H \subset V$ corresponds to an incidence $p \in \Lambda$ of a point p and a line Λ in the projective plane $P(V)$. A configuration C as above corresponds to a labeled triple of point-line incidences $p_i \in \Lambda_i$, $i = 1, 2, 3$, in $P(V)$. The tri-ratio $\mathrm{TR}(C)$ corresponds to a tri-ratio for a generic triple of point-line incidences $\mathrm{TR}(\Lambda_1, \Lambda_2, \Lambda_3, p_1, p_2, p_3)$. Two labeled triples of point-line incidences can be transformed into each other by a projective motion of $P(V)$ if and only if their tri-ratios agree.

Exercise 3.24. Let C be a generic configuration $C = (H_1, H_2, H_3, L_1, L_2, L_3)$ of subspaces in V with $L_i \subset H_i$ of real dimensions 6 for V, 4 for H_i and 2 for L_i. Formulate a computable condition P such that P holds if and only if there exists a complex structure J on V with $J(H_i) = H_i$ and $J(L_i) = L_i$.

3.3 Hyperbolic Trigonometry

We will consider the elements in $\mathbb{H}$ as *points*. A *triangle* is a triple of points (A, B, C) in $\mathbb{H}$ that are not elements of a common line. We will denote by $a, b, c \in \mathbb{R}$ the distances $a = d_{\mathbb{H}}(B, C)$, $b = d_{\mathbb{H}}(C, A)$, $c = d_{\mathbb{H}}(A, B)$. The numbers a, b, c are called the side lengths of the triangle. The angle at A of the triangle (A, B, C) between the lines through A, B and A, C is denoted by $\alpha \in]0, \pi[$. Accordingly, the angles at B and C are denoted by β and γ.

Two triangles (A, B, C) and (A', B', C') are called congruent if there exists an isometry of (H, d_H) mapping A to A', B to B' and C to C'.

A first result claims the existence of triangles, unique up congruence.

Theorem 3.25. *Let $(a, b, c) \in \mathbb{R}$ be three numbers satisfying the inequalities:*

$$0 < a < b + c, \quad 0 < b < a + c, \quad 0 < c < a + b.$$

Then there exists a triangle (A, B, C) in the hyperbolic plane $\mathbb{H}$ having side lengths a, b, c. Moreover, if (A', B', C') is a triangle with side lengths a, b, c, then there exists a unique isometry of $\mathbb{H}$ mapping A to A', B to B' and C to C'.

Proof. Start with three points A, C, B on a line in that order such that the distances AC and CB are equal to b and a. Further assume $a \leq b$. In that position the distance AB is equal to $a + b = c$. By varying the angle $\sphericalangle ACB$ from π to 0, and keeping the points B and C fixed, the point A moves in between C and B and the distance AB varies continuously from $a + b$ to $-a + b < c$. (Use the Hyperbolic Cosine Law, which will be treated later in this section in Theorem 3.31.) By the intermediate value theorem there exists an angle $\sphericalangle ACB$ such that the distance AB equals c. $\quad\square$

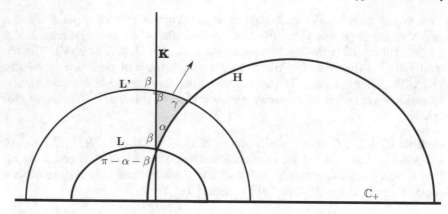

Fig. 3.6 Construction of a triangle from α, β, γ.

Theorem 3.26. *Let* $(\alpha, \beta, \gamma) \in \,]0, \pi[$ *be three numbers satisfying the inequality*

$$\alpha + \beta + \gamma < \pi.$$

Then there exists a triangle (A, B, C) *in the hyperbolic plane* $\mathbb{H}$ *having angles* α, β, γ. *Moreover, if* (A', B', C') *is a triangle with angles* α, β, γ, *then there exists a unique isometry of* $\mathbb{H}$ *mapping A to A', B to B' and C to C'.*

Proof. Again use the intermediate value theorem. Start with three oriented lines L, K and H through a point A such that the angles of the three sectors are $\sphericalangle HK = \alpha, \sphericalangle KL = \beta, \sphericalangle HL = \alpha + \beta$ (Fig. 3.6). Now move the line L continuously while keeping the angle with K equal to β and such that the intersection $L \cap H$ is non-empty. The angles of the triangles bordered by the lines H, K, L are α, β and a third angle that varies continuously from $\pi - \alpha - \beta$ to 0. So there exists a position L' for the moving line L with third angle equal to γ.

The Hyperbolic Cosine Law (Theorem 3.31) permits us to compute the side lengths from the angles. Unicity follows from the previous theorem. $\square$

The first theorem also holds in the Euclidean plane. The second does not. This striking difference was directly, but also indirectly, used to question the existence of the hyperbolic plane. The existence of a planar geometry satisfying the modified fifth postulate V' was already questioned by Euclid,[3] and has remained a hot topic of mathematical research for at least 2000 years, see the book of Roberto Bonola [14] with its supplements by the inventors of hyperbolic geometry, Nikolai Iwanowitsch Lobatschevsky and János Bolyai.

Both theorems together have as a consequence that the side lengths of a triangle in $\mathbb{H}$ determine the angles, and conversely, that the angles of a triangle in $\mathbb{H}$ determine the side lengths, which is rather unexpected to those educated in a Euclidean school.

---

[3] Ευκλείδης από την Αλεξάνδρεια (c.300 BC – 270 BC), Euclid of Alexandria, cited as ὁ στοιχειώτης, author of the Elements, was a Greek mathematician. His axiomatic approach to reality still remains a main method of study after 2300 years. Moreover today, his approach creates candidate realities even before they become visible by our eyes or are detected by our experiments.

In hyperbolic trigonometry the following theorem for right-angled triangles is as central as Pythagoras' Theorem in Euclidean geometry:

Theorem 3.27 (Pythagoras in Hyperbolic Geometry). *Let (A, B, C) be a triangle in the hyperbolic plane $\mathbb{H}$ with right angle at C. The hyperbolic Pythagoras type statement is*

$$\cosh(c) = \cosh(a)\cosh(b).$$

Proof. The half-lines $L^+ = \{q_t \mid q_t = ((1-t)(X^2+1) + t(X-1)^2), \ t \in [0,1[\}$ and $Y^+ = \{r_s \mid r_s = (X^2 + s^2), \ s \in [1, +\infty[\}$ in $\mathbb{H}_I$ meet at $(X^2 + 1)$ with right angles. Indeed the isometry $(P(X)) \mapsto X^2 P(\frac{1}{X})$ fixes pointwise the half-line Y^+ and fixes globally the line L supporting L^+. Every triangle (A, B, C) with right angle at C can be moved by an isometry such that $C = (X^2 + 1)$, $B = q_t \in L$ and $A = r_s \in Y$ hold. Previous computations of $d_{\mathbb{H}}$ (see Section 3.1) yield

$$\cosh(a) = \frac{1}{2\sqrt{t(1-t)}}, \quad \cosh(b) = \frac{s^2+1}{2s}$$

and

$$\cosh(c) = 1 + \frac{(2t-1)^2 + (s - \sqrt{1 - (1-2t)^2})^2}{2s\sqrt{1 - (1-2t)^2}}$$

$$= \frac{s^2+1}{2s\sqrt{1 - (1-2t)^2}} = \frac{1}{2\sqrt{t(1-t)}}\frac{s^2+1}{2s}.$$

The result follows.   □

The Euclidean Pythagoras' Theorem reads $c^2 = a^2 + b^2$, which can be interpreted as an equality of areas. On spheres, Euclidean lengths of intervals on great circles can be measured by numbers in $[0, 2\pi]$. The Spherical Pythagoras' Theorem reads $\cos(c) = \cos(a)\cos(b)$. It is natural to ask for an interpretation of the Pythagorean Theorems in hyperbolic and spherical geometry. See [2] for a more geometric proof and also for how the Euclidean Pythagoras' Theorem is hidden in the hyperbolic and spherical theorems.

In planar Euclidean geometry twice the area $2V$ of a triangle (A, B, C) with right angle at C can be computed in two different ways: $2V = ab$ and $2V = hc$ where h is the length of the height perpendicular to the hypotenuse. So $ab = hc$ holds. The following result provides an analogue in hyperbolic geometry.

Theorem 3.28 (Hyperbolic Sinh Product Theorem). *Let (A, B, C) be a triangle in the hyperbolic plane $\mathbb{H}$ with right angle at C (Fig. 3.7). Let h be the length of the height perpendicular to the hypotenuse. Then*

$$\sinh(a)\sinh(b) = \sinh(h)\sinh(c).$$

Proof. The foot point P of the height to the hypotenuse divides the hypotenuse into two pieces of length c', c''. We choose the piece of length c' to be incident with the vertex B.

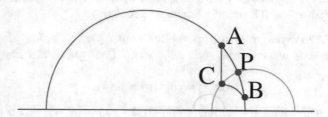

Fig. 3.7 Triangle ABC with height CP and right angle C. Red and yellow lines are symmetric by point involution through C. The green line is the common perpendicular of the red and the yellow line.

The logic of our proof is to use the result in order to have a candidate value for the quantity $h > 0$. Now we build the candidate quadrilateral (A, P, B, C) by first computing candidate values for c', c'' and by gluing the triangles (A, P, C) and (B, P, C) along the common edge PC. The points A, P, B should by aligned since the angle at P is right by construction in both triangles. The punch line of the proof is to check $c' + c'' = c$, which proves that the points A, P, B are indeed aligned. This will be done by checking $\cosh(c' + c'') = \cosh(a)\cosh(b)$.

The candidate value for h is implicitly given by

$$\sinh(h) = \frac{\sinh(a)\sinh(b)}{\sinh(c)}.$$

It follows that

$$\cosh(h) = \sqrt{1 + \sinh^2(h)} = \sqrt{1 + \frac{\sinh^2(a)\sinh^2(b)}{\sinh^2(c)}}.$$

From the hyperbolic Pythagoras' Theorem in the triangles APC and BPC with right angles at P one deduces candidate values for c', c''

$$\cosh(c') = \frac{\cosh(a)}{\sqrt{1 + \frac{\sinh^2(a)\sinh^2(b)}{\sinh^2(c)}}} = \frac{\sinh(c)\cosh(a)}{\sqrt{\sinh^2(a)\sinh^2(b) + \sinh^2(c)}}$$

and

$$\cosh(c'') = \frac{\cosh(b)}{\sqrt{1 + \frac{\sinh^2(a)\sinh^2(b)}{\sinh^2(c)}}} = \frac{\sinh(c)\cosh(b)}{\sqrt{\sinh^2(a)\sinh^2(b) + \sinh^2(c)}}$$

and hence, using $\cosh^2 = 1 + \sinh^2$, $\cosh(c) = \cosh(a)\cosh(b)$, also

$$\sinh(c') = \sqrt{-1 + \frac{\sinh^2(c)\cosh^2(b)}{\sinh^2(c) + \sinh^2(a)\sinh^2(b)}}$$

$$= \sinh(b)\sqrt{\frac{\sinh^2(c) - \sinh^2(a)}{\sinh^2(c) + \sinh^2(a)\sinh^2(b)}}$$

$$= \sinh(b)\sqrt{\frac{\cosh^2(c) - \cosh^2(a)}{\sinh^2(c) + \sinh^2(a)\sinh^2(b)}}$$

$$= \frac{\sinh^2(b)\cosh(a)}{\sqrt{\sinh^2(c) + \sinh^2(a)\sinh^2(b)}}$$

and also

$$\sinh(c'') = \frac{\sinh^2(a)\cosh(b)}{\sqrt{\sinh^2(c) + \sinh^2(a)\sinh^2(b)}}.$$

We prepare the check of $c = c' + c''$ by starting from the cosh addition formula:

$$\cosh(c' + c'') = \cosh(c')\cosh(c'') + \sinh(c')\sinh(c'').$$

First

$$\cosh(c')\cosh(c'') = \frac{\sinh^2(c)\cosh(a)\cosh(b)}{\sinh^2(c) + \sinh^2(a)\sinh^2(b)}.$$

Next

$$\sinh(c')\sinh(c'') = \frac{\sinh^2(a)\sinh^2(b)\cosh(a)\cosh(b)}{\sinh^2(c) + \sinh^2(a)\sinh^2(b)}.$$

Now, clearly, ouf!, $\cosh(c' + c'') = \cosh(a)\cosh(b) = \cosh(c)$. $\qquad\square$

In planar Euclidean geometry the trigonometric functions are defined as ratios of lengths. The following result provides an analogue in hyperbolic geometry.

Theorem 3.29. *Let (A, B, C) be a triangle in the hyperbolic plane $\mathbb{H}$. The measurement of angles is given by the Riemannian metric of $\mathbb{H}$. Assume that the angle at C is right, i.e. $\gamma = \frac{\pi}{2}$. Then $\sin(\alpha) = \frac{\sinh(a)}{\sinh(c)}$ holds.*

Proof. Construct a sequence $(C_n)_{n\in\mathbb{N}}$ as follows: put $C_0 = C$. Let C_1 be the foot point of the orthogonal projection of C_0 to the hypotenuse of (A, B, C). Let C_2 be the foot point of the orthogonal projection of C_1 to the hypotenuse of (A, C_0, C_1). Inductively, let C_{n+1} be the foot point of the orthogonal projection of C_n to the hypotenuse of (A, C_{n-1}, C_n). Claim: $\lim_{n\to\infty} C_n = A$.

Observe, that the point C_{2n+2} is in between the points A and C_{2n}. The sequence of distances $n \mapsto d_{\mathbb{H}}(A, C_{2n})$ is monotonously decreasing, hence convergent. It follows that the sequence of points C_{2n} has a limit, say C'. Analogously, the sequence of points C_{2n+1} has a limit, say B'. Assuming, $C' \neq A$ leads to a contradiction, namely the triangle $AB'C'$ would have right angles at B' and C', contradicting the fact that the sum of the angles of a triangle is strictly less than the sum of two right angles. Hence the claim.

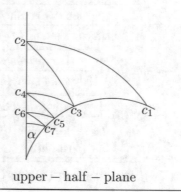

$$upper - half - plane$$

Fig. 3.8 Zigzag of perpendicular segments.

So all lengths of the triangles (A, C_n, C_{n+1}) tend to 0 if $n \to \infty$. The coordinate $z \colon \mathbb{H} = \mathbb{H}_I \to \mathbb{C}_+$ is an isometry into $(\mathbb{C}_+, \rho_{\mathbb{C}_+})$. The Riemannian metric $\rho_{\mathbb{C}_+}$ and the Euclidean metric $\rho_{\mathbb{C}}$ are conformal. Hence,

$$\lim_{n \to \infty} \frac{d_{\mathbb{H}}(C_n, C_{n+1})}{d_{\mathbb{H}}(A, C_n)} = \lim_{n \to \infty} \frac{d_{\mathbb{C}_+}(z(C_n), z(C_{n+1}))}{d_{\mathbb{C}_+}(z(A), z(C_n))}$$

$$= \lim_{n \to \infty} \frac{d_{\mathbb{C}}(z(C_n), z(C_{n+1}))}{d_{\mathbb{C}}(z(A), z(C_n))} = \sin(\alpha),$$

where $\alpha \in \,]0, \frac{\pi}{2}[$ is the angle at A of the triangle (A, B, C) or at $z(A)$ of the triangle $(z(A), z(B), z(C))$ with segments of half circles as sides.

Using the Taylor expansion of the sinh function, one shows

$$\lim_{n \to \infty} \frac{\sinh(d_{\mathbb{H}}(C_n, C_{n+1}))}{\sinh(d_{\mathbb{H}}(A, C_n))} = \lim_{n \to \infty} \frac{d_{\mathbb{H}}(C_n, C_{n+1})}{d_{\mathbb{H}}(A, C_n)} = \sin(\alpha).$$

By the previous Hyperbolic Sinh Product Theorem 3.28 the leftmost limit above is stationary from its first term on. The result follows. $\qquad \square$

As a consequence, by using identities between elementary functions, we have:

Theorem 3.30. *Let (A, B, C) be a triangle in the hyperbolic plane $\mathbb{H}$. Assume that the angle at C is right, i.e $\gamma = \frac{\pi}{2}$. Then*

$$\sin(\alpha) = \frac{\sinh(a)}{\sinh(c)},$$

$$\cos(\alpha) = \frac{\tanh(b)}{\tanh(c)} = \frac{\sinh(b)\cosh(a)}{\sinh(c)},$$

$$\tan(\alpha) = \frac{\tanh(a)}{\sinh(b)} = \frac{\sinh(a)}{\cosh(a)\sinh(b)}.$$

The following theorems describe the three basic couplings of lengths of sides and angles for hyperbolic triangles.

Theorem 3.31 (Hyperbolic Cosine Law). *Let* (A, B, C) *be a triangle in* $\mathbb{H}$. *The Hyperbolic Cosine Law states:*

$$\cosh(a) = \cosh(b)\cosh(c) - \sinh(b)\sinh(c)\cos(\alpha),$$
$$\cosh(b) = \cosh(a)\cosh(c) - \sinh(a)\sinh(c)\cos(\beta),$$
$$\cosh(c) = \cosh(a)\cosh(b) - \sinh(a)\sinh(b)\cos(\gamma).$$

Proof. Let P be the foot point of the height from A. Let a' be the length of BP, and a'' the length PC. Let h be the length of the height AP. Let α' be the angle BAP and α'' the angle PAC.

With the addition formulae

$$\cosh(a) = \cosh(a')\cosh(a'') + \sinh(a')\sinh(a''),$$
$$\cos(\alpha) = \cos(\alpha')\cos(\alpha'') - \sin(\alpha')\sin(\alpha''),$$

the hyperbolic Pythagoras' Theorem and Theorem 3.30 one gets

$$\cosh(b)\cosh(c) - \sinh(b)\sinh(c)\cos(\alpha' + \alpha'')$$
$$= \cosh(a')\cosh(a'')\cosh^2(h)$$
$$- \sinh(b)\sinh(c)\left(\frac{\sinh(h)\cosh(a')}{\sinh(c)}\frac{\sinh(h)\cosh(a'')}{\sinh(b)} - \frac{\sinh(a')}{\sinh(c)}\frac{\sinh(a'')}{\sinh(b)}\right)$$
$$= (\cosh^2(h) - \sinh^2(h))(\cosh(a')\cosh(a'') + \sinh(a')\sinh(a''))$$
$$= \cosh(a). \qquad \square$$

Theorem 3.32 (Hyperbolic Sine-Sinh Law). *Let* (A, B, C) *be a triangle in* $\mathbb{H}$. *The Hyperbolic Sine-Sinh Law states:*

$$\frac{\sin(\alpha)}{\sinh(a)} = \frac{\sin(\beta)}{\sinh(b)} = \frac{\sin(\gamma)}{\sinh(c)}.$$

Proof. The symmetric expression

$$\frac{1 - \cosh^2(a) - \cosh^2(b) - \cosh^2(c) + 2\cosh(a)\cosh(b)\cosh(c)}{\sinh^2(a)\sinh^2(b)\sinh^2(c)}$$

reduces by simplification to $\frac{\sin^2(\alpha)}{\sinh^2(a)}$. $\qquad \square$

Theorem 3.33 (Angular Hyperbolic Cosine Law). *Let* (A, B, C) *be a triangle in* $\mathbb{H}$. *The Angular Hyperbolic Cosine Law states:*

$$-\cos(\alpha) = \cos(\beta)\cos(\gamma) - \sin(\beta)\sin(\gamma)\cosh(a),$$
$$-\cos(\beta) = \cos(\alpha)\cos(\gamma) - \sin(\alpha)\sin(\gamma)\cosh(b),$$
$$-\cos(\gamma) = \cos(\alpha)\cos(\beta) - \sin(\alpha)\sin(\beta)\cosh(c).$$

Proof. Compute $\cos(\beta), \cos(\gamma)$ using the Hyperbolic Cosine Law, next compute $\sin(\beta)$ and $\sin(\gamma)$. The expression $\cos(\beta)\cos(\gamma) - \sin(\beta)\sin(\gamma)\cosh(a)$ will simplify to $-\cos(\alpha)$. $\qquad\square$

We remember that the function $\cos\colon [0, \pi] \to [-1, 1]$ is a strictly monotonously decreasing bijection. The inverse bijection is the function $\arccos\colon [-1, 1] \to [0, \pi]$. Recall that $\arccos(h) = \dfrac{1}{i}\log(h + i\sqrt{1 - h^2})$. From the Angular Hyperbolic Cosine Law we deduce:

Theorem 3.34. *Let* (A, B, C) *be a triangle in* $\mathbb{H}$. *We have*

$$\alpha = \arccos\left(\frac{\cosh(a) - \cosh(b)\cosh(c)}{\sinh(b)\sinh(c)}\right).$$

The angles β, γ *are determined by the same expression after a cyclic permutation.* $\square$

The function $\cosh\colon [0, +\infty[\to [1, +\infty[$ is strictly monotonously increasing. The inverse function is $\mathrm{arcCosh}\colon [1, +\infty[\to [0, +\infty[$. Recall that $\mathrm{arcCosh}(h) = \log(h + \sqrt{h^2 - 1})$. From the Angular Hyperbolic Cosine Law we deduce:

Theorem 3.35. *Let* (A, B, C) *be a triangle in* $\mathbb{H}$. *We have*

$$a = \mathrm{arcCosh}\left(\frac{\cos(\alpha) + \cos(\beta)\cos(\gamma)}{\sin(\beta)\sin(\gamma)}\right).$$

The side lengths b, c *are determined by an analogous expression after a cyclic permutation.* $\qquad\square$

For right hyperbolic triangles simplifications occur:

Theorem 3.36. *Let* (A, B, C) *be a triangle in* $\mathbb{H}$ *with right angle at* C. *The coupling lengths-angles become:*

$$\cosh(a) = \cos(\alpha)/\sin(\beta),$$
$$\cosh(b) = \cos(\beta)/\sin(\alpha),$$
$$\cosh(c) = \coth(\alpha)\coth(\beta). \qquad\square$$

There are many natural questions concerning the trigonometry of more general triangles in the hyperbolic plane $\mathbb{H}$. Instead of encoding a triangle by its vertices, it is more convenient to define a triangle by the triple of lines $(\mathbf{a}, \mathbf{b}, \mathbf{c})$ that carry the sides of the triangle. The line $\mathbf{a}$ carries the side BC etc. If the three lines intersect pairwise in three points, one recovers the notion of a usual triangle. If one or more pairs of lines do not intersect, the triple $(\mathbf{a}, \mathbf{b}, \mathbf{c})$ defines a general hyperbolic triangle.

Pairs of lines $(\mathbf{a}, \mathbf{b})$ in the hyperbolic plane $\mathbb{H}$ that do not intersect are of two kinds. Let $\gamma' \in \mathbb{R}$ be the infimum of the distance $d_{\mathbb{H}}(P, Q)$ with $P \in \mathbf{a}$ and $Q \in \mathbf{b}$. If $\gamma' > 0$ the lines are called *strictly disjoint*, if $\gamma' = 0$ the lines are called *asymptotic*.

Most general triangles are triples $(\mathbf{a}, \mathbf{b}, \mathbf{c})$ of pairwise strictly disjoint lines. Let A be the shortest geodesic segment that joins $\mathbf{b}, \mathbf{c}$, etc. The three geodesic segments A, B, C together with three geodesic segments on the lines $(\mathbf{a}, \mathbf{b}, \mathbf{c})$ build a hexagon

$H(\mathbf{a}, \mathbf{b}, \mathbf{c})$ with right angles. Let a, b, c be the lengths of the segments carried by $(\mathbf{a}, \mathbf{b}, \mathbf{c})$ and let α', β', γ' be the lengths of the segments A, B, C. Observe that all quantities are hyperbolic lengths, and a combinatorial central symmetry exchanges the roles of a, b, c and α', β', γ'.

The basic trigonometric relation for most general triangles $(\mathbf{a}, \mathbf{b}, \mathbf{c})$ and the corresponding right angled hexagons $H(\mathbf{a}, \mathbf{b}, \mathbf{c})$ is given by

Theorem 3.37 (Hyperbolic Cosine Law for $\perp$-hexagons). *With the notation from above the following identities hold.*

$$\cosh(a) = -\cosh(b)\cosh(c) + \sinh(b)\sinh(c)\cosh(\alpha'),$$
$$\cosh(\alpha') = -\cosh(\beta')\cosh(\gamma') + \sinh(\beta')\sinh(\gamma')\cosh(a).$$

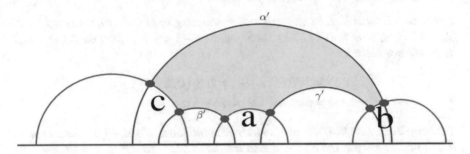

Fig. 3.9 Right-angled hexagon. The red lines are common perpendiculars to the blue lines.

Before giving a proof we need to study pairs of lines in $\mathbb{H}$. In particular we need expressions for the angle if they intersect and for the shortest distance if they do not intersect.

Let K, L be two distinct oriented lines in $\mathbb{H} = \mathbb{H}_I$ that intersect in $\mathbb{H}$. By an isometry we transform for some $s \in]-1, +1[$ the pair (K, L) into the pair (C, L_s), where C is the oriented line $\{t((X-1)^2) + (1-t)((X+1)^2) \mid t \in [0, 1[\}$ and L_s is the oriented line $\{((X-s)^2 + r^2 \mid r > 0\}$. The lines C and L_s are oriented by the parameters $t \in]0, 1[$ and $r \in]0, +\infty[$. At the intersection point of C and L_s appear four sectors oriented by the restriction of the z-coordinate orientation of $\mathbb{H}$: at precisely one of those four oriented sectors the union of boundary legs with the induced orientation form the oriented boundary of this oriented sector. The oriented angle $\alpha(K, L)$ between C and L_s is the angle of this sector. Switching from z-coordinate to $\bar{z}$-coordinate orientation changes this sector to the opposite sector and does not change the angle α. Changing the orientation on one line only changes the angle from α to $\pi - \alpha$.

We define the oriented cross-ratio $CR(K, L)$ of a pair of intersecting oriented lines K, L as

$$CR(K, L) = CR(C, L_s) = cr(-1, 1, s, \infty) = \frac{1+s}{2}.$$

For two lines K, L that are strictly disjoint we orient the lines such that their union becomes the oriented boundary of the z-coordinate oriented region in between K and L. By an isometry we transform the pair (K, L) for some $r \in]1, +\infty[$ to the pair (C, C_r), C as before and $C_r = \{t((X + r)^2) + (1 - t)((X - r)^2) \mid t \in [0, 1[\}$. The shortest distance from C to C_r is realized by the geodesic segment that connects $(X^2 + 1)$ and $(X^2 + r^2)$. Its length is $\log(r)$.

We define the oriented cross-ratio $CR(K, L)$ of a pair of intersecting oriented lines K, L as

$$CR(K, L) = CR(C, C_r) = cr(-1, 1, r, -r) = \frac{(r + 1)^2}{4r}.$$

The pair (K, L) of oriented lines can also be transformed to the pair (C, L_s) with $s = \frac{r^2+1}{2r}$.

From the previous computations we deduce

Theorem 3.38. *Let K, L be two oriented intersecting lines with oriented angle α. If (K, L) is congruent to (C, L_s) as above for some $-1 < s < 1$, then the angle α and its cosine are given by*

$$\cos(\alpha) = 2CR(K, L) - 1 = 2CR(C, L_s) - 1 = s,$$
$$\alpha = \arccos(2CR(K, L) - 1). \qquad \square$$

Theorem 3.39. *Let K, L be two strictly disjoint lines. If (K, L) is congruent to (C, C_r) as above for some $r > 1$, and also to (C, L_s) for $s = \frac{1}{2}(r + \frac{1}{r})$, then the distance from K to L is given by*

$$d_{\mathbb{H}}(K, L) = d_{\mathbb{H}}(C, C_r) = \log(r) = \operatorname{arcCosh}(2CR(K, L) - 1)$$

and

$$\cosh(d_{\mathbb{H}}(K, L)) = s = 2CR(K, L) - 1$$

holds. $\qquad \square$

Proof (Hyperbolic Cosine Law for $\perp$-hexagons). Due to the central combinatorial symmetry of order 6 it suffices to prove only one identity. First we orient the hexagon according to the z-coordinate orientation and name the sides a, C, b, A, c, B. The letters a, b, c also stand for the lengths of the corresponding sides. The lengths of A, B, C are denoted by α', β', γ'. The side a is carried by the oriented line $a = (t(X + a_2)^2 + (1 - t)(X + a_1)^2), t \in]0, 1[$. We introduce a similar notation for the other sides. We get 12 distinct points on $\mathbb{R} \cup \{\infty\}$ which appear in the following cyclic order:

$$a_1, B_2, C_1, a_2, b_1, C_2, A_1, b_2, c_1, A_2, B_1, c_2.$$

For convenience of calculation, assume that all differ from ∞ and that they appear in increasing order according to the order of $\mathbb{R}$. The six right angles

$$\sphericalangle aC, \sphericalangle Cb, \sphericalangle bA, \sphericalangle Ac, \sphericalangle cB, \sphericalangle Ba$$

between oriented lines lead to the cross-ratio equations

$$\frac{1}{2} = CR(aC) = CR(Cb) = CR(bA) = CR(Ac) = CR(cB) = CR(Ba),$$

hence to

$$\begin{aligned}
\frac{1}{2} &= \frac{(C_1 - a_1)(C_2 - a_2)}{(a_2 - a_1)(C_2 - C_1)} = \frac{(b_1 - C_1)(b_2 - C_2)}{(C_2 - C_1)(b_2 - b_1)} = \frac{(A_1 - b_1)(A_2 - b_2)}{(b_2 - b_1)(A_2 - A_1)} \\
&= \frac{(c_1 - A_1)(c_2 - A_2)}{(A_2 - A_1)(c_2 - c_1)} = \frac{(B_1 - c_1)(B_2 - c_2)}{(c_2 - c_1)(B_2 - B_1)} = \frac{(a_1 - B_1)(a_2 - B_2)}{(B_2 - B_1)(a_2 - a_1)}.
\end{aligned}$$

Using the trigonometric identity $\cosh^2 -1 = \sinh^2$, rearranging and squaring, the identity that has to be proved becomes equivalent to the following identity that only uses cosh functions

$$(\cosh(a) - \cosh(b)\cosh(c))^2 - (\cosh^2(b) - 1)(\cosh^2(c) - 1)\cosh^2(\alpha') = 0.$$

One has

$$\cosh(a) = 2CR(BC) - 1 = 2\frac{(C_1 - B_1)(C_2 - B_2)}{(B_2 - B_1)(C_2 - C_1)} - 1,$$

$$\cosh(b) = 2CR(CA) - 1 = 2\frac{(A_1 - C_1)(A_2 - C_2)}{(C_2 - C_1)(A_2 - A_1)} - 1,$$

$$\cosh(c) = 2CR(AB) - 1 = 2\frac{(B_1 - A_1)(B_2 - A_2)}{(A_2 - A_1)(B_2 - B_1)} - 1,$$

$$\cosh(\alpha') = 2CR(bc) - 1 = 2\frac{(c_1 - b_1)(c_2 - b_2)}{(b_2 - b_1)(c_2 - c_1)} - 1.$$

Now it is straightforward to check the identity by a tedious calculation. □

Remark. In spherical geometry no embedded right-angled hexagons exist. However, every spherical triangle Δ can be completed to an immersed spherical right-angled hexagon $H_\Delta = (A, B, C)$. To do this, extend each pair of sides of Δ beyond their intersection point until reaching a pair of points realizing the maximal distance along their supporting lines. Connecting at three places those pairs of points by shortest segments constructs an immersed right-angled spherical hexagon. Here, the lengths of these segments correspond to the angles of the triangle Δ. At the vertex A of Δ one adds a double right-angled spherical triangle. To both sides b, c one adds at A segments of equal length α''. By the spherical Pythagoras' Theorem the length α'' is given by $\cos(\alpha'') = \cos^2(\alpha)$.

Using the notation as above, the spherical length α' of the added segment between the lines supporting the sides b, c is equal to the angle α of the triangle Δ. Using consistent notation, the lengths of the sides of the hexagon H_Δ are in cyclic order $\alpha', \alpha'' + b + \gamma'', \gamma', \gamma'' + a + \beta'', \beta', \beta'' + c + \alpha''$.

A Spherical Cosine Law holds for the $\perp$-hexagon H_Δ which in fact reduces to the Spherical Cosine Law for the triangle Δ.

The triangle Δ can be retrieved from the immersed hexagon H_Δ as the triangle spanned by its double points.

Exercise 3.40. Let $p = (p_i)_I$ be a countable system of points in $\mathbb{H}$, weighted by a weight function $w \colon I \to \mathbb{R}_{\geq 0}$ with $\sum_I w_i = 1$. Assume that the set $\{p_i\} \subset \mathbb{H}$ is of finite diameter. Define the energy $E(q; p, w)$ of (p, w) with respect to a point $q \in \mathbb{H}$ by $E(q; p, w) := -1 + \sum_I w_i \cosh(d_{\mathbb{H}}(q, p_i))$. Show that the map $q \in \mathbb{H} \mapsto E(q; p, w) \in \mathbb{R}$ achieves its minimum at a unique point $b(p, w)$.

Hint: Show that the restriction of this map to a geodesic in $\mathbb{H}$ is strictly convex.

Exercise 3.41. Show, in the case that the cardinality of I is finite, that $b(p, w)$ depends differentiably upon the multi-point p and the weights w. We call the point $b(p, w)$ the energy-barycenter of the weighted multi-point (p, w).

Exercise 3.42. Construct hexagons with 6 equal angles in $\mathbb{H}$. Let a, b, c the lengths of 3 sides that pairwise do not have a vertex in common and let $\alpha \in [0, \frac{2\pi}{3}]$ be the common angle. Compute the three other lengths.

Exercise 3.43. Let $h = (A_1, A_2, \ldots, A_6)$ and $h' = (A'_1, A'_2, \ldots, A'_6)$ be hyperbolic hexagons with oriented numberings of the vertices. Assume h to be right angled, h' to have angles $> \frac{\pi}{2}$ and that the sides of h' have smaller length than the corresponding side in h. Let $B = b(A_1, A_2, \ldots, A_6), B' = b(A'_1, A'_2, \ldots, A'_6)$ be the energy-barycenters. Show that there exists a unique isometric inclusion of h' in h such that $B = B'$ and the points $B = B', A'_1, A_1$ are aligned. We call such an inclusion a rigid inclusion. If the other pairs of vertices are required to be aligned, then h' admits six rigid inclusions in h.

Exercise 3.44 (Busemann's axiom for negative curvature). [4] In a hyperbolic triangle ABC the length of the segment joining the mid-points of the sides AB and AC is strictly less than half the length of the side BC. See [17].

Together with right angled triangles, the Khayyám–Saccheri[5] and Lambert quadrilaterals[6] are basic building blocks in hyperbolic and spherical trigonometry.

A Lambert quadrilateral is a quadrilateral $AOBC$ with three right angles, say at A, O and B. A Khayyám–Saccheri quadrilateral is a quadrilateral $ABCD$ with right angles at B, C and sides AB and CD of equal length.

A generalized Lambert quadrilateral is a quadrilateral with three right angles, say at A, O and B such that the sides opposite to the sides OA and OB do not necessarily meet. A generalized Lambert quadrilateral is up to isometry characterized by the lengths $|AO|, |BO|$. We distinguish three cases:

 (i) The (ordinary) ones: the opposite sides meet in a point C.
 (ii) The ideal ones: the opposite sides do not meet, but have asymptotic ends.
 (iii) The trans-ideal ones: the opposite sides do not meet, and the ends are not asymptotic.

---

[4] Named after German-American mathematician Herbert Busemann (1905–1994)
[5] Named after Persian mathematician, astronomer, philosopher, and poet Omar Khayyám (1048–1131) and Italian philosopher and mathematician Giovanni Girolamo Saccheri (1667–1733).
[6] Named after Swiss mathematician and physician Johann Heinrich Lambert (1728–1777).

One can think of the rays $[O, A[$ and $[O, B[$ as orthogonal coordinate axes X_+, Y_+ and the points $A \in X_+$, $B \in Y_+$ as the orthogonal projections of the existing, or non-existing point C of the quadrilateral. Think of the numbers $x = \tanh(|OA|)$, $y = \tanh(|OB|)$ as "coordinates of the point C". If $A \notin X_+$ but $A \in X$, where X is the line carrying X_+, then define the coordinate number by $x = -\tanh(|OA|)$. Extend the definition of the coordinate y accordingly.

Now, the hyperbolic plane $\mathbb{H}$ has coordinates $p \in \mathbb{H} \mapsto (x(p), y(p)) \in \mathbb{R}^2$. Not every pair (a, b) of numbers is realized as coordinates of a point $p \in \mathbb{H}$. Let $\mathbb{K}$ be the open unit disk in $\mathbb{R}^2$.

Theorem 3.45 (Kleinian Model $\mathbb{K}$). [7] *The map $p \in \mathbb{H} \mapsto (x(p), y(p)) \in \mathbb{K}$ is a bijection. Images of hyperbolic lines are chords in $\mathbb{K}$. The hyperbolic metric $d_{\mathbb{H}}$ is transformed to the Hilbert metric $d_{\mathbb{K}}$. The space $(\mathbb{K}, d_{\mathbb{K}})$ is the so-called Kleinian model of the hyperbolic plane.*

Proof. Let X, Y be lines in $\mathbb{H}$ that intersect each other with a right angle at the point O. Think of X, Y as coordinate axes. For $p \in \mathbb{H}$ let p_X, O, p_Y, p be the Lambert quadrilateral where p_X, p_Y are the projections of p on the axes. Denote by α, β the angles pOp_X, pOp_Y. Observe $\cos(\alpha)^2 + \cos(\beta)^2 = 1$ since $\alpha + \beta = \frac{\pi}{2}$. From $\cos(\alpha) = \frac{x(p)}{\tanh(|Op|)}$ and $\cos(\beta) = \frac{y(p)}{\tanh(|Op|)}$ follows $\tanh(|Op|)^2 = x(p)^2 + y(p)^2$. By varying p over $\mathbb{H}$ the quantity $\tanh(|Op|)$ varies over $[0, 1[$. Now it follows that the map $p \in \mathbb{H} \mapsto (x(p), y(p)) \in \mathbb{K}$ is a bijection. We leave the rest of the proof as an exercise. Hint: the coordinate map is equivariant with respect to rotation of $\mathbb{H}$ with center O and rotations of $\mathbb{R}^2$ with center $(0, 0)$. $\qquad\square$

From this proof also follows

Theorem 3.46. *A generalized Lambert quadrilateral AOB with $a = \tanh(|AO|)$, $b = \tanh(|BO|)$ is ordinary if $a^2 + b^2 < 1$, ideal if $a^2 + b^2 = 1$, and trans-ideal if $a^2 + b^2 > 1$.* $\qquad\square$

In the forthcoming proof of the Uniformization Theorem we will use that the lengths u, v of two adjacent edges of a right-angled hyperbolic hexagon satisfy $\tanh(u)^2 + \tanh(v)^2 > 1$. Indeed, four consecutive edges define a trans-ideal Lambert quadrilateral. Consequently

Theorem 3.47 (Degeneration of hyperbolic $\perp$-hexagons). *Let $(H_n)_{n \in \mathbb{N}}$ be a sequence of labeled hyperbolic $\perp$-hexagons. Let d_n be the distance between the lines that support the edge U_n and the opposite edge U'_n. If the edge length u_n of the edge U_n tends to 0 for $n \to \infty$, then the edge length v_n of an adjacent edge V_n tends to ∞ with*

$$v_n \geq \operatorname{arctanh}(\sqrt{1 - \tanh(u_n)^2}).$$

The distance d_n tends to ∞ with

$$d_n \geq \operatorname{arctanh}\left(\sqrt{1 - \tanh(u_n)^2 - \tanh(\frac{u_n}{2})^2}\right).$$

---

[7] Named after German mathematician Felix Christian Klein (1849–1925).

Proof. The first inequality follows from $\tanh(u_n)^2 + \tanh(v_n)^2 > 1$. For the second inequality observe that a Saccheri quadrilateral S with bottom edge U_n and two perpendicular sides of length $A = \operatorname{arctanh}(\sqrt{1 - \tanh(u_n)^2})$ can be placed inside H_n. Let h be the height of S, i.e. the distance between U_n and the opposite edge U_n'' in S. Observe $d_n \geq h$. The Saccheri quadrilateral S is the union of two Lambert quadrilaterals L, L'. Let δ be the length of the diagonal in L' that connects the midpoint of U_n with the acute angle of L'. Now observe, by using "Cartesian coordinates" in L' and the hyperbolic Pythagoras' Theorem,

$$\tanh(h)^2 + \tanh\left(\frac{u_n}{2}\right)^2 = \tanh(\delta)^2 \geq \tanh(A)^2.$$

Hence,

$$d_n \geq h \geq \operatorname{arctanh}\left(\sqrt{1 - \tanh(u_n)^2 - \tanh\left(\frac{u_n}{2}\right)^2}\right). \qquad \square$$

3.4 Hyperbolic Area

The angular excess $\epsilon(ABC) = \alpha + \beta + \gamma - \pi$ of a spherical triangle $\delta(ABC)$ was discovered as a measure of area by Albert Girard[8] and Thomas Harriot[9] [8]. A direct inspection shows that the angular excess behaves in an additive way under geodesic subdivision.

A similar behavior is known in hyperbolic geometry for the angular defect $\delta(ABC) = \pi - \alpha - \beta - \gamma$ that was introduced by Lambert. See [3].

Let $\mu_{\mathbb{H}}$ be the Lebesgue measure on $\mathbb{H}$ induced by the Riemannian metric $g_{\mathbb{H}}$ of Gaussian curvature -1. Similarly on the Euclidean unit sphere $\mathbb{S}$ let $\mu_{\mathbb{S}}$ be the Lebesgue measure induced by the Riemannian metric $g_{\mathbb{S}}$ of Gaussian curvature $+1$.

In both geometries by integration we get an area function defined on triangles that behaves in an additive way under geodesic subdivision:

$$A_{\mathbb{H}}(ABC) = \int_{\Delta(ABC)} d\mu_{\mathbb{H}}, \quad A_{\mathbb{S}}(ABC) = \int_{\Delta(ABC)} d\mu_{\mathbb{S}}.$$

The defect/excess functions and the area functions are moreover invariant under isometries.

It follows that the angular defect/excess functions and the area functions are proportional. So there exist gauge constants $\lambda_{\mathbb{H}}, \lambda_{\mathbb{S}}$ such that for triangles ABC in $\mathbb{H}$ or $\mathbb{S}$ the following formulas hold

$$A_{\mathbb{H}}(ABC) = \lambda_{\mathbb{H}}\, \delta(ABC), \quad A_{\mathbb{S}}(ABC) = \lambda_{\mathbb{S}}\, \epsilon(ABC).$$

In order to calibrate these gauge constants, an example will suffice.

---

[8] Albert Girard (1595–1632), French-Dutch mathematician.

[9] Thomas Harriot (1560–1621), English mathematician and astronomer.

For the spherical geometry the half sphere considered as a triangle ABC with vertices on the equator gives

$$A_{\mathbb{S}}(ABC) = 2\pi = \pi + \pi + \pi - \pi = \epsilon(ABC).$$

This shows $\lambda_{\mathbb{S}} = 1$.

We have used that the total area of the unit two-sphere $\mathbb{S}$ is 4π, as computed by Archimedes of Syracuse[10] [7].

Let ABC in $\mathbb{H}$ be the triangle with image $z(\Delta(ABC))$ under the coordinate z in $\mathbb{C}_+$ that is limited by the lines $L_{-1,\infty}, L_{-1,1}, L_{1,\infty}$. The angles are $\alpha = \beta = \gamma = 0$. One computes

$$\delta(ABC) = \pi - 0 - 0 - 0 = \pi,$$

$$A_{\mathbb{H}}(ABC) = \int_{z(\Delta(ABC))} \frac{dx \wedge dy}{y^2} = \int_{-1}^{+1} dx \int_{\sqrt{1-x^2}}^{+\infty} \frac{dy}{y^2} = \int_{-1}^{+1} \frac{dx}{\sqrt{1-x^2}} = \pi$$

and concludes $\lambda_{\mathbb{H}} = 1$.

As an application, a regular N-gon P_N with right angles in $\mathbb{H}$, $N \geq 5$, has area $(N-4)\frac{\pi}{2}$. Indeed, P_N is the union of $2N$ triangles with angles $\frac{\pi}{N}, \frac{\pi}{2}, \frac{\pi}{4}$. This union is not disjoint, but the union of their interiors is. Even more generally, the area of an N-gon P_N with angles $\alpha_1, \alpha_2, \ldots, \alpha_N$ is $(N-2)\pi - \sum_i \alpha_i$.

Exercise 3.48. The area of a hyperbolic right-angled hexagon equals π. The side lengths of a regular right-angled hyperbolic hexagon equal $\log(2 + \sqrt{3})$. Compute the area of a right-angled hyperbolic tetragon.

Exercise 3.49. Let H be a hyperbolic right-angled hexagon with three sides of length $< \log(2+\sqrt{3})$. Show that the union of those sides is not connected. Hint: assuming the contrary, construct a regular right-angled hyperbolic hexagon containing H. Show that the other sides have length $> \log(2 + \sqrt{3})$.

3.5 A Compact Hyperbolic Surface of Genus $g \geq 2$

With the help of hyperbolic trigonometry we can construct a compact hyperbolic surface for every genus $g \geq 2$.

Theorem 3.50. *For any $g \geq 2$, there exists a compact hyperbolic surface S_g of genus g. The hyperbolic area of S_g is $-2\pi\chi(S_g)$. The surface is a $4g$ fold branched cover of the sphere $\Sigma_{1,1/4g,1/4g}$ with 3 orbifold angles $\dfrac{\pi}{1}, \dfrac{\pi}{4g}, \dfrac{\pi}{4g}$. The covering map is a Belyi map.*

A hyperbolic surface S is a manifold given by families $(U_\alpha \subset \mathbb{H})_{\alpha \in A}, (U_{\beta\alpha} \subset U_\alpha)$ and $(\phi_{\alpha\beta} : U_{\alpha\beta} \to U_{\beta\alpha})_{(\alpha,\beta) \in A \times A}$. Note that the subsets U_α are open subsets of $\mathbb{H}$. Moreover, the maps $\phi_{\alpha\beta} : U_{\alpha\beta} \to U_{\beta\alpha}$ preserve all the structure that we have

---

[10] Archimedes of Syracuse (287 BC–212 BC), Greek mathematician, physicist, engineer, and astronomer

in the hyperbolic plane $\mathbb{H}$. This means that the map $\phi_{\alpha\beta}\colon U_{\alpha\beta} \to U_{\beta\alpha}$ is the restriction of an automorphism $\Phi_{\alpha\beta}\colon \mathbb{H} \to \mathbb{H}$. In particular, $\phi_{\alpha\beta}$ is differentiable, so S is a differentiable manifold. But $\Phi_{\alpha\beta}$ preserves the fields $p \in \mathbb{H} \mapsto \mathrm{Hess}_p$ and $p \in \mathbb{H} \mapsto J_p$, so S is equipped with a Riemannian metric and also with a complex structure. Since the local geometry is the local geometry inherited from the hyperbolic plane, we say that S is a hyperbolic surface.

Proof. Let $g \in \mathbb{N}$, $g > 1$. First we construct a hyperbolic triangle T, see Fig. 3.10, with angles $\dfrac{\pi}{2}, \dfrac{\pi}{4g}, \dfrac{\pi}{4g}$. This is possible, since $\dfrac{\pi}{2} + \dfrac{\pi}{4g} + \dfrac{\pi}{4g} < \pi$. It is a triangle with a

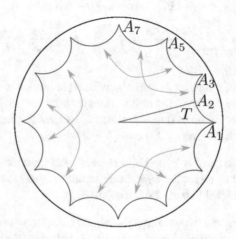

Fig. 3.10 Hyperbolic 12-gon in the Poincaré disk with identifications.

right angle. The hypotenuse has length $\mathrm{arcCosh}\left(\dfrac{\cos(\pi/4g)^2}{\sin(\pi/4g)^2}\right)$ and both catheti are of equal length $\mathrm{arcCosh}\left(\dfrac{\cos(\pi/4g)}{\sin(\pi/4g)}\right)$. We put $8g$ such triangles together and construct a regular $4g$-gon P in the hyperbolic plane $\mathbb{H}$ with angles $\dfrac{2\pi}{4g}$ and side length $2\,\mathrm{arcCosh}\left(\dfrac{\cos(\pi/4g)}{\sin(\pi/4g)}\right)$. The $4g$-gon P is the tile of a tessellation of the hyperbolic plane $\mathbb{H}$. We will describe the manifold S with $8g + 1$ charts $U_0, U_1, U_2, \ldots, U_{8g}$. The open set U_0 is the interior of the $4g$-gon P. We number its vertices following the boundary of P and the midpoints of the edges by $A_1, A_2, \ldots, A_{8g}$. Numberings are cyclic, i.e. $A_{8g+j} = A_j$. Let $E_1, E_2, \ldots, E_{8g}$ be the (half) edges: $E_i = A_i A_{i+1}$, $1 \le i \le 8g$. Let U_i be an open ϵ-tubular neighborhood of the edge E_i. Here $0 < \epsilon \ll \mathrm{arccos}\left(\dfrac{\cos(\pi/4g)}{\sin(\pi/4g)}\right)$. In order to describe the manifold S we have to define subsets $U_{ji} \subset U_i$ and a family $\phi_{ji}\colon U_{ji} \to U_{ij}$ of gluing maps.

If $U_i \cap U_j \ne \emptyset$ we put $U_{ij} = U_i \cap U_j = U_{ji}$ and $\phi_{ij} = \mathrm{Id}_{U_j \cap U_i}\colon U_{ij} \to U_{ji}$.

For $k = 0, 1, 2, \ldots, g - 1$ define

$$U_{4k+6,4k+1} := U_{4k+1}, \quad U_{4k+5,4k+2} := U_{4k+2},$$

$$U_{4k+8,4k+3} := U_{4k+3}, \quad U_{4k+7,4k+4} := U_{4k+4}.$$

We orient the edges E_i from A_i to A_{i+1} if i is odd, otherwise from A_{i+1} to A_i. For $k = 0, 1, 2, \ldots, g - 1$, let

$$\Phi_{4k+6,4k+1} \colon \mathbb{H} \to \mathbb{H}$$

be the isometry with $\Phi_{4k+6,4k+1}(U_0) \cap U_0 = \emptyset$ that maps the oriented edge E_{4k+1} to the oriented edge E_{4k+6}. Let

$$\Phi_{4k+5,4k+2} \colon \mathbb{H} \to \mathbb{H}$$

be the isometry with $\Phi_{4k+5,4k+2}(U_0) \cap U_0 = \emptyset$ that maps the oriented edge E_{4k+2} to the oriented edge E_{4k+5}. The isometries $\Phi_{4k+6,4k+1}^{\pm 1}$, $\Phi_{4k+5,4k+2}^{\pm 1}$ correspond to the green arrows in the figure. Let $\phi_{4k+6,4k+1}, \phi_{4k+5,4k+2}$ be the restrictions to $U_{4k+6,4k+1}$, $U_{4k+5,4k+2}$. Let $\sim$ be the equivalence relation on $\dot{\bigcup}_{0 \leq i \leq 8g} U_i$ generated by the gluings

$$p \sim \phi_{i,j}(p), \; p \in U_{ij}, \; U_i \cap U_j \neq \emptyset,$$

$$p \sim \phi_{4k+6,4k+1}(p), \; p \in U_{4k+6,4k+1},$$

$$p \sim \phi_{4k+5,4k+2}(p), \; p \in U_{4k+5,4k+2}.$$

The manifold $S_g = S := \dot{\bigcup}_{0 \leq i \leq 8g} U_i / \sim$ is now defined.

The $4g$-gon P is the union of $8g$ triangles with angles $\dfrac{\pi}{2}, \dfrac{\pi}{4g}, \dfrac{\pi}{4g}$. Each triangle maps by $q \colon \dot{\bigcup}_{0 \leq i \leq 8g} U_i \to S := \dot{\bigcup}_{0 \leq i \leq 8g} U_i / \sim$ injectively to S. In S we will have $(2+2g)$ vertices, $12g$ edges and $8g$ triangles. Hence $\chi(S) = (2+2g) - 12g + 8g = 2 - 2g$ is the Euler characteristic and the genus of S equals g. For Belyi maps, and Belyi's Theorem, see the book by Ernesto Girondo and Gabino Gonzales-Diáz [40]. □

The same construction works for $g = 1$. Let P be a square in the Euclidean plane. We introduce vertices $A_1, A_2, \ldots, A_8$ in the plane as before, also open sets $U_0, U_1, \ldots, U_8$ and corresponding gluings ϕ_{ij}. The resulting surface will be a surface of genus 1, i.e. a torus, but with a Euclidean structure.

A genus 0 surface can be obtained by starting out with two discs. No Euclidean motion will allow us to glue them in order to make a manifold. But don't worry, a surface with spherical geometry is already present, the unit two-sphere S^2 in Euclidean three space.

Exercise 3.51. A topological pair of pants is a compact surface of genus 0 with three boundary components. Construct a hyperbolic pair of pants P by gluing three pairs of corresponding edges of two copies H_1, H_2 of a hyperbolic polygon with 6 edges and right angles. The boundary components are closed geodesics. Up to isometry the lengths of the 3 boundary components determine the hyperbolic pair of pants P. Show that each isometry of P preserves the decomposition into a pair of hexagons.

Exercise 3.52. Study the topology of the surface $\{p \in \mathbb{C}^2 \mid w(p)^2 = z(p)^5 - 1\}$.

Exercise 3.53. Compute the genus of the Fermat surface $\{p \in \mathbb{P}^2(\mathbb{C}) \mid x(p)^d + y(p)^d + z(p)^d = 0\}$ of degree d.

In spherical, Euclidean and hyperbolic geometry a triangle is determined up to isometry by its side lengths. Hence the area is a function of the side lengths. It is interesting to have explicit expressions for the area.

Theorem 3.54 (Heron, Huilier). *Suppose we have a triangle ABC with sides of length a, b, and c, and semi-perimeter s. In spherical geometry, the angular excess $\epsilon(ABC)$ is given by Huilier's formula*

$$\tan\left(\frac{\epsilon(ABC)}{4}\right) = \sqrt{\left(\tan\left(\frac{s}{2}\right)\tan\left(\frac{s-a}{2}\right)\tan\left(\frac{s-b}{2}\right)\tan\left(\frac{s-c}{2}\right)\right)}.$$

In hyperbolic geometry, the angular defect $\delta(ABC)$ is given by

$$\tan\left(\frac{\delta(ABC)}{4}\right) = \sqrt{\left(\tanh\left(\frac{s}{2}\right)\tanh\left(\frac{s-a}{2}\right)\tanh\left(\frac{s-b}{2}\right)\tanh\left(\frac{s-c}{2}\right)\right)}.$$

In Euclidean geometry, the area A is given by Heron's formula

$$A = \sqrt{s(s-a)(s-b)(s-c)}.$$

3.6 The Riemann Sphere $\mathbb{C} \cup \{\infty\}$

In the above construction, the hyperbolic plane $\mathbb{H} = \mathbb{H}_I$ was a subset of the quotient space $S/\sim$ where we started with the set

$$S = \{\phi \colon \mathbb{R}[X] \to F \mid \phi \text{ surjective ring homomorphism to a field}\}.$$

Up to isomorphism of fields, the only fields occurring as the image of a surjective ring homomorphism $\phi \colon \mathbb{R}[X] \to F$ are the three subfields $F_1 = \{0\}, \mathbb{R}, \mathbb{C}$ of $\mathbb{C}$. So, we may replace the set S by

$$S = \{\phi \colon \mathbb{R}[X] \to \mathbb{C} \mid \phi \text{ surjective ring homomorphism to } F_1, \mathbb{R} \text{ or } \mathbb{C}\}$$

or by

$$S = \{\phi \colon \mathbb{R}[X] \to \mathbb{C} \mid \phi \ \mathbb{R}\text{-algebra homomorphism to } \mathbb{C}\}.$$

In order to have a better understanding of the set S we use a coordinate Ev_X on this set. Here consider $\mathbb{R}$ as a subfield of $\mathbb{C}$ and denote the ring homomorphism to the less common field F_1 by ∞. Here "∞" is just a symbol with the property: $\infty \notin \mathbb{C}$. We define the bijective coordinate $\mathrm{Ev}_X \colon S \to \mathbb{C} \cup \{\infty\}$ by $\mathrm{Ev}_X(\infty) = \infty \in \mathbb{C} \cup \{\infty\}$ and by the evaluation $\mathrm{Ev}_X(\phi) = \phi(X) \in \mathbb{C} \cup \{\infty\}$ if $\phi \neq \infty$. Now identify S and $\mathbb{C} \cup \{\infty\}$ via Ev_X. As a natural topology T on $\mathbb{C} \cup \{\infty\}$, hence on S, choose the only compact topology that on $\mathbb{C}$ induces the usual norm topology. The topological space (S, T) is homeomorphic to the 2-sphere S^2.

Bernhard Riemann[11], by using the coordinates $z, \frac{1}{z}$ on $\mathbb{C}$ and on $(\mathbb{C} \setminus \{0\}) \cup \{\infty\}$ respectively, made $\mathbb{C} \cup \{\infty\}$ into a Riemann surface of genus 0. See Chapter 6.

Coordinates $z, \frac{1}{z} : \mathbb{P}^1(\mathbb{C}) \to \mathbb{C} \cup \{\infty\}$ allow us to identify the space $\mathbb{P}^1(\mathbb{C})$ of vector subspaces of dimension 1 in $\mathbb{C}^2$ with $\mathbb{C} \cup \{\infty\}$. Namely define $z(L)$, $L \in \mathbb{P}^1(\mathbb{C})$, by $z(L) = \frac{a}{b}$ with $(a, b) \in L \setminus \{0\}$.

More generally, similar coordinates allow for any field K to identify the projective space $\mathbb{P}^1(K)$ of 1-dimensional subspaces of K^2 with $K \cup \{\infty\}$.

It is a surprise to us that the set S of all ring homomorphisms of $\mathbb{R}[X]$ to a field becomes, in a natural way, a manifold diffeomorphic to the two-sphere in three space together with a J-field and a rich panoply of geometries. The study of the two-sphere has a long history: for instance, by Archimedes, as early as 225 BC, see [7], and Menelaus, as early as 75 BC, see the book of Roshdi Rashed and Athanase Papadopoulos [100]. The remarkable study of Menelaus introduces spherical geometry axiomatically without using the sphere in Euclidean 3-space as a model.

---

[11] Bernhard Riemann (1826–1866), German mathematician, was Professor in Göttingen

Chapter 4
Some Examples and Sources of Geometry

Life takes place in the Universe. Humanity is doomed to study "bon gré, mal gré" the geometry of the Universe. Explore the book *The Road to Reality. A Complete Guide to the Laws of the Universe* by Roger Penrose [91]. It is hard to imagine a definition of Geometry. The understanding of RNA is a challenge by which Geometry, in particular Topology, will grow. See the work of Robert Penner [89, 90].

4.1 The Space of Norms

In the above presentation of the hyperbolic plane $\mathbb{H}_I$ we first introduced a set E of points, where each point represented a mathematical object of some kind. In a next step one studies how these points interact. In this way the set obtains a rich structure and becomes a space E with geometry.

The next example is based on a private communication with Bernard Mischler [79], who taught me the following.

Let V be a vector space. Let $\mathbb{B} = \mathbb{B}(V)$ be the set of norms on V. The structure will be an internal composition law

$$\mu \colon \mathbb{B} \times \mathbb{B} \to \mathbb{B}.$$

In his first step Mischler introduces a map $\mathrm{Diagr} \colon \mathbb{B} \times \mathbb{B} \times V \to \mathbb{D}(\mathbb{P}_+)$ to the space of Newton diagrams $\mathbb{D}(\mathbb{P}_+)$ in $\mathbb{P}_+ := \mathbb{R}_+ \times \mathbb{R}_+$. A Newton diagram in $\mathbb{P}_+$ is a closed convex subset D in $\mathbb{P}_+$ such that the intersections of D with each axis $\{(a,b) \in \mathbb{P}_+ | a = 0\}$ or $\{(a,b) \in \mathbb{P}_+ | b = 0\}$ is non-bounded. The set of Newton diagrams itself has an interesting geometry, which we will not explain further here, except that in particular, for each Newton diagram D one has a number $\alpha(D) \in \mathbb{R}$, namely the area of $\mathbb{P}_+ \setminus D$.

Given two norms N, M on V and a vector $u \in V$, the Newton diagram $\mathrm{Diagr}(N, M)(u)$ attached to N, M, u is the set

$$\mathrm{Diagr}(N, M)(u) := \{(r, s) \in \mathbb{P}_+ \, | \, B^N(0, r) \cap B^M(u, s) \neq \emptyset\}.$$

© The Author(s), under exclusive license to Springer Nature Switzerland AG 2021
N. A'Campo, *Topological, Differential and Conformal Geometry of Surfaces*, Universitext,
https://doi.org/10.1007/978-3-030-89032-2_4

95

In other words, $\mathrm{Diagr}(N, M)(u)$ is the set of pairs of radii $(r, s) \in \mathbb{P}_+$ such that the closed N-ball with center $0 \in V$ and radius r intersects the closed M-ball with center $u \in V$ and radius s. The binary composition μ of norms $(N, M) \mapsto \mu(N, M)$ is defined as follows: the square of the $\mu(N, M)$-norm of $u \in V$ is given by

$$\mu(N, M)(u)^2 := 2\alpha(\mathrm{Diagr}(N, M)(u)).$$

Theorem 4.1 (Bernard Mischler). *Let V be a real vector space with basis $(e_1, e_2, \ldots, e_n)$. Let N be the l^1-norm*

$$N(u) = \|u\|_{l^1} = \sum_{1 \le i \le n} |x_i(u)|$$

and M the l^∞-norm

$$M(u) = \|u\|_{l^\infty} = \mathrm{Max}_{1 \le i \le n} |x_i(u)|.$$

Then $\mu(N, M)$ is the l^2-norm

$$\mu(N, M)(u) = \|u\|_{l^2} = \left(\sum_{1 \le i \le n} |x_i(u)|^2 \right)^{1/2}.$$

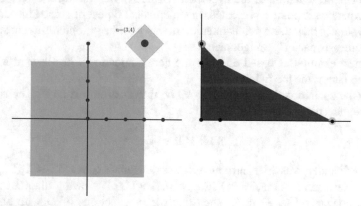

Fig. 4.1 Interpolation of l^1- and l^∞-norms. Twice the red area is the l^2-norm of $u = (3, 4) = (3, 3) + (0, 1)$.

An immediate consequence is

Theorem 4.2. *If a linear map $A \colon V \to V$ on a based vector space contracts the l^1- and the l^∞-norms, then A contracts the l^2-norm.*

We have learned once more that a set of mathematical objects of some given type has a rich structure and geometry.

Exercise 4.3. Work out the inequalities that are hidden in the above theorem, say for cases $\mathrm{Dim}(V) = 2, 3, 10$.

Exercise 4.4. Let $\| \cdot \|$ be a norm on $\mathbb{R}^n$ that takes integral values at vectors with integral coordinates. Show that the unit ball of $\| \cdot \|$ is the convex hull of a finite set.

The following example was explained at the *Topology Seminar 1977* in Orsay by A. Douady[1] and shows that the conclusion of the above exercise does not hold if one assumes that the norm takes rational values at vectors with integral coordinates. Douady's norm on $\mathbb{R}^2$ has as unit ball the convex hull of the union of the two discs of Euclidean radius 1 and centers at $(-1, 0)$ and $(1, 0)$.

The above exercise is the key step in the study of the Thurston norm on the cohomology group $H^1(M, \mathbb{Z}) \otimes_{\mathbb{Z}} \mathbb{R}$ of 3-manifolds M [120].

4.2 Combinatorial Geometry

As an example, let $E = \mathrm{Grass}_{4,2}(F_2)$ be the set of 2-dimensional linear subspaces of the 4-dimensional vector space F_2^4 over the field F_2 with two elements. We want to study structures on the set E.

The 2-point function $(a, b) \in E \times E \mapsto \delta(a, b) = 2 - \mathrm{Dim}_{F_2}(a \cap b)$ is a metric on E. The group $\mathrm{GL}_4(F_2)$ acts transitively on E, but does not act transitively on the set $E \times E \setminus E$ of distinct pairs. Indeed, the metric δ takes only the values $0, 1, 2$. The group $\mathrm{GL}_4(F_2)$ acts by isometries on (E, δ).

Exercise 4.5. Is $\mathrm{GL}_4(F_2)$ the full group of isometries of (E, δ)?

A ball $B^\delta(a, 1)$ has 19 elements. If $\delta(a, b) = 2$, then $\#B^\delta(a, 1) \cap B^\delta(b, 1) = 9$ and if $\delta(a, b) = 1$ then $\#B^\delta(a, 1) \cap B^\delta(b, 1) = 14$. In particular, (E, δ) is a homogeneous metric space with integral metric, of diameter 2, with balls of radius 1 having 19 elements and mutual intersections having 9 or 14 elements. In short, we say it is a $(19, 14, 9)$-triple space.

Many questions arise: For which triples (a, b, c) does there exist an (a, b, c)-triple space? Are $\mathrm{Grass}_{4,2}(F_q)$ with the metric $\delta(a, b) = 2 - \mathrm{Dim}_{F_q}(a \cap b)$ the only such spaces?

4.3 Spaces of Involutions

Recall that the spaces of involutions acting fixed-point-free on $\mathbb{P}^1(\mathbb{R})$ or $\mathbb{P}^1(\mathbb{C})$ that preserve the 4-point function given by cross-ratio are models of the hyperbolic spaces $\mathbb{H}^2$ and $\mathbb{H}^3$. The Busemann boundary of $\mathbb{H}^2$ can be identified with the real projective line $\mathbb{P}^1(\mathbb{R})$. Switching the endpoints of the geodesics passing through a point $p \in \mathbb{H}^2$ is a fixed-point-free involution I_p of $\mathbb{P}^1(\mathbb{R})$ that preserves the cross-ratio. A similar construction identifies points $p \in \mathbb{H}^3$ with cross-ratio preserving fixed-point-free involutions I_p of $\mathbb{P}^1(\mathbb{C})$. The involutions I_p extend to isometric involutions of $\mathbb{H}^2$ and $\mathbb{H}^3$. In short, points p are identified with isometric involutions that have p as the

---

[1] Adrien Douady (1935–2006), French mathematician.

only fixed point. At the fixed point p the differential of the involution I_p is the $-\mathrm{Id}$ involution of the tangent space at p.

This last formulation was used by Élie Cartan as an axiom in the definition of an important class of geometric spaces [20, 21]. A *Riemannian symmetric space* is a connected Riemannian manifold X such that each point $p \in X$ is a fixed point of an isometric involution I_p with $\mathrm{D}_p I_p = -\mathrm{Id}_{T_p X}$.

The first examples are the Euclidean spaces, spheres, projective spaces and hyperbolic spaces. There are many more. A basic example is the space of Euclidean norms on $\mathbb{R}^n$ with unit balls of volume normalized by 1. This space can be parametrized by the homogeneous space $\mathrm{SL}(n, \mathbb{R})/\mathrm{SO}(n)$.

Symmetric spaces and Involutions are central in Geometry, Mathematics and Physics etc., see the books by Helgason [50] and by Knus, Merkurjev, Rost and Tignol [60].

4.4 Conflicts and Dynamics

Consider the map $T: \mathbb{P}^1(\mathbb{C}) \to \mathbb{P}^1(\mathbb{C})$ given by $z \mapsto z^2 + 1/16$. For most points $z \in \mathbb{C}$ the dynamics of T produces by iteration a sequence $z, T(z), T^2(z), \ldots$ that converges to one of the locally contracting fixed points ∞, $a = \frac{1}{2} - \frac{1}{4}\sqrt{3}$ of T. So, $\mathbb{P}^1(\mathbb{C})$ is partitioned into three parts, the open subsets U_a, U_∞ consisting of those points that converge by iteration to a or ∞ and the closed subset U_u of those points that produce sequences that do not converge to a or ∞. Think of U_a and U_∞ as countries with one capital source of inspiration and truth and the dynamics of T as a way to quickly reach both inspiration and truth. It turns out that the third set U_u, where the two ideal sources are in conflict, has a very interesting self-similar geometry: the more one zooms in, geometric properties repeat at finer and finer scales!

Chapter 5
Differential Topology of Surfaces

5.1 0- and 1-de Rham Cohomology of Surfaces

Cohomology appears as soon as one wishes to understand globally problems that one can solve locally. The cohomology will be a bookkeeping device for the obstructions preventing the existence of global solutions, provided that there are local solutions. This is very philosophical and not a definition of cohomology! There exist many different types of cohomology theories and it is far from being true that all the global problems with local solutions can be understood through an appropriate cohomology theory.

Let us move to an example to explain this idea. Let S be a compact, connected, oriented surface of genus g. Let ω be a 1-differential form on S. We want to investigate the possibility of finding a function f on S with $df = \omega$. We already know the answer locally: the answer is locally yes if and only if the form ω is closed, i.e. if $d\omega = 0$ holds. So, the answer to the question whether there is a function on S with $df = \omega$ has a chance to be yes only if $d\omega = 0$ holds. Given two 1-forms ω_1, ω_2 on S, the problem of solving $df_1 = \omega_1$ is equivalent to the problem of solving $df_2 = \omega_2$ as soon as the difference 1-form $\omega_1 - \omega_2$ is *exact*, i.e. if $dg = \omega_1 - \omega_2$ for a function g holds. So the following quotient vector space measures how impossible it is to solve the equation $df = \omega$ with unknown function f:

$$\{\omega \in \Omega^1(S) \mid d\omega = 0\}/\{df \mid f \in \Omega^0(S)\},$$

where we have used the notation $\Omega^1(S)$ for the vector space of all 1-differential forms on S and the notation $\Omega^0(S)$ for the vector space of all functions on S. This quotient vector space, denoted by $H^1_{dR}(S)$, of the space of closed 1-forms modulo the space of exact 1-forms on S is called the *first de Rham cohomology group* of S.[1]

More generally, let M be a differentiable manifold. The quotient vector spaces

$$\{\omega \in \Omega^k(M) \mid d\omega = 0\}/\{d\eta \mid \eta \in \Omega^{k-1}(M)\}, \quad k = 0, 1, 2, \ldots$$

---

[1] In honor of the Swiss mathematician Georges de Rham (1903–1990).

measure how possible (better impossible) it is to solve the equation $d\eta = \omega$ with $\eta \in \Omega^{k-1}(M)$ for given $\omega \in \Omega^k(M)$ satisfying the necessary condition $d\omega = 0$. This quotient vector space is denoted by $H_{dR}^k(M)$ and is called the kth de Rham cohomology group of M. Its dimension is called the kth Betti number of M.[2] If $H_{dR}^k(M) = 0$ the equation $d\eta = \omega$ will have global solutions for those $\omega \in \Omega^k(M)$ with $d\omega = 0$. If $\text{Dim}(H_{dR}^k(M)) = n$ there will be n linearly independent "reasons", called *obstructions*, for the impossibility of solving the equation $d\eta = \omega$.

Theorem 5.1. *Let S be a compact, oriented, connected surface of genus g. Then* $\text{Dim}(H_{dR}^1(S)) = 2g$.

Proof. We only consider the case $g \geq 2$. Remember that the surface S can be obtained as $S := \dot{\cup}_{0 \leq i \leq 8g} U_i / \sim$. We map the open sets $U_i \subset \mathbb{H}$ by the coordinate $w : \mathbb{H}^2 \to \mathbb{D}$ to the Poincaré disk. The open set $w(U_i)$ in $\mathbb{D}$ is still denoted by U_i. Let $U := \cup_{0 \leq i \leq 8g} U_i \subset \mathbb{D}$. It is an open star-shaped subset of $\mathbb{D} \subset \mathbb{C} = V$. Let ω be a closed 1-differential form on S given by a family of forms $\omega_i : U_i \to V^*$. This family defines a closed form ω_U on U. Since U is star-shaped there exists a function $f_U : U \to \mathbb{R}$ with $d f_U = \omega_U$. Each time we use a gluing transformation $\phi_{ij} : U_{ij} \to U_{ji}$, we have to check that the solution f_U is invariant under this gluing. This means for the function f_U that for all $p \in U_{ij}$ the values $f_U(p)$ and $f_U(\phi_{ij}(p))$ coincide. This leads to necessary conditions for being able to define the function f_U on S, namely the equations

$$f_U(\phi_{ij}(p)) - f_U(p) = \int_{[p, \phi_{ij}(p)]} \omega_U = 0, \quad p \in U_{ij}.$$

Here $[p, \phi_{ij}(p)]$ is the segment in $\mathbb{D}$ on the hyperbolic line through the points $p, \phi_{ij}(p)$. The open sets $U_{i,j}$ are path connected, so if the above condition is fulfilled for one $p \in U_{i,j}$ then the conditions are fulfilled for all $p \in U_{i,j}$. The above necessary conditions reduce to a finite list of conditions.

We take for points p the midpoints A_{2h}, $h = 1, 2, \ldots, 4g$, of the edges of P. The obstructions t_k, s_k are the numbers

$$t_k = \int_{[A_{8k-6}, A_{8k-2}]} \omega_U, \quad s_k = \int_{[A_{8k-4}, A_{8k}]} \omega_U$$

and the above necessary conditions ask for the vanishing of the numbers t_k, s_k.

The oriented segment $[A_{8k-6}, A_{8k-2}]$ closes in the surface S up to a cycle a_k. Also the oriented segment $[A_{8k-4}, A_{8k}]$ closes in S to a cycle b_k. We can also express the obstructions as

$$t_k = \int_{a_k} \omega, \quad s_k = \int_{b_k} \omega,$$

which are in fact linear maps $t_k, s_k : \Omega^1(S) \to \mathbb{R}$. If all obstructions vanish, the function f_U becomes a function f on S solving $df = \omega$. We have proved $\text{Dim}(H_{dR}^1(S)) \leq 2g$.

---

[2] In honor of the Italian mathematician Enrico Betti (1823–1892).

The cycles a_k, b_h are pairwise disjoint, except for the pairs (a_k, b_k) that intersect on S in one point like coordinate axes intersect in the plane.

We proceed to describe a construction of a list $\alpha_k, \beta_k, 1 \leq k \leq g$, of closed 1-forms with $t_k(\beta_h) = -s_k(\alpha_h) = \delta_h^k, 1 \leq k, h \leq g$. We have used here the Kronecker symbol δ_h^k with values $\delta_h^k = 0$ if $k \neq h$ and $\delta_h^k = 1$ if $k = h$.

The circle a_k has a cylindrical tubular neighborhood C_k on S since S is orientable. Let $g_k \colon C_k \to \mathbb{R}$ be a function that is identically 1 near one boundary component of C_k and identically 0 near the other boundary component. Let dg_k be its differential. It is crucial that g_k does not extend to a function on S with vanishing differential on $S \setminus C_k$ but dg_k extends (by 0 outside of C_k) to a smooth 1-differential form on S. Let α_k be the extension of dg or of $-dg$ by 0 on S such that all obstructions evaluate to 0 except $\int_{b_k} \alpha_k = -1$. The construction of β_k in a tubular neighborhood of b_k is similar with $\int_{a_k} \beta_k = 1$.

This proves that the obstructions are linearly independent. Now the equality $\mathrm{Dim}(H_{\mathrm{dR}}^1(S)) = 2g$ follows. □

A system of oriented cycles a_i, b_i as above is called a *symplectic marking*. The corresponding system of 1-differential forms $\alpha_i, \beta_i, 1 \leq i \leq g$, as in the above proof provide a basis $([\alpha_i], [\beta_i]), 1 \leq i \leq g$, for the real vector space $H_{\mathrm{dR}}^1(S, \mathbb{R})$ as for the complex vector space $H_{\mathrm{dR}}^1(S, \mathbb{C})$. For each closed 1-differential form ω its cohomology class $[\omega]$ expands as

$$[\omega] = \sum_{i=1}^{g} \int_{a_i} \omega \ [\beta_i] - \int_{b_i} \omega \ [\alpha_i].$$

Observe that $\int_S \alpha_i \wedge \beta_j = \delta_i^j$ and $\int_S \alpha_i \wedge \alpha_j = \int_S \beta_i \wedge \beta_j = 0$.

Theorem 5.2 (Green–Riemann relation). *For closed 1-differential forms ω_1, ω_2 on a closed oriented surface S of genus $g \geq 1$ with a symplectic marking of cycles a, b the following identity holds:*

$$\int_S \omega_1 \wedge \omega_2 = \sum_{i=1}^{g} \int_{a_i} \omega_1 \int_{b_i} \omega_2 - \int_{b_i} \omega_1 \int_{a_i} \omega_2.$$

Proof. Both sides are alternating bilinear in the arguments ω_1, ω_2. Replacing an argument ω by $\omega + df$ does not change the values of both sides. Replacing the arguments by expansions

$$\omega = \sum_{i=1}^{g} \int_{a_i} \omega \ \beta_i - \int_{b_i} \omega \ \alpha_i$$

does not change the values. Clearly the identity holds for pairs with entries from the set $\{\alpha_i, \beta_j\}$. □

The differential df of a function f measures the local behavior of a function. Rolle's Theorem[3] implies that a function f with $df = 0$ is constant along every

---

[3] Named after French mathematician Michel Rolle (1652–1719).

differentiable path. A connected differentiable manifold is also connected by differentiable paths. It follows that the restrictions of a function f with $df = 0$ to the connected components of a differentiable manifold are constant functions. In particular:

Theorem 5.3. *For a connected non-empty manifold M the space $H^0_{dR}(M)$ is the space of constant functions and $\mathrm{Dim}(H^0_{dR}(M)) = 1$.*

Let a be a simply closed regularly parametrized curve on a compact orientable surface S of genus $g > 0$. We can extend the parametrization to an embedding of a cylinder $\Gamma\colon S^1 \times [-1, +1] \to S$ such that $t \in S^1 \mapsto \Gamma(t, 0) \in S$ parametrizes the curve a. The image of Γ is called a collar neighborhood of the curve a. Let $\chi\colon [-1, +1] \to \mathbb{R}$ be a function such that $\chi'(s) \geq 0$, $\chi(+1) - \chi(-1) = 1$ and $\chi'(s) = 0$ near ± 1. The push-forward 1-differential form $\Gamma_*(d\chi)$ on the image of Γ extends by 0 to a 1-differential form $\alpha(a)$ on S. Let b be a closed parametrized curve in S that intersects a in one point such that the speeds of a and b at this point are linearly independent. One has $\int_b \alpha(a) = \pm 1$. More generally, $\int_b \alpha(a) \in \mathbb{Z}$ counts with sign how many times a curve b goes through the collar neighborhood of the curve a.

Exercise 5.4. Do the computation of $H^1_{dR}(S_g)$ for the remaining cases of genus $g = 0, 1$.

Exercise 5.5. Compute $H^1_{dR}(K)$ for the Klein bottle K.

Exercise 5.6. Compute $H^1_{dR}(\mathbb{P}^2(\mathbb{R}))$.

Exercise 5.7. Let $A\colon \mathbb{R}^2 \to S^1 \times S^1$ be the map $p \in \mathbb{R}^2 \mapsto (e^{2\pi i x(p)}, e^{2\pi i y(p)}) \in S^1 \times S^1$. Let $\phi\colon S^1 \times S^1 \to S^1 \times S^1$ be a diffeomorphism of the torus. Construct a diffeomorphism $\Phi\colon \mathbb{R}^2 \to \mathbb{R}^2$ satisfying $\Phi(0) = 0$ and $A \circ \Phi = \phi \circ A$. Construct a linear map $\Psi\colon \mathbb{R}^2 \to \mathbb{R}^2$ such that $\Psi - \Phi$ is bounded. Show that Ψ maps $A^{-1}(1, 1)$ into $A^{-1}(1, 1)$. Identify $H^1_{dR}(S^1 \times S^1)$ and $\mathbb{R}^2$ such that Ψ becomes the induced map by Φ on $H^1_{dR}(S^1 \times S^1)$. Show that Ψ induces a diffeomorphism ψ of $S^1 \times S^1$, satisfying $A \circ \Psi = \psi \circ A$, that is homotopic, even isotopic, to ϕ.

Exercise 5.8. The augmented mapping class group of the torus $T^2 = S^1 \times S^1$ is the quotient group $\mathrm{AMod}(T^2) = \mathrm{Diff}(T^2)/\mathrm{Diff}_0(T^2)$, where $\mathrm{Diff}_0(T^2)$ is the group of diffeomorphisms that are homotopic or isotopic to the identity. Show that the groups $\mathrm{AMod}(T^2)$ and $\mathrm{GL}(2, \mathbb{Z})$ are isomorphic.

5.2 The Hyperbolic Plane Again, Now $\mathbb{H} = \mathbb{H}_J$

We work with different personalities. We define the hyperbolic plane as a set by

$$\mathbb{H}_J = \{J \in \mathrm{Hom}(\mathbb{R}^2, \mathbb{R}^2) \mid J \circ J = J^2 = -\mathrm{Id}_{\mathbb{R}^2}, J \text{ left oriented}\}.$$

Remember we use $\mathbb{H}$ for hyperbolic and J since the personalities are linear complex structures on $\mathbb{R}^2$. Let $J \in \mathbb{H}_J$. From $J \circ J = J^2 = -\mathrm{Id}_{\mathbb{R}^2}$ it follows that for all

$u \in \mathbb{R}^2, u \neq 0$, the vectors $u, J(u)$ are linearly independent. The imposed condition, namely J *left oriented*, means that the pair of vectors $(u, J(u))$ and the pair (e_1, e_2) of standard basis vectors give equal orientations to $\mathbb{R}^2$. We can represent J by a real two-by-two matrix $\left(\begin{smallmatrix} h & x \\ k & y \end{smallmatrix}\right)$. The condition of left orientation implies $k > 0$. The condition $J \circ J = J^2 = -\mathrm{Id}_{\mathbb{R}^2}$ implies that the eigenvalues are $i, -i$, so $\mathrm{Trace}(J) = 0$, meaning $y = -h$. The condition also implies $\mathrm{Det}(J) = 1$, so $x = \dfrac{-h^2 - 1}{k}$. Representing elements of $\mathbb{H}$ by matrices we conclude

$$\mathbb{H}_J = \left\{ \left(\begin{smallmatrix} h & * \\ k & -h \end{smallmatrix}\right) \mid h, k \in \mathbb{R}, k > 0, * = \frac{-h^2 - 1}{k} \right\}.$$

What are the symmetries, the interactions in $\mathbb{H}$? Is there a geometry defined by lines or by a two-point function?

A base change $f_1 = P(e_1), f_2 = P(e_2), P \in \mathrm{GL}^+(2, \mathbb{R})$ acts on $\mathbb{H}_J$ by $(P, J) \mapsto P \circ J \circ P^{-1}$. Here $\mathrm{GL}^+(2, \mathbb{R})$ is the group of linear automorphisms of $\mathbb{R}^2$ that preserve orientation.

We define as a line in $\mathbb{H}_J$ a subset L that is maximal for the following property: given $p, q, p', q' \in L \subset \mathbb{H}_J$ the compositions $p \circ q$ and $p' \circ q'$ commute in $\mathrm{GL}^+(2, \mathbb{R})$. This definition is based on the composition of elements in $\mathbb{H}_J$ with values in $\mathrm{GL}^+(2, \mathbb{R})$. Think of this composition as an interaction. The effective group of symmetries is the group $\mathrm{PSL}(2, \mathbb{R})$. We get a symmetric two-point function $T \colon \mathbb{H}_J \times \mathbb{H}_J \to \mathbb{R}$ with positive values by putting $T(p, q) := -\dfrac{1}{2}\mathrm{Trace}(p \circ q)$. The two-point function $d_{\mathbb{H}_J} \colon \mathbb{H}_J \times \mathbb{H}_J \to \mathbb{R}$ with

$$d_{\mathbb{H}_J}(p, q) := \log\!\left(T(p, q) + \sqrt{T(p, q)^2 - 1}\right)$$

is a symmetry-invariant distance function.

Recall that the group $\mathrm{PSL}(2, \mathbb{R})$ acts by homographic transformations on $\mathbb{C}_+$. The coordinate Fix is defined by

$$\mathrm{Fix} \colon \mathbb{H}_J \to \mathbb{C}_+, \ \left(\begin{smallmatrix} h & * \\ k & -h \end{smallmatrix}\right) \mapsto a + bi \in \mathbb{C}_+,$$

where $a + bi \in \mathbb{C}_+$ is the fixed point of the homographic transformation

$$u \in \mathbb{C}_+ \mapsto \frac{hu + *}{ku - h} \in \mathbb{C}_+, \quad * = \frac{-h^2 - 1}{k}.$$

We have $a + bi = \dfrac{h}{k} + \dfrac{i}{k}$. The coordinate Fix is an equivariant bijection with respect to the symmetries by $\mathrm{PSL}(2, \mathbb{R})$. One should not be surprised that Fix maps lines in $\mathbb{H}_J$ to half circles perpendicular to the real axes in $\mathbb{C}_+$, and that the composition $\Pi \circ \mathrm{Fix}$ maps lines in $\mathbb{H} = \mathbb{H}_J$ to lines in $\mathbb{H}_I$. The inverse map

$$\mathrm{Xif} \colon \mathbb{C}_+ \to \mathbb{H}_J$$

is given by

$$a + bi \in \mathbb{C}_+ \mapsto \mathrm{Xif}(a + bi) = \begin{pmatrix} a/b & * \\ 1/b & -a/b \end{pmatrix} \in \mathbb{H}_J.$$

Theorem 5.9. *The tangent space $T_J(\mathbb{H}_J)$ at $J \in \mathbb{H}_J$ is the vector space of all endomorphisms h of $\mathbb{R}^2$ that anti-commute with J.*

Proof. Up to first order an endomorphism h deforms J to an almost complex structure $J + h$ on $\mathbb{R}^2$ if $J \circ h + h \circ J = 0$ since

$$(J + h)^2 = J^2 + J \circ h + h \circ J + h \circ h$$

and $J^2 = -\mathrm{Id}$.   □

The following theorem constructs a natural J-field on the manifold $\mathbb{H}_J$. Thus the hyperbolic plane has a natural complex structure.

Theorem 5.10. *The map $J_J \colon h \in T_J \mathbb{H}_J \mapsto h \circ J \in T_J \mathbb{H}_J$ defines a J-field on $\mathbb{H}_J$.*

Proof. For $h \in T_J \mathbb{H}_J$ the endomorphism $k = h \circ J$ anti-commutes with J. Indeed,

$$k \circ J + J \circ k = h \circ J \circ J + J \circ h \circ J = h \circ J \circ J - h \circ J \circ J = 0.$$

Clearly $J_J \circ J_J = -\mathrm{Id}_{T_J \mathbb{H}_J}$.   □

Theorem 5.11. *The coordinate* $\mathrm{Fix} \colon \mathbb{H}_J \to \mathbb{C}_+ \subset \mathbb{C}$ *is* (J_J, m_i)-*holomorphic.*

Proof. The map Fix is equivariant for the transitive holomorphic actions of $\mathrm{PGL}^+(2, \mathbb{R})$ on $(\mathbb{H}_J, J_J)$ and on $(\mathbb{C}_+, m_i)$. The map Fix is (J_J, m_i)-holomorphic at the point $\begin{pmatrix} 0 & -1 \\ 1 & 0 \end{pmatrix} \in \mathbb{H}_J$.   □

Exercise 5.12. Let $J_1, J_2, \ldots, J_n$ be n different points in $\mathbb{H}_J$. Put $J_{n+1} = J_1$. We assume that $J_{k+1} = J_{k-1} \circ J_k \circ J_{k-1}$ for $k = 1, 2, \ldots, n$ hold. Show that the images $\mathrm{Fix}(J_k)$ are the vertices of a regular right-angled n-gon, $n \geq 5$, in $\mathbb{C}_+$.

Exercise 5.13. Given $J \in \mathbb{H}_J$ and $\theta \in \mathbb{R}/2\pi\mathbb{Z}$ define $J(\theta) \colon \mathbb{R}^2 \to \mathbb{R}^2$ by $J(\theta)(u) := \sin(\theta)J(u) + \cos(\theta)u$. Show that $(\theta, u) \mapsto J(\theta)(u)$ defines a faithful linear action of the circle group $\mathbb{R}/2\pi\mathbb{Z}$ on $\mathbb{R}^2$. Define a hyperbolic plane $\mathbb{H} = \mathbb{H}_\rho$ as the set of faithful linear actions of the group $\mathbb{R}/2\pi\mathbb{Z}$ on $\mathbb{R}^2$.

Exercise 5.14. In $\mathbb{H}_I$ we have a notion of energy-barycenter. How about $\mathbb{H}_J$? Let (p, w) be a weighted multi-point in $\mathbb{H}_J$. For $q \in \mathbb{H}_J$ let R_q be the rotation of $\mathbb{H}_J$ with center q and constant angular speed 1. Compute according to Newton's law the kinetic energy $K(q; p, w)$ of the motion of the multi-point (p, w). Think of the points p_i, w_i as a point mass. Does the function $q \mapsto K(q; p, w)$ achieve a unique minimum at a point $k(p, w)$? Do we have $k(p, w) = b(p, w)$?

The group $\mathrm{SL}(2, \mathbb{R})$ acts by $(P, J) \in \mathbb{H}_J \mapsto P^{-1}JP \in \mathbb{H}_J$ transitively on the space $\mathbb{H}_J$. The stabilizer $\mathrm{Stab}_{\mathrm{SL}(2,\mathbb{R})}(J)$ of J in $\mathrm{SL}(2, \mathbb{R})$ is the fixed point set of the involution $Q \in \mathrm{SL}(2, \mathbb{R}) \mapsto -JQJ \in \mathrm{SL}(2, \mathbb{R})$. Hence $\mathbb{H}_J$ is identified as the symmetric space $\mathrm{SL}(2, \mathbb{R})/\mathrm{Stab}_{\mathrm{SL}(2,\mathbb{R})}(J)$, which is a model of the hyperbolic plane. See the book of Helgason [50] for a study of symmetric spaces.

The hyperbolic geometry of $\mathbb{H}_J$ was used in a seminal way by Dennis Sullivan in [116]. Dennis has orally explained to me his construction.

Let $\Gamma \subset \mathrm{Diff}^+(\mathbb{C} \cup \{\infty\})$ be a quasi-conformal group of orientation-preserving diffeomorphisms of the Riemann Sphere (S^2, J), i.e. there exists a constant $k > 0$ such that for all $p \in S^2 = \mathbb{C} \cup \{\infty\}$ the subset $B_p := \{\phi_*^* J_{\phi(p)}\}$ of the hyperbolic plane $J(T_p S^2)$ has diameter less than k. The field $p \in S^2 \mapsto B_p \subset J(T_p(S^2))$ is Γ-equivariant by construction. Likewise for similar fields, such as the field of closures $p \mapsto \bar{B}_p$, the field of hyperbolic convex hulls C_p of $\bar{B}_p$ and the field $N_p = N_p(R)$ of tubular neighborhoods of radius $R > 0$. The boundary ∂N_p is a C^1-curve with a well-defined probability measure μ_p and continuously varying normals. The measure μ_p can be smeared out along geodesics normal to ∂N_p to a probability measure σ_p on $J(T_p(S^2))$ by thinking of the pointed normal geodesic $(G_{p,q}, q), q \in \partial N_p$ as the real line with its standard Gaussian measure. One obtains a Γ-equivariant field $p \mapsto \sigma_p$. There are several ways of concentrating above Γ-equivariant set or measure-valued sections to a Γ-equivariant section J_Γ of the bundle $J(TS^2)$. For instance, define $J'_\Gamma(p)$ to be the center of the smallest hyperbolic disk that contains B_p. The measure σ_p can be concentrated as follows: put $E_p = \int_{J(T_p S^2)} J \, d\sigma_p \in \mathrm{End}(T_p S^2)$, which is a traceless endomorphism without real eigenvectors. So $J''_p := \frac{1}{\sqrt{\det(E_p)}} E_p \in J(T_p S^2)$.

The section J'_p is measurable and we expect that the section J''_p is continuous. Let $u', u'': (S^2, J', J'') \to (S^2, J)$ be homeomorphisms, both uniformizing J' or J''. Sullivan's Theorem follows now from the measurable Riemann mapping theorem of Alfors and Bers [5]:

Theorem 5.15 (Dennis Sullivan). *Let Γ be a quasi-conformal group of diffeomorphism of $\mathbb{C} \cup \{\infty\}$. Then Γ is conjugated by a homeomorphism to a subgroup of $\mathrm{PSL}(2, \mathbb{C})$.* $\qquad\square$

5.3 Reminder: Multi-Linear Algebra

Let V be a real vector space of finite dimension and let $J: V \to V$ be a linear map with $J \circ J = -\mathrm{Id}_V$. Using J we can multiply vectors u in V by complex numbers $\lambda = a + bi$ in such a way that V together with this multiplication is a complex vector space. The multiplication λu is defined for $T\lambda = a + bi \in \mathbb{C}$ by

$$\lambda u := au + bJ(u).$$

We denote by $\Lambda^k(V, \mathbb{C})$ the complex vector space of all maps:

$$\omega: V^k \to \mathbb{C}$$

that are $\mathbb{R}$-multi-linear and alternating. Remember, $\mathbb{R}$-*multi-linear* means that the map

$$(u_1, u_2, \ldots, u_k) \in V^k \mapsto \omega(u_1, u_2, \ldots, u_k) \in \mathbb{C}$$

is $\mathbb{R}$-linear in each of its arguments. *Alternating* means that if in ω one exchanges two arguments $u_i, u_j, 1 \le i < j \le k$, the value of

$$\omega(u_1, u_2, \ldots, u_k) \in \mathbb{C}$$

gets multiplied by -1. Since the target space of $\omega\colon V^k \to \mathbb{C}$ is the $\mathbb{C}$-vector space $\mathbb{C}$ the space $\Lambda^k(V, \mathbb{C})$ becomes a $\mathbb{C}$-vector space: the addition $\omega_1 + \omega_2$ is defined as

$$(\omega_1 + \omega_2)(u_1, u_2, \ldots, u_k) := \omega_1(u_1, u_2, \ldots, u_k) + \omega_2(u_1, u_2, \ldots, u_k)$$

and the multiplication $\lambda\omega$ is defined as

$$(\lambda\omega)(u_1, u_2, \ldots, u_k) := \lambda\omega(u_1, u_2, \ldots, u_k).$$

It is important to observe that the map

$$(u_1, u_2, \ldots, u_k) \in V^k \mapsto \omega(u_1, u_2, \ldots, u_k) \in \mathbb{C}$$

is only $\mathbb{R}$-linear and not $\mathbb{C}$-linear in each of its arguments.

Exercise 5.16. Let $\sigma\colon E \to E$ be a permutation of a finite set E. Minimal non-empty subsets $A \subset E$ that are stable under σ are called σ-orbits. Define orb(σ) as the partition of E into σ-orbits. Let #orb(σ) be the length of the partition, i.e. the number of σ-orbits. Define the signature of the permutation σ by

$$\text{sign}(\sigma) = (-1)^{\#E - \#\text{orb}(\sigma)}.$$

Prove that the signature of a transposition equals -1. Prove that the signature is multiplicative: $\text{sign}(\sigma_1 \circ \sigma_2) = \text{sign}(\sigma_1)\text{sign}(\sigma_2)$. Show that the above definition of *alternating* is consistent.

Let $\Lambda_J^{k-l,l}(V, \mathbb{C}), l = 0, 1, \ldots, k$, be the subspace of $\Lambda^k(V, \mathbb{C})$ of all maps $\omega\colon V^k \to \mathbb{C}$ that satisfy

$$\omega(\lambda u_1, \lambda u_2, \ldots, \lambda u_k) = \lambda^{k-l}\bar{\lambda}^l \omega_1(u_1, u_2, \ldots, u_k), \lambda \in \mathbb{C}.$$

The spaces $\Lambda_J^{k-l,l}(V, \mathbb{C}), l = 0, 1, \ldots, k$, are $\mathbb{C}$-sub-vector spaces.

Theorem 5.17. *The space $\Lambda^k(V, \mathbb{C})$ is the direct sum of its sub-vector spaces $\Lambda_J^{k-l,l}(V, \mathbb{C}), l = 0, 1, \ldots, k$:*

$$\Lambda^k(V, \mathbb{C}) = \bigoplus_{l=0}^{l=k} \Lambda_J^{k-l,l}(V, \mathbb{C}).$$

Proof. For $v = (u_1, u_2, \ldots, u_k) \in V^k, \omega \in \Lambda_J^{k-l,l}(V, \mathbb{C}, \lambda = e^{it}$ we have $\omega(e^{it}v) = e^{(k-2l)it}\omega(v)$. We define a projector $\P_J^{k-l,l}\colon \Lambda^k(V, \mathbb{C}) \to \Lambda^k(V, \mathbb{C})$ with image $\Lambda_J^{k-l,l}(V, \mathbb{C})$ by

$$(\P_J^{k-l,l}\omega)(v) = \frac{1}{2\pi} \int_0^{2\pi} e^{(2l-k)it}\omega(e^{it}v)dt.$$

It suffices to observe

$$\P_J^{k-l,l} = 0,\ l \notin \{0, 1, 2, \ldots, k\},\quad \sum_{l\in\mathbb{Z}} \P_J^{k-l,l} = \text{Id}_{\Lambda^k(V,\mathbb{C})}.\qquad \square$$

Exercise 5.18. Compute the dimension of $\Lambda^k(V, \mathbb{C})$ as a function of $n = \text{Dim}(V)$ and k.

5.4 Reminder: Holomorphic Functions in One Complex Variable

Let U be an open subset in $\mathbb{C}$. A local path in U is a path $\gamma\colon [0, 1] \to U$ taking its values in a disk $B(p, r) = \{q \in \mathbb{C} \mid |p - q| < r\}$ which is contained in U.

Holomorphic functions on U, open in $\mathbb{C}$, are complex-valued continuous functions for which all path integrals $\int_\gamma f\,dz$ over local piece-wise smooth closed paths in U vanish.

Holomorphic functions have many properties. The following theorem provides three more ways of characterizing holomorphic functions. The first two (i) and (ii) apply to a continuous function.

Theorem 5.19. *Let U be an open subset in $\mathbb{C}$ and let f be a complex-valued contin-uous function on U. The following statements are equivalent:*

(i) *f is holomorphic,*
(ii) *the 1-differential form $f\,dz$ is closed,*
(iii) *the function f is continuously differentiable and its differential df is $\mathbb{C}$-linear,*
(iv) *in each subdisc $B(p, r) \subset U$ the function f is given by a convergent power series.*

Proof. (i) implies (ii): remember, the value of $d(f\,dz)$ at $q \in U$ on vectors $u, v \in \mathbb{C}$ is given by

$$(d(f\,dz))_q(u, v) = \lim_{s\to 0, s>0} \frac{1}{s^2} \int_{\partial P_q(u,v,s)} f\,dz,$$

a scaled limit of path integrals over boundaries of parallelograms. By (i)

$$\int_{\partial P_q(u,v,s)} f\,dz = 0$$

holds for $q \in U$, $u, v \in \mathbb{C}$ and $s > 0$ with $P_q(u, v, s) \subset U$. Hence $d(f\,dz) = 0$.

(ii) implies (iii): let $D = B(p, r) \subset U$ be a subdisc in U. On D the closed form $f\,dz$ is exact. Let $F\colon D \to \mathbb{C}$ be a function with $dF = f\,dz$.

Remember that dg for a complex-valued function g on U splits as $dg = d'g + d''g$ into $\mathbb{C}$-linear and anti-$\mathbb{C}$-linear parts.

Observe $dF = d'F$ and $d''F = 0$. The function F being continuously differentiable, from $d(dF) = 0$ follows $d''d'F = 0$. Hence F is harmonic, and in particular smooth.

From $dF = f\,dz$ it follows that f is smooth, in particular continuously differentiable on D and also on U. Since f is smooth, the formula $d(f\,dz) = d''f \wedge dz = 0$ applies, and $d''f = 0$ follows. From $d''f = 0$ follows $df = d'f$, showing that the differential of f is $\mathbb{C}$-linear.

(iii) implies (iv): let $B(p,r) \subset U$ be a subdisc in U such that its closure is also a subset of U. For $q \in B(p,r)$ and $0 < \rho < r - |q - p|$ let $R(p,q,r,\rho)$ be the closure of $B(p,r) \setminus B(q,\rho)$ in U. The function $g = \frac{f}{z(q)-z}$ is smooth and $d(g\,dz) = 0$. It follows by Stokes' Theorem that the path integral of $g\,dz$ along the oriented boundary of $R(p,q,r,\rho)$ vanishes.

Evaluating separately on each boundary component yields

$$\int_{\partial^+ B(q,\rho)} \frac{f\,dz}{z(q) - z} = \int_{\partial^+ B(p,r)} \frac{f\,dz}{z(q) - z}.$$

The left-hand side is independent of ρ since no ρ appears in the right-hand side. Its value is $2\pi i f(q)$, which can be seen by taking the limit for $\rho \to 0$. In conclusion

$$f(q) = \frac{1}{2\pi i} \int_{\partial^+ B(p,r)} \frac{f\,dz}{z(q) - z},$$

which shows the interesting and fundamental fact that the value of the function f at $q \in B(p,r) \subset U$ is determined by the restriction of f to the boundary of the disk $B(p,r)$.

The integral expression for $f(q)$ expands into a power series.

$$\begin{aligned}
f(q) &= \frac{1}{2\pi i} \int_{\partial^+ B(p,r)} \frac{f\,dz}{z(q) - z} \\
&= \frac{1}{2\pi i} \int_{\partial^+ B(p,r)} \frac{f\,dz}{(z(q) - z(p)) - (z - z(p))} \\
&= \frac{1}{2\pi i} \int_{\partial^+ B(p,r)} \frac{f\,dz}{\frac{z(q)-z(p)}{z-z(p)} - 1}.
\end{aligned}$$

Along the path of integration $\left|\frac{z(q)-z(p)}{z-z(p)}\right| = \frac{|q-p|}{r} < 1$ holds, which permits an expansion using the geometric series

$$\frac{1}{2\pi i} \int_{\partial^+ B(p,r)} \sum_{n=0}^{\infty} (z(q) - z(p))^n \frac{f\,dz}{(z - z(q))^{n+1}}$$

$$= \sum_{n=0}^{\infty} a_n (z(q) - z(p))^n,$$

with

$$a_n = \frac{1}{2\pi i} \int_{\partial^+ B(p,r)} \frac{f\,dz}{(z - z(q))^{n+1}}.$$

If using the coordinate function $u = z - z(p)$ centered at p one has $du = dz$ and more simply the power series expression in u for f

$$f = \sum_{n=0}^{\infty} a_n u^n$$

with

$$a_n = \frac{1}{2\pi i} \int_{\partial^+ B(p,r)} \frac{f du}{u^{n+1}}.$$

(iv) implies (i): path integration and summation commute for convergent power series. For a closed local path $\gamma: [0,1] \to U$ with values in a disk $B(p,r) \subset U$ and coordinate $u = z - z(p)$ centered at p one gets

$$\int_\gamma f dz = \int_\gamma f du = \int_\gamma \sum_{n=0}^{\infty} a_n u^n du$$

$$= \sum_{n=0}^{\infty} \int_\gamma a_n u^n du = \sum_{n=1}^{\infty} a_n (u(\gamma(0)))^{n+1} - a_n (u(\gamma(1)))^{n+1} = 0.$$

(iv) implies (ii): differentiation and summation commute for convergent power series. The property $d(f dz) = 0$ is local, hence it suffices to work in discs $B(p,r) \subset U$. So working with the coordinate $u = z - z(p)$ at p, one gets $d(f dz) = d(f du) = (\sum_{n=1}^{\infty} n a_n u^{n-1}) du \wedge du = 0$. $\qquad \square$

Let $f: U \to \mathbb{C}$ be a holomorphic function on an open subset U of $\mathbb{C}$. From (ii) it follows that the differential df can be written in a unique way as $df = g dz$ for a function g on U. The function g is called the complex derivative of f and is sometimes denoted by f' or by $\frac{\partial f}{\partial z}$. From (iv) it follows that the complex derivative f' of a holomorphic function f is holomorphic too.

Let $(f_n)_{n \in \mathbb{N}}$ be a sequence of holomorphic functions in an open subset U of $\mathbb{C}$. Suppose that the sequence converges uniformly on each compact subset of U. It is well known that the limit $f = \lim_{n \to \infty} f_n$ is a continuous function on U. From (i) it follows that the function f is holomorphic. Indeed, for a local closed path $\gamma: [0,1] \to U$ the image of γ is compact and path integration commutes with taking uniform limits, hence

$$0 = \lim_{n \to \infty} \int_\gamma f_n dz = \int_\gamma \lim_{n \to \infty} f_n dz = \int_\gamma f dz,$$

which proves

Theorem 5.20. *The limit function for uniform convergence on compact subsets of a sequence of holomorphic functions on an open subset of $\mathbb{C}$ is holomorphic.*

Proof. See above. $\qquad \square$

The characterization (i), due to G. Morera,[4] is very useful, since it is hereditary with respect to taking limits for the uniform convergence on compacta in the domain of functions.

The characterization (iv) allows us to define interesting functions, for example the exponential function $e^z = \sum_{n=0}^{\infty} \frac{z^n}{n!}$, trigonometric functions $\cos(z) = \sum_{k=0}^{\infty} (-1)^k \frac{z^{2k}}{(2k)!}$ and many more.

Exercise 5.21. The function $w = \sum_{n=0}^{\infty} z^{n!}$ is holomorphic in the open unit disk D. Let U be open and connected in $\mathbb{C}$. Assume $D \subset U$ and $D \neq U$. Prove that w does not extend to a holomorphic function on U. Hint: try to define $w(\mu)$ for a root of unity μ.

Exercise 5.22. The function $\Theta = \sum_{n=0}^{\infty} z^{n^2}$ is holomorphic in the open unit disk D. Does Θ extend to an open neighborhood of -1?

A zoom in an open subset U of $\mathbb{C}$ is a pair of nested discs $(B(p,r) \subset B(p,2r))$, $p \in U$, $r > 0$, such that the closure in $\mathbb{C}$ of $B(p,2r)$ is a subset of U. A zoom-cover of U is a family $(B(p_j,r_j) \subset B(p_j,2r_j))_{j \in J}$ of zooms in U such that the inner discs $B(p_j,r_j)$ cover U. Every open subset U in $\mathbb{C}$ admits a countable zoom-cover. For a continuous function f on U and subset A in U denote by $\|f\|_A$ the least upper bound of $|f|$ on A. So,

$$\|f\|_A = \mathrm{Sup}_{a \in A} |f(a)|.$$

Exercise 5.23. For every zoom-cover $(B(p_j,r_j) \subset B(p_j,2r_j))_{j \in J}$ of an open subset U in $\mathbb{C}$ there exists a countable subset J_0 in J such that $(B(p_j,r_j) \subset B(p_j,2r_j))_{j \in J_0}$ is still a zoom-cover of U.

Exercise 5.24. Let $C^0(U, \mathbb{C})$ be the space of complex continuous functions on an open subset U of $\mathbb{C}$ equipped with the topology of uniform convergence on compacta in U. Let $Z = (B(p_j,r_j) \subset B(p_j,2r_j))_{j \in \mathbb{N}}$ be a countable zoom-cover of U indexed by the natural numbers. For $f, g \in C^0(U, \mathbb{C})$ define

$$d_Z(f,g) = \sum_{j=0}^{\infty} \frac{1}{2^j} \frac{\|f-g\|_{B(p_j,2r_j)}}{1 + \|f-g\|_{B(p_j,2r_j)}}$$

and show that d_Z is a metric on $C^0(U, \mathbb{C})$ with the topology of uniform convergence on compacta in U.

Exercise 5.25. Balls for d_Z in $C^0(U, \mathbb{C})$ are convex. The function $n(f) = d_Z(f, 0)$ is not a norm, but satisfies $n(f + g) \leq n(f) + n(g)$.

The following theorems and their consequences were discovered by P. Montel.[5]

---

[4] Giacinto Morera (1856–1909), Italian mathematician.

[5] Paul Antoine Aristide Montel (1876–1975), French mathematician.

Theorem 5.26. *Let* $(B(p,r) \subset B(p,2r))$ *be a zoom in an open subset U of $\mathbb{C}$. For a holomorphic function f on U the inequality*

$$\|f'\|_{B(p,r)} \leq \frac{1}{r}\|f\|_{B(p,2r)}$$

holds. The restriction of f to $B(p,r)$ is Lipschitz continuous with Lipschitz constant $\frac{1}{r}\|f\|_{B(p,2r)}$.

Proof. For $q \in B(p,r)$ the disk $B(q,r)$ is contained in $B(p,2r)$. Hence

$$f'(q) = \frac{1}{2\pi i} \int_{\partial^+ B(q,r)} \frac{f\,du}{u^2}\, f\,du,$$

where u is the coordinate $u = z - z(q)$ centered at q. For $q \in B(p,r)$ it follows that

$$|f'(q)| \leq \frac{1}{2\pi} \frac{\|f\|_{B(q,r)}}{r^2} 2\pi r = \frac{1}{r}\|f\|_{B(p,2r)}$$

and hence $\|f'\|_{B(p,r)} \leq \frac{1}{r}\|f\|_{B(p,2r)}$.

For $q_0, q_1 \in B(p,r)$ one has $f(q_1) - f(q_0) = \int_{[q_0,q_1]} f'\,dz$. Hence, from the Lipschitz continuity on $B(p,r)$ with the claimed constant we get

$$|f(q_1) - f(q_0)| \leq \frac{1}{r}\|f\|_{B(p,2r)}\,|q_1 - q_0|, \quad q_0, q_1 \in B(p,r). \qquad \square$$

Theorem 5.27. *Let $(f_n)_{n \in \mathbb{N}}$ be a sequence of holomorphic functions on $B(p,2r)$ such that for $M \in \mathbb{R}$ the estimates $\|f_n\|_{B(p,2r)} \leq M$, $n \in \mathbb{N}$, hold. Then there exists a subsequence $(f_{n_k})_{k \in \mathbb{N}}$ and a holomorphic function g on $B(p,r)$ such that*

$$\lim_{k \to \infty} \|f_{n_k} - g\|_{B(p,r)} = 0.$$

Proof. Let p_i, $i \in \mathbb{N}$, be an enumeration of the points in $B(p,2r)$ with rational real and imaginary parts. For each i the values $f_n(p_i)$, $n \in \mathbb{N}$, belong to the closed disk centered at $0 \in \mathbb{C}$ of radius M. Hence for each i there exists a subsequence $(f_{n_k^i})_{k \in \mathbb{N}}$ such that the sequence of values $(f_{n_k^i}(p_i))_{k \in \mathbb{N}}$ converges. For each i the subsequence $(f_{n_k^{i+1}})_{k \in \mathbb{N}}$ can be chosen as a subsequence of $(f_{n_k^i})_{k \in \mathbb{N}}$. The diagonal subsequence $(f_{n_k^k})_{k \in \mathbb{N}}$ is then such that at each point p_i the sequence of its values converges to a value $g(p_i)$.

The Lipschitz inequalities

$$|g(p_i) - g(p_j)| \leq \frac{M}{r}\,|p_i - p_j|, \quad i, j \in \mathbb{N},$$

are preserved by taking limits.

The partially defined Lipschitz continuous function g can be extended in a unique way to a Lipschitz continuous function $g\colon B(p,r) \to \mathbb{C}$ with Lipschitz constant $\frac{M}{r}$.

One has

$$\lim_{k \to \infty} \|f_{n_k^k} - g\|_{B(p,r)} = 0.$$

Indeed, assume the contrary, i.e. the existence of $\epsilon > 0$, of a subsequence $(f_{m_h})_{h \in \mathbb{N}}$ of $(f_{n_k^k})_{k \in \mathbb{N}}$ and of a sequence of points $(q_h)_{h \in \mathbb{N}}$ in $B(p,r)$ such that

$$\|f_{m_h}(q_h) - g(q_h)\|_{B(p,r)} \ge \epsilon.$$

By taking a subsequence we may achieve that the points q_h converge to a point q in the closure of $B(p,r)$. Now choose $p_i \in B(p,r)$ with $|q - p_i| < \frac{r}{8M}\epsilon$. There exists an H such that $|q_h - q| < \frac{r}{8M}\epsilon$, $h \ge H$. Observe $|q_h - p_i| < \frac{r}{4M}\epsilon$, hence $\|f_{m_h}(q_h) - f_{m_h}(p_i)\|_{B(p,r)} \ge \frac{1}{4}\epsilon$ and $\|g(q_h) - g(p_i)\|_{B(p,r)} \ge \frac{1}{4}\epsilon$. It follows that $\|f_{m_h}(p_i) - g(p_i)\| \ge \frac{1}{2}\epsilon$, $h \ge H$, contradicting $\lim_{h \to \infty} f_{m_h}(p_i) = g(p_i)$.

The function g is holomorphic, since it is a uniform limit of holomorphic functions. $\qquad\square$

In a finite-dimensional normed real or complex vector space $(V, \|\cdot\|_V)$ compacta are precisely the closed and bounded subsets. For instance closed balls $\{v \in V \mid \|v\|_V \le R\}$ are compact. This property is characteristic of finite-dimensional spaces by a theorem of F. Riesz.[6]

Theorem 5.28. *Let $(V, \|\cdot\|_V)$ be a real or complex normed vector space. Then the following are equivalent:*

(i) *V is finite-dimensional.*
(ii) *The closed unit ball in V is compact.*

Proof. Clearly $(i) \Rightarrow (ii)$. The following proves the implication $(ii) \Rightarrow (i)$. The closed unit ball $B_1 \subset V$, being compact, is covered by finitely many balls

$$B(v_1, \frac{1}{2}), B(v_2, \frac{1}{2}), \ldots, B(v_k, \frac{1}{2}).$$

For $v \in V$, $v \ne 0$, put $w = \frac{1}{\|v\|_V} v \in B_1$. The vector w belongs to $B(v_{i_1}, \frac{1}{2})$ for some $1 \le i_1 \le k$. Hence $w = v_{i_1} + w_1$ with $2w_1 \in B_1$. Again the vector $2w_1$ belongs to $B(v_{i_2}, \frac{1}{2})$ for some $1 \le i_2 \le k$. Hence $w = v_{i_1} + \frac{1}{2}v_{i_2} + w_2$ with $4w_2 \in B_1$. Repeating one gets for $n \in \mathbb{N}$ an expansion

$$w = v_{i_1} + \frac{1}{2}v_{i_2} + \cdots + \frac{1}{2^{n-1}}v_{i_n} + w_n$$

with $2^n w_n \in B_1$. This expansion can be rewritten as a linear combination with positive coefficients of the vectors $v_1, v_2, \ldots, v_k$ yielding

$$w = a_1(n)v_1 + \cdots + a_k(n)v_k + w_n$$

with $a_i(n) \in [0, 2]$ and $2^n w_n \in B_1$. Observe $\lim_{n \to \infty} w_n = 0$. Moreover the coefficients $a_i(n)$ satisfy $a_i(n) \le a_i(n+1) \le 2$. Putting $a_i = \lim_{n \to \infty} a_i(n)$ one gets

---

[6] Frigyes Riesz (1880–1956), Hungarian mathematician.

$$w = a_1 v_1 + \cdots + a_k v_k, \ a_i \geq 0,$$

showing that $v = \|v\|_V w$ is in the positive cone generated by the vectors v_i. One concludes that the vector space V is of finite dimension $< k$. $\qquad\square$

We turn back to the space $C^0(U, \mathbb{C})$ with metric d_Z given by a countable zoom-cover $Z = (B(p_j, r_j) \subset B(p_j, 2r_j))_{j \in J}$. The space $\hbar(U, \mathbb{C})$ of holomorphic functions is a closed subspace of $C^0(U, \mathbb{C})$ by Theorem 5.20. The following theorem, due to Montel, describes the compacta in $\hbar(U, \mathbb{C})$.

Theorem 5.29. *Let* $M = (M_j)_{j \in J}$ *be a system of real numbers* > 0. *The set*

$$K_M = \{ f \in \hbar(U, \mathbb{C}) \mid \|f\|_{B(p_j, 2r_j)} \leq M_j, \ j \in J \}$$

is compact. Moreover, for every compact set K in $\hbar(U, \mathbb{C})$ there exists a system $M = (M_j)_{j \in J}$ *of numbers with* $K \subset K_M$.

Proof. The topology on K_M is given by the metric d_Z. In order to prove that K_M is compact, it suffices to show that every sequence $(f_n)_{n \in \mathbb{N}}$ in K_M has a subsequence that converges to an element g in K_M. First identify J, the index set of the zoom Z, with $\mathbb{N}$. So $j \in J = \mathbb{N}$. By the previous theorem there exist subsequences $(f_{n_k^j})_{k \in \mathbb{N}}$ that converge for the norm $\| \cdot \|_{B(p_j, r_j)}$ to a holomorphic function g_j on $B(p_j, r_j)$. The subsequences can be chosen such that for each $j \in \mathbb{N}$ the subsequence $(f_{n_k^{j+1}})_{k \in \mathbb{N}}$ is a subsequence of $(f_{n_k^j})_{k \in \mathbb{N}}$.

The diagonal subsequence $(f_{n_k^k})_{k \in \mathbb{N}}$ converges for the distance d_Z to a function g that agrees for every $j \in J$ with the holomorphic function g_j on $B(p_j, r_j)$. The function g is holomorphic on U since the balls $B(p_j, r_j)$ cover U. Finally $g \in K_M$ since inequalities like $\|f_n\|_{B(p_j, 2r_j)} \leq M_j$ are preserved after taking limits. $\qquad\square$

5.5 J-Laplace Operator and Metric

Let (S, J) be a connected compact surface with J-field J. We assume that the surface S is defined by charts $U_\alpha \subset V$. The first Laplace operator $\Delta_J : \Omega^0(S, \mathbb{R}) \to \Omega^2(S, \mathbb{R})$ is the map $f \in \Omega^0(S, \mathbb{R}) \mapsto d(df \circ J) \in \Omega^2(S, \mathbb{R})$ that transforms a function to a differential 2-form.

Let $\omega \in \Omega^2(S, \mathbb{R})$ be a volume form on S such that for all $p \in S$ and all $u \in T_p S$, $u \neq 0$, we have $\omega_p(u, Ju) > 0$. The given J-field orients the surface S. A volume form ω also orients the surface S. The inequalities $\omega_p(u, Ju) > 0$ just mean that both orientations agree.

From the pair of fields (J, ω) we deduce a new field of bilinear maps $g(J, \omega)$ that in a chart $U_\alpha \subset V$ is given by $p \in U_\alpha \mapsto g_{\alpha,p}$, where $g_{\alpha,p} : V \times V \to \mathbb{R}$ is the bilinear map

$$(u, v) \in V \times V \mapsto \omega_{\alpha,p}(u, J_{\alpha,p}(v)) \in \mathbb{R}.$$

The field $g(J, \omega)$ on S defines a Riemannian metric $g = g(J, \omega)$ on S which depends on the data of the J-field J and the volume form ω, both with the same orientation. The endomorphism J_p fixes the form ω_p since $\mathrm{Det}(J_p) = 1$. It follows that

$$g_p(u, v) = \omega_p(u, Jv) = \omega_p(Ju, J^2v)$$
$$= \omega_p(Ju, -v) = \omega_p(v, Ju) = g_p(v, u).$$

Hence g_p is symmetric, also clearly bilinear, and positive by construction, so a scalar product. The pointwise action of J is isometric and the volume form of the Riemannian metric $g(J, \omega)$ coincides with the form ω.

From a volume form ω on S we get an operator

$$\%_\omega : \Omega^2(S, \mathbb{R}) \to \Omega^0(S, \mathbb{R})$$

as follows. For $\eta \in \Omega^2(S, \mathbb{R})$ the function $\%_\omega(\eta)$ is the unique function f satisfying $\eta = f\omega$. We define the second Laplace operator

$$\Delta_{J,\omega} : \Omega^0(S, \mathbb{R}) \to \Omega^0(S, \mathbb{R})$$

by $\Delta_{J,\omega} := \%_\omega \circ \Delta_J$. The operator $\Delta_{J,\omega}$ transforms a function to a function.

The choice of a J-field J and a volume form ω on a differential surface S opens the possibility to apply a wide range of mathematical and physical methods to study the triple (S, J, ω). Given a triple (S, J, ω), we can measure the length of curves, compute curvature, and make many more measurements since on S we have the Riemannian metric $g_{J,\omega}$. We can study the eigenfunctions and eigenvalues of the operator $\Delta_{J,\omega}$ on functions, also on the space $\mathrm{Bal}(S, \omega)$ of balanced functions f with $\int_S f\omega = 0$. The restriction of $\Delta_{J,\omega}$ to $\mathrm{Bal}(S, \omega)$ has a regularized determinant $Z(S, J, \omega) \in \mathbb{R}$. So one can optimize the quantity $\omega \mapsto Z(S, J, \omega)$ keeping (S, J) fixed. This was done and initiated by Alexander Markowitsch Polyakov, yielding spectacular results. We will not explain these results, but refer to [96].

5.6 J-Surfaces

A J-surface is a pair (S, J) consisting of a manifold S with charts in $\mathbb{R}^2$ and a J-field on it. Examples are $(\mathbb{R}^2, J)$ where the J-field is constant with value $\left(\begin{smallmatrix} 0 & -1 \\ 1 & 0 \end{smallmatrix}\right)$. This example is very important, it corresponds to $\mathbb{C}$ with the multiplication m_i by i. The torus is the surface $T^2 := \mathbb{R}^2/2\pi\mathbb{Z}^2$. The previous J-field m_i is constant, hence $2\pi\mathbb{Z}^2$-periodic and descends to the torus T^2, which then as a J-surface is denoted by $\mathbb{C}/\Gamma$, Γ being the standard lattice $\Gamma := 2\pi\mathbb{Z} + 2\pi i\mathbb{Z}$. Stretching by the factor 2π transforms the lattice $\Gamma := \mathbb{Z} + i\mathbb{Z}$ into $\Gamma := 2\pi\mathbb{Z} + 2\pi i\mathbb{Z}$. The J-tori $\mathbb{R}^2/2\pi\mathbb{Z}^2$ and $\mathbb{R}^2/\mathbb{Z}^2$ are diffeomorphic as J-surfaces. We can vary the lattice by putting $\Gamma_\tau := \mathbb{Z} + \tau\mathbb{Z}$, $\tau \in \mathbb{C}_+$, and in this way obtain many J-tori $\mathbb{C}/\Gamma_\tau$.

A more general construction of a J-field on an oriented surface uses the notion of energy-barycenter in $\mathbb{H} = \mathbb{H}_J$. Let S be an oriented differential surface given by a finite family of charts $U_\alpha \subset \mathbb{R}^2$. Let $J_\alpha : U_\alpha \to \mathbb{H}_J = \mathbb{H}_J(\mathbb{R}^2)$ be a family of oriented differentiable J-fields. We want to construct a J-field on the surface S. The construction uses a differentiable partition of unity χ_α subordinated to the given system of charts. For a point $p \in S$ with representatives $p_1, p_2, \ldots, p_k$ in charts $U_{\alpha_1}, U_{\alpha_2}, \ldots, U_{\alpha_k}$ we define $J(p)$ as the barycenter of

$J_{\alpha_1}(p_1), J_{\alpha_2}(p_2), \ldots, J_{\alpha_k}(p_k)$ with weights $\chi(p_1), \chi(p_2), \ldots, \chi(p_k)$. This construction provides many *J*-surfaces.

For an oriented real vector space V of dimension two, let $H_J(V)$ be the space of $J \in \text{End}(V)$ with $J^2 = -\text{Id}_V$ and matching orientation. We have seen that $H_J(V)$ carries the geometry of a hyperbolic plane.

Let S be an oriented, compact and connected differentiable surface. Let $\mathbb{J}(TS)$ be the space of *J*-fields on S matching the orientation of S. The space $\mathbb{J}(TS)$ is equipped with a natural metric, namely the following. Let J_0, J_1 be two *J*-fields on S. For $p \in S$ the values $J_{0,p}, J_{1,p}$ are in $\mathbb{H}_J(T_p(S))$, which is naturally a model of the hyperbolic plane. The natural distance on $\mathbb{J}(TS)$ is

$$\text{dist}_J(J_0, J_1) := \text{Supremum}_{p \in S} \text{dist}_{\mathbb{H}_J(T_p(S))}(J_{0,p}, J_{1,p}).$$

The topological space $\mathbb{J}(TS)$ with its distance dist_J is naturally path connected by geodesic paths, namely for $p \in S$ let $t \in [0, 1] \mapsto J_{p,t} \in \mathbb{H}_J(T_p(S))$ be the geodesic in $\mathbb{H}_J(V)$ with constant speed from $J_{0,p}$ to $J_{1,p}$, V being the oriented 2-dimensional vector space $T_p(S)$.

We have repeatedly used the word natural. Here it just means that the distance dist_J and the above geodesic segments $[J_0, J_1]$ are preserved by the action of the group of orientation-preserving diffeomorphisms of the surface S.

In real dimension two a *J*-field $J: S \to \text{End}(TS)$ on a differentiable surface defines a canonical structure of a holomorphic manifold on S. In order to obtain this structure it suffices to locally construct *J*-holomorphic functions. This is addressed in the next theorem, which is in fact a local rigidity theorem for *J*-fields in dimension 2.

Theorem 5.30 (Local rigidity of *J*-fields in real dimension two). *Let V be a real two-dimensional vector space. Let $J: V \to \text{End}(V)$ be a J-field. For all $p \in V$ there exists an open neighborhood U of p and a smooth injective map $z: U \to \mathbb{C}$ such that at all $q \in U$ its differential $(\text{D}z)_q$ is of real rank 2 and satisfies $(\text{D}z)_q(J_q(h)) = i(\text{D}z)_q(h)$, $h \in V$.*

In other words, for $p \in V$ there exists a complex coordinate function $z: U \to \mathbb{C}$ on an open neighborhood of p that is (J, m_i)-holomorphic. The local coordinate functions of the theorem show that locally near each point $p \in V$ the *J*-field J is isomorphic to the standard *J*-field m_i on $\mathbb{C}$ near $z(p)$.

The proof uses the Toric Laplace Inversion Theorem:

Theorem 5.31 (Toric Laplace Inversion). *The Laplace operator*

$$f \mapsto \left(-\frac{\partial^2 f}{\partial^2 x} - \frac{\partial^2 f}{\partial^2 y} \right) \text{d}x \wedge \text{d}y$$

is an isomorphism between the Fréchet spaces of smooth functions f and smooth 2-differential forms ω on the 2-torus $T^2 = \mathbb{R}^2/\mathbb{Z}^2$ with $\int_{T^2} f \, \text{d}x \wedge \text{d}y = 0$ and $\int_{T^2} \omega = 0$ respectively.

Proof. A continuous function f on the torus T^2 is represented by a Fourier series

$$\sum_{(n,m) \in \mathbb{Z}^2} a_{n,m} e^{2\pi i (nx+my)}$$

with

$$a_{n,m} = \int_0^1 \int_0^1 f(x,y) e^{-2\pi i (nx+my)} \, dx dy.$$

The coefficients $a_{n,m}$ are bounded by the supremum norm $\|f\|_{\sup}$ of f. More is true, the coefficients tend to 0 if (n,m) tends to infinity. For $2k$-times continuously differential functions f the coefficients tend faster to 0, more precisely $(n^2 + m^2)^k a_{n,m}$ tend to 0 if (n,m) tends to infinity. A converse holds: a function f on T^2 is smooth if and only if for all $k \in \mathbb{N}$ the Fourier coefficients $a_{n,m}$ tend to 0 such that $(n^2 + m^2)^k a_{n,m}$ also tend to 0. Since $a_{n,m}(-\frac{\partial^2 f}{\partial^2 x} - \frac{\partial^2 f}{\partial^2 y}) = 4\pi^2 (n^2 + m^2) a_{n,m}(f)$ the operator $f \mapsto (-\frac{\partial^2 f}{\partial^2 x} - \frac{\partial^2 f}{\partial^2 y}) dx \wedge dy$ is an isomorphism of Fréchet spaces between the spaces of functions with $a_{0,0}(f) = 0$ and of forms ω with integral 0. □

Proof (Local Rigidity of J-fields in real dimension 2). Assume p to be the origin of V and let e_1, e_2 be a basis for V with $J_p(e_1) = e_2$. Let x, y be the corresponding linear coordinates on V. Let $\chi : V \to [0,1]$ be a smooth function with $\chi(q) = 1$ if $|x(q)|, |y(q)| \leq \frac{1}{4}$ and $\chi(q) = 0$ if $|x(q)| \geq \frac{1}{2}$ or $|y(q)| \geq \frac{1}{2}$. For $s \in]0,1]$ put $\chi_s(q) = \chi(\frac{1}{s} q)$. Let J_0 on $P = \{q \in V \mid |x(q)|, |y(q)| \leq 1\}$ be the constant J-field $J_q = J_p$. For $q \in P$, let $t \in [0,1] \mapsto J(t,q) \in H_J(V)$ be the geodesic from $J_{0,q}$ to J_q. For $s \in]0,1]$, let J_s be the J-field on P defined by $J_{s,q} := J(\chi_s(q), q)$. Observe $\lim_{s \to 0} J_s = J_0$. For $s \in]0,1]$, define $U_s = \{q \in P \mid |x(q)|, |y(q)| < \frac{s}{4}\}$ and observe that on the open neighborhood U_s of p the fields J_s and J agree.

The J-fields J_s correspond to J-fields, again denoted by J_s, on the torus $T = \mathbb{R}^2/\mathbb{Z}^2$. The 1-differential form dx corresponds to a 1-form, again and with a slight abuse of notation denoted by dx on T. For smooth functions f with $\int_T f dx \wedge dy = 0$ on T we consider the equation

$$d((dx + df) \circ J_s) = 0,$$

which can be rewritten as

$$\Delta_{J_s}(f) = -d(dx \circ J_s).$$

By the Toric Laplace Inversion, the equation for $s = 0$ has only the solution $f = 0$. The operator Δ_{J_s} for s close to 0 is close to the operator $\Delta_{J_0} = \Delta$ of the Toric Laplace Inversion theorem, so is also invertible. Hence for $s > 0$ but small enough, the above equation has a solution f_s that in the C^∞-Fréchet topology is close to the function 0. Observe that $dx + df_s \neq 0$ at p. Fix such an s.

Put $\omega_s = (dx + df_s) - i(dx + df_s) \circ J_s \in \Omega^1(U_s, \mathbb{C})$. Observe that $d\omega_s = 0$. Note that $\omega_s \circ J_s = i\omega_s$. Let $z : U_s \to \mathbb{C}$ be the function

$$z(q) = \int_{[p,q]} \omega.$$

Observe that the function z is (J, m_i)-holomorphic since it is (J_s, m_i)-holomorphic and $J_s = J$ on U_s. The function z is a (J, m_i)-holomorphic coordinate at p on an open neighborhood $U' \subset U_s$ of p since $(Dz)_p \neq 0$. $\qquad\square$

We introduce the term z-structure. We call a system of (U_a, z_a), $a \in A$, of charts on a differentiable manifold M with values in $\mathbb{C}^d$ a *z-structure* on M if the transition maps are holomorphic. Such a system of charts is called a holomorphic atlas for brevity. Two z-structures (U_a, z_a), $a \in A$, and (U_b, z_b), $b \in B$, are equivalent if all transition maps between charts U_a and U_b are holomorphic. A holomorphic structure is a z-structure up to equivalence of holomorphic atlases.

The main result of this section affirms that in real dimension 2 every J-structure determines a z-structure and a unique compatible holomorphic structure. In fact, the proof of the local rigidity theorem yields more: In real dimension 2 every smooth family J_t of J-structures determines a smooth family z_t of z-structures.

Chapter 6
Riemann Surfaces

6.1 Riemann Surfaces as z- and as J-Surfaces

A *Riemann surface* is a differentiable manifold with charts in $\mathbb{C}$ and holomorphic transition maps. In short, we can say that a Riemann surface is a surface with a z-structure. It follows from the Local Rigidity of J-Fields (Theorem 5.30), that a Riemann surface can also be defined as a differentiable manifold S of dimension two, together with a J-field on it. Having these two equivalent definitions implies:

Theorem 6.1. *Let S be a differentiable surface. The set $Z(S)$ of holomorphic structures on S has a natural topology, namely the compact-open topology on the space of J-fields. On a compact surface, the topology is induced by the metric $d_{Z(S)}$ of maximal pointwise distance in the hyperbolic plane and the action of diffeomorphisms on $Z(S)$ is by isometries.* ☐

Since the pointwise values of J-fields in real dimension two are in the hyperbolic plane we conclude:

Theorem 6.2. *Let S be an oriented differentiable surface. The space $Z(S)$ of holomorphic structures on S is path connected. Moreover between two points there exists the path that traces pointwise geodesics in the hyperbolic plane.* ☐

As we will see later, a little more effort, not using the two definitions of Riemann surface, but merely the proof of the local rigidity of J-fields will give:

Theorem 6.3. *Let Σ be a Riemann surface of genus 1. Then Σ is bi-holomorphic to $\mathbb{C}/\Gamma_\tau$, $\tau \in \mathbb{C}_+$, for a suitable lattice Γ_τ in $\mathbb{C}$.*

© The Author(s), under exclusive license to Springer Nature Switzerland AG 2021
N. A'Campo, *Topological, Differential and Conformal Geometry of Surfaces*, Universitext,
https://doi.org/10.1007/978-3-030-89032-2_6

6.2 Natural Structures on the Space $\mathbb{J}(TS)$

Let S be a connected compact oriented surface. Let $\mathbb{J}(TS)$ be the space of all smooth J-fields on S that agree with the orientation of S. The space $\mathbb{J}(TS)$ is a manifold of infinite dimension, it has charts in a complex Fréchet space of complex smooth functions.

More explicitly, let (U_α, ϕ_α) be an atlas for the surface S. A $J \in \mathbb{J}(TS)$ field is given on $\phi_\alpha(U_\alpha) = U'_\alpha \subset \mathbb{R}^2$ by a smooth map $J_\alpha \colon U'_\alpha \to \mathbb{H}_J(\mathbb{R}^2) = \mathbb{H}_J$. Denote by $z(J_\alpha)$ the post composition of J_α with the coordinate $z \colon \mathbb{H}_J \to \mathbb{C}_+ \subset \mathbb{C}$. Further, denote by $\mathbb{U}'_\alpha$ all smooth maps $J \colon U'_\alpha \to \mathbb{H}_J$, by $z(\mathbb{U}'_\alpha)$ their post compositions with the coordinate z and by $\mathbb{U}'_{\beta\alpha}$, $z(\mathbb{U}'_{\beta\alpha})$ the restriction to $\phi_\alpha(U_\beta \cap U_\alpha)$. The space $z(\mathbb{U}'_\alpha)$ is an open set in the complex Fréchet space of smooth functions on U'_α. The coordinate change maps $\phi_{\beta\alpha}$ induce holomorphic maps $\Phi_{\beta\alpha} \colon z(\mathbb{U}'_{\beta\alpha}) \to z(\mathbb{U}'_{\alpha\beta})$ which allows us to build the manifold $\mathbb{J}(TS)$ as a Fréchet manifold modelled on a complex function space $C^\infty(U, \mathbb{C})$. Indeed, it suffices to choose the atlas (U_α, ϕ_α) such that all chart images U'_α are equal to some open subset $U' \subset \mathbb{R}^2$.

The group of orientation-preserving diffeomorphisms $\mathrm{Diff}^+(S)$ of the oriented surface S acts on $\mathbb{J}(TS)$. Define a *natural structure* on the space $\mathbb{J}(TS)$ to be a $\mathrm{Diff}^+(S)$-invariant structure.

An important example of a natural structure is given by the pointwise geodesic paths $t \in \mathbb{R} \mapsto J_t \in \mathbb{J}(TS)$ that we have already encountered. This structure assigns to a pair of points $J_0, J_1 \in \mathbb{J}(TS)$ a parametrized segment $[J_0, J_1] \subset \mathbb{J}(TS)$ consisting of a map $t \in [0, 1] \mapsto J_t \in \mathbb{J}(TS)$ that traces at every point $p \in S$ a geodesic of constant speed in the hyperbolic plane $J_{t,p} \in \mathbb{H}_J(T_pS)$.

Let us denote a point in $\mathbb{J}(TS)$ by (S, J) or simply by J. The tangent space $T_J\mathbb{J}(TS)$ at $J \in \mathbb{J}(TS)$ is the space of smooth fields $H \colon S \to \mathrm{End}(TS)$ of endomorphisms of the tangent space TS, that pointwise anti-commute with the J-field J. Indeed, from

$$(J + H)^2 = -\mathrm{Id} + H \circ J + J \circ H + H \circ H$$

it follows that $J + H$ satisfies $(J + H)^2 = -\mathrm{Id}$ up to order 1 if and only if H anti-commutes with J pointwise. At a point $p \in S$, the value H_p is a linear map $H_p \colon T_pS \to T_pS$ with $H_p \circ J_p + J_p \circ H_p = 0$. Hence, almost clearly:

Theorem 6.4. *The map* $J^\circ \colon H \in T_J(\mathbb{J}(TS)) \mapsto J \circ H \in T_J(\mathbb{J}(TS))$ *is a natural complex structure* J° *on the manifold* $\mathbb{J}(TS)$. *In other words, the group of orientation-preserving diffeomorphisms of S acts by* J°-*holomorphic automorphisms of* $\mathbb{J}(TS)$.

Proof. If H anti-commutes with J, then $J^\circ(H) = J \circ H$ also anti-commutes with J, so J° is an endomorphism of the tangent bundle $T(\mathbb{J}(TS))$. Clearly, $J^\circ \circ J^\circ = -\mathrm{Id}_{T\mathbb{J}(TS)}$. So, J° is an almost complex structure on $\mathbb{J}(TS)$.

The structure J° is integrable. Indeed, we will construct a J°-holomorphic atlas on the Fréchet manifold $\mathbb{J}(TS)$.

Let the oriented surface S be presented by an atlas (U_α, ϕ_α) such that the charts U_α have compact closure in S and such that the images $\phi_\alpha(U_\alpha) \subset \mathbb{R}^2$ are a fixed open disk $U \subset \mathbb{R}^2$.

Let $F = \mathbb{J}(TU)$ be the Fréchet manifold of smooth bounded fields $J \colon U \to \mathbb{H}_J(\mathbb{R}^2) = \mathbb{H}_J$. Bounded here means that $J(U) \subset \mathbb{H}_J$ is bounded in the hyperbolic plane and that all partial derivatives are 2×2-matrices with bounded coefficients. The Fréchet topology is given by the family of C^k-norms. Postcomposing $J \in F$ with the complex coordinate Fix: $\mathbb{H}_J \to \mathbb{C}_+$ shows that the Fréchet manifold F is identified with an open subset $\mathrm{Fix} F = \{\mathrm{Fix} \circ J \mid J \in F\}$ in the complex linear Fréchet space $C_b^\infty(U, \mathbb{C})$ of all bounded smooth maps from U to $\mathbb{C}$, that admit moreover a smooth extension to the closure $\bar{U}$.

The (integrable almost) complex structure on $\mathrm{Fix} F$ is the linear structure m_i. The post composition with Fix is (J°, m_i)-holomorphic, so the almost complex structure J° on F is integrable too.

The coordinate map $\phi_\alpha : U_\alpha \to U$ induces $\phi_\alpha^* : \mathbb{J}(U_\alpha) \to F$, which is J°-biholomorphic. So, J° on $\mathbb{J}(U_\alpha)$ is integrable too.

The restriction maps $R_\alpha : \mathbb{J}(TS) \to \mathbb{J}(U_\alpha)$ are J°-holomorphic and define the complex Fréchet manifold structure. It follows that J° on $\mathbb{J}(TS)$ is integrable too. $\square$

Our next aim is the construction of a natural (pre-)symplectic structure ω° on $\mathbb{J}(TS)$. We need some preparations.

Given $J \in \mathbb{J}(TS)$ on a surface S, let (U_α) be an open cover of S such that each U_α admits a (bijective) holomorphic coordinate $z_\alpha : U_\alpha \to U \subset \mathbb{C}$. A tangent vector $H \in T_J \mathbb{J}(TS)$ is given by a compatible system $H_\alpha \in T_{m_i} \mathbb{J}(U)$. Each H_α is a field on U of 2×2-matrices that anti-commute with m_i, or equivalently the field $p \in U \mapsto H_{\alpha,p}$ is a field of traceless symmetric matrices. So $H_{\alpha,p} = \begin{pmatrix} a(p) & b(p) \\ b(p) & -a(p) \end{pmatrix}$. Observe that

$$m_i \circ H_{\alpha,p} = \begin{pmatrix} b(p) & -a(p) \\ -a(p) & -b(p) \end{pmatrix}, \quad H_{\alpha,p} \circ m_i = \begin{pmatrix} -b(p) & a(p) \\ a(p) & b(p) \end{pmatrix}.$$

Define the norm of a matrix $\begin{pmatrix} a & b \\ b & -a \end{pmatrix}$ as $\sqrt{a^2 + b^2}$. Let $D \subset T_{m_i} \mathbb{J}(TU)$ be the space of smooth traceless symmetric fields of 2×2-matrices

$$p \in U \mapsto d(p) = \begin{pmatrix} a(p) & b(p) \\ b(p) & -a(p) \end{pmatrix}$$

of norm < 1. The map

$$E : D \to \mathbb{J}(TU)$$

$$d \in D \mapsto \left(p \mapsto \left(E(d)\right)_p = \frac{1}{\sqrt{1 - a(p)^2 - b(p)^2}} \begin{pmatrix} a(p) & -1+b(p) \\ 1+b(p) & -a(p) \end{pmatrix}\right)$$

smoothly parametrizes an open neighborhood N_{m_i} of m_i in $\mathbb{J}(TU)$.

Note that the anti-commutator $[A, B]_+ := AB + BA$ of traceless symmetric 2×2-matrices evaluates to a multiple of the identity matrix.

Keeping the notation $U, \mathbb{J}(TU), m_i \in \mathbb{J}(TU), N_{m_i}, E, D$ as before, we obtain:

Theorem 6.5. *For $H \in T_{m_i} \mathbb{J}(TU)$ the field*

$$E(d) \in N_{m_i} \mapsto \tilde{H} := H + \frac{1}{4} \mathrm{Trace}([H, E(d) - m_i]_+) E(d)$$

is a smooth tangent vector field $\tilde{H}$ on N_{m_i} that extends H.

Proof. Putting $X_{E(d)} = \frac{1}{4}\text{Trace}([H, E(d) - m_i]_+)E(d)$, observe $X_{E(0)} = 0$ since $E(0) = m_i$. So the field of 2×2-matrices $p \in U \mapsto H_p + (X_{E(d)})_p$ extends the field $p \in U \mapsto H_p$. Observe that

$$[H + X_{E(d)}, E(d)]_+ = [H, E(d)]_+ + [X_{E(d)}, E(d)]_+$$
$$= [H, E(d) - m_i]_+ + [X_{E(d)}, E(d)]_+$$

and

$$[H, E(d) - m_i]_+ = \begin{pmatrix} \lambda & 0 \\ 0 & \lambda \end{pmatrix},$$

$$[E(d), E(d)]_+ = \begin{pmatrix} -2 & 0 \\ 0 & -2 \end{pmatrix}.$$

So, by putting $X = \frac{1}{4}\text{Trace}([H, E(d) - m_i]_+)E(d)$ makes $H + X$ and $E(d)$ anti-commute, hence $H + X \in T_{E(d)}\mathbb{J}(TU)$. $\square$

The above proof even shows a little more, namely that the tangent vectors $H \in T_J\mathbb{J}(TS)$ extend explicitly smoothly to a tangent vector field over an open neighborhood N_J of J in $\mathbb{J}(TS)$. Put $D_J = \{d \in T_J\mathbb{J}(TS) \mid \text{Det}(d_p + J_p) > 0, \, p \in S\}$. The map

$$E: d \in D_J \to \mathbb{J}(TS), \, d \mapsto \frac{1}{\sqrt{\text{Det}(d + J)}}(d + J)$$

is a parametrization of an open neighborhood N_J of J in $\mathbb{J}(TS)$. The expression

$$\tilde{H}_{E(d)} := H + \frac{1}{4}\text{Trace}([H, E(d) - J]_+E(d)$$

extends $H \in T_J\mathbb{J}(TS)$ to a vector field over N_J.

Theorem 6.6. *The expression*

$$\omega_J^\circ(H, K) := -\text{Trace}HJK, \, J \in \mathbb{J}(TS), \, H, K \in T_J(\mathbb{J}(TS))$$

defines a natural non-degenerate closed 2-differential form ω° on the Fréchet manifold $\mathbb{J}(TS)$ with values in the space $C^\infty(S, \mathbb{R})$ of real functions on S. For $H \in T_J(\mathbb{J}(TS))$, the function $\omega_J^\circ(H, J^\circ H)$ is non-negative, and moreover positive at $p \in S$ if $H_p \neq 0$.

Proof. Clearly $\omega_J^\circ(H, K)$ is linear in its arguments H and K. From $KJ = -JK$ and the property $\text{Trace}AB = \text{Trace}BA$ its anti-symmetry follows:

$$\omega_J^\circ(H, K) = -\text{Trace}HJK = \text{Trace}HKJ = \text{Trace}KJH = -\omega_J^\circ(K, H)$$

so the expression defines a 2-differential form ω° on $\mathbb{J}(TS)$.

The 2-form ω° is closed. Indeed, for $J \in \mathbb{J}(TS)$, $H, K, L \in T_J(\mathbb{J}(TS))$ let $\tilde{H}, \tilde{K}, \tilde{L}$ be extensions over N_J. In order to prove $d\omega^\circ = 0$ we evaluate at $J \in N_J$

$$d\omega^\circ(\tilde{H}, \tilde{K}, \tilde{L}) = \tilde{H}\omega^\circ(\tilde{K}, \tilde{L}) - \tilde{K}\omega^\circ(\tilde{H}, \tilde{L}) + \tilde{L}\omega^\circ(\tilde{H}, \tilde{K})$$
$$- \omega^\circ([\tilde{H}, \tilde{K}]_{\text{Lie}}, \tilde{L}) + \omega^\circ([\tilde{H}, \tilde{L}]_{\text{Lie}}, \tilde{K}) - \omega^\circ([\tilde{K}, \tilde{L}]_{\text{Lie}}, \tilde{H}).$$

The first term at J is the directional derivative in the direction $H \in T_J \mathbb{J}(TS)$ of the function $J' \in N_J \mapsto \omega^\circ_{J'}(\tilde{K}_{J'}, \tilde{L}_{J'}) = -\mathrm{Trace}\tilde{K}_{J'} J' \tilde{L}_{J'}$. One contribution, the partial derivative with respect to J', is $-\mathrm{Trace} KHL$. So for all three first terms we collect the contributions

$$-\mathrm{Trace}(KHL - HKL + HLK).$$

The remaining six contributions of the two partial derivatives from the first three terms cancel against the last three terms. So finally

$$(\mathrm{d}\omega^\circ(\tilde{H}, \tilde{K}, \tilde{L}))_J = -\mathrm{Trace}(KHL - HKL + HLK).$$

Products with three factors like KHL anti-commute with J, so $\mathrm{Trace} KHL = 0$. We conclude that ω° is a closed 2-form on $\mathbb{J}(TS)$.

Locally, expressed in suitable coordinates, a field J is the constant field $\left(\begin{smallmatrix} 0 & -1 \\ 1 & 0 \end{smallmatrix}\right)$ and a field H that anti-commutes with J is of the form $\left(\begin{smallmatrix} a(p) & b(p) \\ b(p) & -a(p) \end{smallmatrix}\right)$. The function $p \mapsto (\omega^\circ(H, J^\circ H))_p$ on S is the non-negative function $p \mapsto 2(a(p)^2 + b(p)^2)$, which is positive at p if $H_p \neq 0$. $\qquad\square$

Much later (see Chapter 14) we will study the *moduli spaces* $\mathbb{M}_g$ and the *Teichmüller spaces* $\mathbb{T}_g$. At this point we can already give definitions, but many justifications have to be postponed at this stage of the study.

Let S_g be a compact, oriented, connected surface of genus $g \geq 0$. The space $\mathbb{M}_g$ is the quotient as metric space of $\mathbb{J}(TS_g)$ by the group $\mathrm{Diff}^+(S_g)$ of oriented diffeomorphisms of S_g. The Teichmüller space[1] $\mathbb{T}_g$ is the quotient of $\mathbb{J}(TS_g)$ by the group $\mathrm{Diff}_0(S_g)$ of diffeomorphisms of S_g that are homotopic to the identity. More precisely, the group Φ being $\mathrm{Diff}^+(S_g)$ or $\mathrm{Diff}_0(S_g)$, the distance between orbits $\Phi^* a, \Phi^* b \in \mathbb{J}(S_g)/\Phi$, $a, b \in \mathbb{J}(TS_g)$ is the infimum of the distances in $\mathbb{J}(TS_g)$ between $(S_g, \phi^* a)$, $(S_g, \psi^* b)$, $\phi, \psi \in \Phi$.

It turns out that $\mathbb{T}_g$ is a smooth manifold of real dimension $6g - 6$ if $g \geq 2$, of real dimension 2 if $g = 1$ and of dimension 0 if $g = 0$. The complex structure J° on $\mathbb{J}(TS_g)$ is invariant under the action of diffeomorphisms, hence will induce a complex structure on $\mathbb{T}_g$, provided that $\mathbb{T}_g$ is naturally a smooth manifold and that the orbits of the group of diffeomorphisms have J°-invariant tangent spaces. Later in Section 14.1 we will prove that $\mathbb{T}_g$ is indeed a smooth complex manifold.

Also the Uniformization Theorem for compact Riemann surfaces will be proved. It provides a Riemann surface (S, J) of genus $g \geq 1$ with a natural Riemannian metric of constant curvature. Let μ_J be the corresponding natural measure. The form ω° on $\mathbb{J}(TS)$ together with the measure μ define a natural 2-differential form $\omega^{\circ,\mu}$ on $\mathbb{J}(TS)$ by

$$\omega^{\circ,\mu}(H, K) := \int_S \omega^\circ(H, K) \mathrm{d}\mu_J$$

The form $\omega^{\circ,\mu}$ is non-degenerate. Indeed, $\omega^{\circ,\mu}(H, J^\circ H) > 0$ for $0 \neq H \in T_J \mathbb{J}(TS)$.

Theorem 6.7. *If S is compact, of genus $g \geq 1$, then the pair $(J^\circ, \omega^{\circ,\mu})$ induces a natural Kähler structure on the Teichmüller space $\mathbb{T}_g$ and on the moduli space $\mathbb{M}_g$.*

---

[1] Named after German mathematician Paul Julius Oswald Teichmüller (1913–1943).

6.3 J-Fields and Integrability in Higher Dimensions

The above decomposition as a direct sum can be done pointwise in the case of a
J-manifold. Let M be a manifold with charts U_α in a vector space V and let J be a J-
field on M given locally by fields $J_\alpha : U_\alpha \to J(V)$. At each point $p \in M$, the tangent
space at p is identified with V. Fields on M that are locally maps $U_\alpha \to \Lambda(V, \mathbb{C})$
are k-differential forms with values in $\mathbb{C}$. The infinite-dimensional $\mathbb{C}$-vector space
of these forms is denoted by $\Omega(M, \mathbb{C})$. The previous theorem applies pointwise and
gives a direct sum decomposition

$$\Omega^k(M, \mathbb{C}) = \bigoplus_{l=0}^{l=k} \Omega_J^{k-l,l}(M, \mathbb{C}).$$

The previously pointwise defined projectors $\P_J^{k-l,l}$ give projectors

$$\P_J^{k-l,l} : \Omega^k(V, \mathbb{C}) \to \Omega^k(V, \mathbb{C})$$

with images $\Omega_J^{k-l,l}(V, \mathbb{C})$. It is interesting to see how the exterior differentials
$d^k : \Omega^k(V, \mathbb{C}) \to \Omega^{k+1}(V, \mathbb{C})$ intertwine with the projectors $\P_J^{k-l,l}$ and $\P_J^{k+1-h,h}$.

In the case of a complex manifold, i.e. a manifold M with charts U_α in $\mathbb{C}^n$ and
holomorphic transition maps $\phi_{\beta\alpha}$, the underlying J-field intertwines in a special way
with the projectors $\P^{*,*}$. In a chart U_α the forms $\omega \in \Omega^{k-l,l}(M, \mathbb{C})$ can be written
in terms of the coordinate differentials $dz_i, d\bar{z}_j$, namely, as a linear combination of
forms

$$f \, dz_{a_1} \wedge \cdots \wedge dz_{a_{k-l}} \wedge d\bar{z}_{b_1} \wedge \cdots \wedge d\bar{z}_{b_l}$$

with $1 \leq a_1 < a_2 < \cdots < a_{k-l} \leq n$ and $1 \leq b_1 < b_2 < \cdots < b_l \leq n$. It follows that
$d^k : \Omega^k(M, \mathbb{C}) \to \Omega^{k+1}(M, \mathbb{C})$ maps the space $\Omega^{k-l,l}(M, \mathbb{C})$ into the sum of two
summands of the decomposition:

$$d^k(\Omega^{k-l,l}(M, \mathbb{C})) \subset \Omega^{k+1-l,l}(M, \mathbb{C}) \oplus \Omega^{k-l,l+1}(M, \mathbb{C}).$$

With the notation $\Omega^{p,q}(M, \mathbb{C}) = \Omega^{k-l,l}(M, \mathbb{C})$ we have $d^{p+q}(\Omega^{p,q}(M, \mathbb{C})) \subset$
$\Omega^{p+1,q}(M, \mathbb{C}) \oplus \Omega^{p,q+1}(M, \mathbb{C})$. Recall that for spaces $\Omega_J^{k-l,l}(V, \mathbb{C})$ or projectors
$\P_J^{k-l,l}$ on a complex manifold associated to the underlying J-field we have omitted
the index J in the notations. In particular, for a J-field that underlies a complex
manifold structure it follows that the composition

$$\P^{0,2} \circ d^1 : \Omega^{1,0}(M, \mathbb{C}) \to \Omega^{0,2}(M, \mathbb{C})$$

evaluates to 0.

What happens for J-fields that do not necessary underlie a holomorphic structure
on the manifold M? It was observed by A. Nijenhuis[2] that the map $\P_J^{0,2} \circ d^1$ is
tensorial [84], which means that for each point $p \in M$ and for each $\omega \in \Omega_J^{1,0}(M, \mathbb{C})$
the value $(\P_J^{0,2} \circ d^1(\omega))_p$ of $\P_J^{0,2} \circ d^1(\omega)$ at the point p only depends on the value

---

[2] Albert Nijenhuis (1926–2015), Dutch-American mathematician.

ω_p of ω at the point p. The map $\P_J^{0,2} \circ d^1$ is the Nijenhuis torsion, now called the Nijenhuis tensor N_J.

A J-field on a manifold M is called integrable if the field underlies a holomorphic structure on M. We have already proved the following result of Nijenhuis [84]:

Theorem 6.8. $N_J = 0$ *for an integrable J-field J.* $\qquad\qquad\qquad\qquad\qquad\square$

The converse statement is true according to a deep result of A. Newlander[3] and L. Nirenberg[4] [83]:

Theorem 6.9 (Newlander–Nirenberg Theorem). *A J-field J on a manifold is integrable if* $N_J = 0$ *holds.*

We already gave a proof for J-fields on 2-dimensional manifolds. For the higher-dimensional cases we give references to the literature. See the more elementary proof by Bernard Malgrange for C^∞-fields [72] and the even more elementary proof by André Weil[5] for real analytic fields.

In real even dimension $4, 6, \ldots$ a J-structure cannot be upgraded in general to a z-structure.

Exercise 6.10. First, show that the map $\P^{0,2} \circ d^1 : \Omega^{1,0}(M, \mathbb{C}) \to \Omega^{0,2}(M, \mathbb{C})$ is smoothly linear and deduce that the $\P^{0,2} \circ d^1$ map is tensorial.

6.4 Integrability of Fibred J-Fields

Let $E = B \times S$ be the product of a holomorphic manifold B and a smooth surface S. Let $\sigma : B \to \mathbb{J}(TS)$ be a smooth map from B to the complex Banach manifold $\mathbb{J}(TS)$ of J-fields on the surface S. The map σ defines a J-field J_σ on E as follows. The tangent space T_pE at $p = (b, s)$ decomposes as a direct sum of the tangent spaces $T_pE = T_bB \oplus T_sS$ and $J_{\sigma,p} = J_b \oplus \sigma(b)_s$.

The J-field J_σ is partially integrable. On B its J-field $b \mapsto J_{B,b}$ is integrable by assumption and on the fibers $\{b\} \times S$ the fields $\sigma(b)$ are integrable by Theorem 5.30. The following theorem simplifies the integrability condition of Nijenhuis for the field J_σ.

Theorem 6.11 (Fibered Integrability Criterion). *Let $E = B \times S$ be the product of a holomorphic manifold B and a smooth surface S. Let $\sigma : B \to \mathbb{J}(TS)$ be a smooth map from B to the complex Banach manifold $\mathbb{J}(TS)$ of J-fields on the surface S. The field J_σ on E is integrable if and only if the map σ is holomorphic.*

Proof. Let $\frac{\partial}{\partial t_1}, \frac{\partial}{\partial t_2}, \ldots, \frac{\partial}{\partial t_{2n}}$ be local coordinate vector fields on B such that $J(\frac{\partial}{\partial t_k}) = \frac{\partial}{\partial t_{k+n}}$, $k = 1, \ldots n$, holds. This is possible to achieve since B is a holomorphic manifold. Let $\frac{\partial}{\partial s_1}, \frac{\partial}{\partial s_2}$ be locally smooth coordinate vector fields on S. It suffices to

---

[3] August Newlander, American mathematician.

[4] Louis Nirenberg (1925–2020), Canadian-American mathematician.

[5] André Weil (1906–1998), French mathematician.

compute the Nijenhuis tensor $N_{J_\sigma}(X, Y)$ for X a coordinate vector field on B and Y a coordinate vector field on S.

Using the expression of Nijenhuis and Woolf [85] for the Nijenhuis tensor

$$N_J(X, Y) = [JX, JY]_{\text{Lie}} - [X, Y]_{\text{Lie}} - J[JX, Y]_{\text{Lie}} - J[X, JY]_{\text{Lie}}$$

it simplifies here for the field J_σ on $B \times S$ to

$$N_{J_\sigma}(X, Y) = [J_\sigma X, J_\sigma Y]_{\text{Lie}} - J_\sigma[X, J_\sigma Y]_{\text{Lie}}$$

since $X, J_\sigma X, Y$ are coordinate vector fields on $B \times S$. Notice

$$[X, J_\sigma Y]_{\text{Lie}} = ((D\sigma)(X))(Y) \ , \ [J_\sigma X, J_\sigma Y]_{\text{Lie}} = ((D\sigma)(J_\sigma X))(Y)$$

and at $(b, s) \in B \times S$

$$(J_\sigma)_{b,s} Y_s = \sigma(b) Y_s, \ (J_\sigma)_{b,s} X_b = J_{B,b} X_b.$$

So, $N_{J_\sigma} = 0$ if and only if for all coordinate vector fields X, Y as above

$$((D\sigma)(J_\sigma X))(Y) = J_\sigma((D\sigma)(X))(Y)$$

or more expanded at $(b, s) \in B \times S$

$$((D\sigma)_b(J_{B,b} X_b))(Y_s) = (((D\sigma)(J_\sigma X))(Y))_{b,s}$$
$$= (J_\sigma((D\sigma)(X))(Y)))_{b,s} = (\sigma(b))_s(((D\sigma)_b(X_b))(Y_s))$$

hold. This formula (which is written with many brackets) means precisely that the map $\sigma \colon B \to \mathbb{J}(TS)$ is $(J_B, J_{J(S)})$-holomorphic. With the Newlander–Nirenberg Theorem (Theorem 6.9) it follows that the field J_σ on $B \times S$ is integrable if and only if the map $\sigma \colon B \to \mathbb{J}(TS)$ is holomorphic. □

6.5 Analysis of Laplace Operators on J-Surfaces

In the proofs of the two main theorems, namely the Uniformization Theorem and the Analytic Characterization of Genus (Theorem 12.2), the following property of the Laplace operator $\Delta_J \colon \Omega^0(S, \mathbb{R}) \to \Omega^2(S, \mathbb{R})$ on a J-surface (S, J) will be of crucial importance. Remember that the spaces $\Omega^0(S, \mathbb{R})$ and $\Omega^2(S, \mathbb{R})$ are spaces of smooth, i.e. C^∞, sections.

Theorem 6.12 (Laplace Inversion Theorem, version 1). *Let (S, J) be a compact connected J-surface. Let ω be a smooth volume form on S. The kernel and co-kernel of the continuous operators on Fréchet spaces $\Delta_J \colon \Omega^0(S, \mathbb{R}) \to \Omega^2(S, \mathbb{R})$ and $\Delta_{J,\omega} \colon \Omega^0(S, \mathbb{R}) \to \Omega^0(S, \mathbb{R})$ are of dimension one. The kernel of Δ_J is the space of constant functions. The image of Δ_J is the space of 2-forms ω with $\int_S \omega = 0$.*

Theorem 6.13 (Laplace Inversion Theorem, version 2)**.** *Let* (S, J) *be a compact connected J-surface. Let ω_0 be a smooth volume form on S of total volume 1. The operator*

$$\Delta_J^+ : \Omega^0(S, \mathbb{R}) \to \Omega^2(S, \mathbb{R})$$

defined by

$$f \in \Omega^0(S, \mathbb{R}) \mapsto \Delta_J(f) + \left(\int_S f\omega_0 \right) \omega_0 \in \Omega^2(S, \mathbb{R})$$

is an isomorphism of Fréchet spaces.

These theorems imply each other. There are several methods of proof. Since the operator Δ_J is an elliptic partial differential operator, the theory of Fredholm operators applies.

For version 2 we supply a proof that uses Potential Theory. Its origins are in the work of Peter Gustav Lejeune Dirichlet, Axel Harnack [49], Robert Brown,[6] Tibor Radó [99] and Oskar Perron [92]. The mathematical treatment of Brownian motion was initiated in the work of Albert Einstein,[7] Norbert Wiener,[8] and Paul Pierre Lévy[9] [68]. The proof of the Laplace Inversion Theorem needs a preparation that uses Brownian motion and constructs meromorphic differentials with only two first order poles.

Theorem 6.14 (Two-point Green function). *Let* (S, J) *be a compact connected J-surface. Given two distinct points $p, q \in S$ there exists a Δ_J-harmonic real function $h_{p,q}$ on $S \setminus \{p, q\}$ such that the J-holomorphic $(1,0)$-form $\omega_{p,q} - dh_{p,q} - idh_{p,q} \circ J$ has poles at p and q with residues $+1$ and -1 respectively.*

Proof. Let ω_0 be a J-oriented volume form on S. The proof uses Brownian motion on the Riemannian surface $(S, g(J, \omega_0))$, see [131, 132]. Choose local z-coordinates z_p, z_q centred at p, q respectively. Put $b_p(r) = \{a \in S \mid |z_p(a)| < r\}$ and $b_q(r) = \{a \in S \mid |z_q(a)| < r\}$. For $r > 0$ and small enough, use Brownian motion on $S_{p,q,r} = S \setminus b_p(r) \cup b_q(r)$ in order to construct a real function $h_{p,q,r}$ on $S_{p,q,r}$ that is harmonic in the interior of $S_{p,q,r}$ and with boundary values $\frac{\pm 1}{2\pi} \log(r)$. The function $h_{p,q}$ is the limit function for $r \to +0$ of the functions $h_{p,q,r}$. The function $h_{p,q}$ is a Green function with poles p, q and moreover in the sense of distribution or Cauchy principal values solves the equation $\Delta_J(h_{p,q}) = \delta_p - \delta_q$. The J-holomorphic $(1,0)$-form $\omega_{p,q} = dh_{p,q} - idh_{p,q} \circ J$ has poles at p and q with residues $+1$ and -1 respectively. $\qquad\square$

For the further study of Brownian Motion, see the books of Yor [131, 132] and Revuz and Yor[10] [101] and for Distributions the book [111] of L. Schwartz.[11] Cauchy principal values give the first examples of distributions.

---

[6] Robert Brown (1773–1858) Scottish botanist who observed the so-called Brownian motion.

[7] Albert Einstein (1879–1955), Swiss-American physicist.

[8] Norbert Wiener (1894–1964), American mathematician.

[9] Paul Pierre Lévy (1886–1971), French mathematician.

[10] Marc Yor (1949–2014), French mathematician.

[11] Laurent Schwartz (1915–2002), French mathematician.

Proof (of the Laplace Inversion Theorem, version 2). Let $M_0^2(S)$ be the set of Radon measures of total mass 2 and integral 0. It is a compact convex set in the dual space, with its weak topology, of the Banach space $C^0(S, \mathbb{R})$ of real continuous functions on S. Its extremal points are the measures $m_{p,q} = \delta_p - \delta_q$, $p, q \in S$, $p \neq q$, where δ_r is the Dirac measure concentrated at $r \in S$. The integrable function $h_{p,q}$ of the previous theorem satisfies $\Delta_J(h_{p,q}) = m_{p,q}$ in the sense of distributions of Laurent Schwartz or principal values of Augustin-Louis Cauchy [111]. The Banach space is separable and therefore the space $M_0^2(S)$ is metrizable. The Krein–Milman Theorem of Mark Grigorjewitsch Krein and David Milman, see [63], represents a point $\omega \in M_0^2(S)$ as a limit of convex combinations of points in the closure of the set of extremal points $EM_0^2(S)$. Since the space $M_0^2(S)$ is metrizable, even Gustave Choquet's strengthening[12] of the Krein–Milman theorem (see [95]) applies, and a probability measure μ_ω on $M_0^2(S)$ may be chosen with support in the set of extremal points such that the representation

$$\omega = \int_{EM_0^1(S)} m_{p,q} \mathrm{d}\mu(p, q)$$

holds. It follows that for every $\omega \in M_0^2(S)$ the integrable function

$$h_\omega = \int_{EM_0^1(S)} h_{p,q} \mathrm{d}\mu(p, q)$$

satisfies $\Delta_J(h_\omega) = \omega$. The function h_ω is unique up to a constant function by the maximum principle. If ω is smooth, the function h_ω is smooth too by standard regularity theory.

If $\omega \in \Omega^2(S)$ is not in $M_0^2(S)$ we first replace ω by $\omega - (\int_S \omega)\omega_0$. After scaling, ω will be in $M_0^2(S)$ and we apply the above.

The operator Δ_J^+ being continuous, injective and surjective is an isomorphism of Fréchet spaces by the Banach–Schauder Isomorphism Theorem (see Theorem 12.14 later). □

Spectral Theory provides another proof of version 1. The operator $\Delta_{J,\omega}$ maps functions to functions, so the notion of eigenvalue and eigenvector make sense. A function $f \neq 0$ on S is an eigenvector if the functions f and $\Delta_{J,\omega}(f)$ are proportional, i.e.

$$\Delta_{J,\omega}(f) = \lambda f.$$

The factors λ that appear are the eigenvalues of the operator $\Delta_{J,\omega}$.

The following "work horse" theorem states essential results from spectral theory.

Theorem 6.15 (Spectral work horse). *The operator $\Delta_{J,\omega}$ has countably many eigenvalues $(\lambda_n)_{n \in \mathbb{N}}$. The eigenvalues λ_n are real non-negative numbers, which can be labeled in increasing order such that*

$$0 = \lambda_0 < \lambda_1 \leq \lambda_2 \leq \cdots$$

---

[12] Gastave Choquet, French mathematician, was Professor in Paris.

holds. Moreover, for every $T \in \mathbb{R}$ the number $\Lambda(T)$ of eigenvalues in $[0,T]$ is finite. The asymptotics of the growth of $\Lambda(T)$ for $T \to +\infty$ is deduced from the law of Weyl [13] [124] and Karamata [14] [58]:

$$\lim_{n \to \infty} \frac{\lambda_n}{n} = \frac{1}{4\pi} \int_S \omega.$$

There exists a system of eigenvectors $(e_n)_{n \in \mathbb{N}}$ which are smooth functions on S with

$$\int_S e_n^2 \omega = 1, \ \int_S e_n e_m \omega = 0, \ 0 \leq n < m.$$

Every smooth function f on S admits a unique expansion in eigenfunctions

$$f = \sum_0^\infty a_n e_n$$

which is uniformly convergent in the Fréchet space of smooth functions. One has for every $k \in \mathbb{N}$

$$\lim_{n \to \infty} n^k a_n = 0.$$

Conversely, a series $\sum_0^\infty a_n e_n$ with $\lim_{n \to \infty} n^k a_n = 0$, $k \in \mathbb{N}$, defines a smooth function by summation.

In short, one can say that the properties given by Fourier's theory for smooth functions and the operator $-\frac{d^2}{d\theta^2}$ on the circle $S^1 = \mathbb{R}/2\pi\mathbb{Z}$ hold for smooth functions and the operator $\Delta_{J,\omega}$ on a J-surface S. Even more generally, these properties hold for the Laplace operator

$$\Delta_M := *d*d \colon \Omega^0(M) \to \Omega^0(M)$$

on a compact Riemannian manifold (M, g) of dimension $d \geq 1$. The law concerning the growth of the eigenvalues depends on the dimension. For $T > 0$ let $\Lambda_M(T)$ be the dimension of the space spanned by eigenfunctions for eigenvalues $\lambda \in [0,T]$. Then the law of Weyl and Karamata becomes

$$\lim_{T \to \infty} \frac{\Lambda_M(T)}{T^{d/2}} = \frac{\beta(d)}{(2\pi)^d} \operatorname{Vol}_g(M),$$

where $\beta(d)$ is the volume of the unit ball in Euclidean space of dimension d. As an illustration, the cases of the circle S^1 of length 2π and the round unit two-sphere S^2 in $\mathbb{R}^3$ of area 4π match the law.

The eigenfunctions on the circle S^1 are

$$\frac{1}{2\pi}, \frac{1}{\sqrt{\pi}} \sin(\theta), \frac{1}{\sqrt{\pi}} \cos(\theta), \frac{1}{\sqrt{\pi}} \sin(2\theta), \frac{1}{\sqrt{\pi}} \cos(2\theta), \frac{1}{\sqrt{\pi}} \sin(3\theta), \frac{1}{\sqrt{\pi}} \cos(3\theta), \dots$$

---

[13] Hermann Klaus Hugo Weyl (1885–1955), German-Swiss mathematician, physician philosopher, Professor at ETH in Zürich and IAS in Princeton.

[14] Jovan Karamata (1902–1967), Serbian mathematician and Professor at the University of Geneva.

and the corresponding eigenvalues $0, 1, 1, 4, 4, 9, 9, \ldots$. Weyl's Law in dimension 1 is

$$\lim_{T \to \infty} \frac{\Lambda_{S^1}(T)}{\sqrt{T}} = \lim_{T \to \infty} \frac{2([\sqrt{T}] + 1)}{\sqrt{T}} = 2 = \frac{\beta(1)}{2\pi} \text{Length}(S^1).$$

The eigenvalues and eigenspaces of the Laplace operator Δ_{S^2} on the round sphere S^2 of radius 1 in Euclidean 3-space $\mathbb{R}^3$ are explicitly known. The space of restrictions of homogeneous harmonic polynomials $F(x, y, z)$ on Euclidean 3-space of degree d to the sphere S^2 is the kernel of the operator $\Delta_{S^2} - d(d + 1)\text{Id}$. The eigenvalues are $0, 2, 6, 12, 20, \ldots, d(d + 1), \ldots$ with multiplicity $2d + 1$ for the eigenvalue $d(d + 1)$, $d \geq 0$. Observe that

$$\lim_{T \to \infty} \frac{\Lambda_{S^2}(T)}{T} = \lim_{d \to \infty} \frac{\Lambda_{S^2}(d(d + 1))}{d(d + 1)}$$

$$= \lim_{d \to \infty} \frac{(d + 1)^2}{d(d + 1)} = 1 = \frac{\beta(2)}{(2\pi)^2} \text{Area}(S^2).$$

For more about Riemannian Geometry and in particular about Weyl and Karamata's Law, study Marcel Berger's[15] impressive book [13].

Proof (of Laplace Inversion Theorem, version 1). Let $f = \sum_0^\infty a_n e_n$ be in the kernel of $\Delta_{J,\omega}$. We have

$$0 = \Delta_{J,\omega}(f) = \sum_1^\infty \lambda_n a_n e_n.$$

From the uniqueness of the expansion in eigenfunctions of the function 0 it follows that $\lambda_n a_n = 0$, $n = 1, 2, \ldots$. Hence $a_n = 0$, $n > 0$, since $\lambda_n > 0$, $n > 0$. We conclude $f = a_0 e_0$ and $\text{Dim}(\text{Ker}(\Delta_{J,\omega})) = 1$. The space of constant functions is of dimension one and a subspace of the kernel, hence equal to the kernel of $\Delta_{J,\omega}$.

All smooth functions $g = \sum_0^\infty b_n e_n$ with $b_0 = 0$ are contained in the image of $\Delta_{J,\omega}$. Indeed, put $f = \sum_1^\infty \frac{b_n}{\lambda_n} e_n$ and we have $\Delta_{J,\omega}(f) = g$. It follows that the dimension of the co-kernel equals one.

The operator $\%{\omega} : \Omega^2(S, \mathbb{R}) \to \Omega^0(S, \mathbb{R})$ is a linear isomorphism, hence the dimensions of the kernel and co-kernel of Δ_J are one. Moreover, $\text{Ker}(\Delta_J) = \text{Ker}(\Delta_{J,\omega})$. One has $\int_S \Delta_J(f) = \int_S d(df \circ J) = 0$ by Stokes' Theorem. Comparing co-dimensions shows that the image of Δ_J coincides with the kernel of the integration $\omega \in \Omega^2(S, \mathbb{R}) \mapsto \int_S \omega \in \mathbb{R}$. $\square$

The following theorem will be needed. Let $G : \mathbb{C} \to \mathbb{R}$ be the function defined by $G(p) := \frac{-1}{2\pi} \log(|p|)$ and $G(0) := 0$. The function G is an elementary solution of the Poisson equation[16] $\Delta G = \delta$, where δ is the Dirac atomic unit measure[17] at the origin $0 \in \mathbb{C}$. The idea of using an elementary solution for the Poisson equation is rooted in the early work of George Green.[18] For every C^∞-function $u : \mathbb{C} \to \mathbb{C}$ with compact support the function $f = u * G$ obtained by convolution

---

[15] Marcel Berger (1927–2016), French mathematician.

[16] Named after French mathematician and physicist Siméon Denis Poisson (1781–1840).

[17] Named after British physicist and mathematician Paul Adrien Maurice Dirac (1902–1984).

[18] George Green (1793–1841), British miller, physicist and mathematician.

$$f(p) = \int_{\mathbb{C}} G(q)u(p-q)dx \wedge dy$$

is a C^∞-function that satisfies the Poisson equation $\Delta f = u$ with $\Delta = \Delta_J = -(\frac{\partial^2}{\partial x^2} + \frac{\partial^2}{\partial y^2})$.

Theorem 6.16 (Local inversion of Laplace operator). *Let $u\colon \mathbb{C} \to \mathbb{C}$ be a C^∞-function with support in the unit disk $D_1 \subset \mathbb{C}$. Then the function $f = u * G$ satisfies $\Delta f = u$. Let M be the integral*

$$M = \int_{D_3} |G(p)|\,d\mathrm{Vol}(p)$$

over the disk of radius 3. For every monomial $P = \frac{\partial^a}{\partial x^a}\frac{\partial^b}{\partial y^b}$ of partial differential operators the inequality

$$\mathrm{Sup}_{p \in D_2}|Pf(p)| \le M\,\mathrm{Sup}_{p \in D_1}|Pu(p)|$$

holds.

Proof. The Laplace operator Δ and the partial differential operator P commute, so $Pf = G * Pu$ solves $\Delta Pf = Pu$. Hence for $p \in D_2$ one estimates $|Pf(p)|$ by $M\,\mathrm{Sup}_{p \in D_1}|Pu(p)|$. $\qquad\square$

It is worth mentioning that the computation of the de Rham cohomology of oriented compact connected surfaces of genus g can now be completed.

Theorem 6.17. *The second de Rham cohomology group $H^2_{dR}(S)$ of a compact, oriented connected surface S is isomorphic to $\mathbb{R}$.*

Proof. First equip S with a J-field J. Since $H^2_{dR}(S) = \Omega^2(S, \mathbb{R})/d\Omega^1(S, \mathbb{R})$ it suffices to show that the space of exact 2-forms $d\Omega^1(S, \mathbb{R})$ is the kernel of the map

$$\int_S : \omega \in \Omega^2(S, \mathbb{R}) \mapsto \int_S \omega \in \mathbb{R}.$$

By Stokes' Theorem we have that $d\Omega^1(S, \mathbb{R}) \subset \mathrm{Ker}(\int_S)$. For $\omega \in \mathrm{Ker}(\int_S)$ by the Laplace Inversion Theorem there exists a function f with $\Delta_J(f) = \omega$. Put $\alpha := df \circ J \in \Omega^1(S, \mathbb{R})$, hence $d\alpha = \omega$ and $d\Omega^1(S, \mathbb{R}) = \mathrm{Ker}(\int_S)$. $\qquad\square$

For a compact, oriented connected surface S let $\mathrm{Vol}(S)$ be the space of volume forms $\omega \in \Omega^2(S, \mathbb{R})$ that are pointwise positive, i.e. $\omega_p(u, v) > 0$ for each oriented basis (u, v) of $T_p(S)$, $p \in S$. Given $v \in \mathbb{R}$, $v > 0$, one constructs using a partition of unity volume forms $\omega \in \mathrm{Vol}(S)$ with $\int_{[S]} \omega = v$. There is a lot of freedom in this construction. The following theorem of J. Moser[19] [81] is a global rigidity theorem.

Theorem 6.18 (J. Moser, Global Volume Rigidity). *Two volume forms ω_0, ω_1 of equal total volume on a compact, connected oriented manifold M are isotopic.*

---

[19] Jürgen Moser (1928–1999), German-American-Swiss mathematician.

Proof. The path of volume forms $\omega_t = (1 - t)\omega_0 + t\omega_1$, $t \in [0, 1]$, connects the forms ω_0, ω_1 with constant speed $\dot{\omega}_t = \omega_1 - \omega_0 \in \Omega^n(M, \mathbb{R})$ with $\int_{[M]} \dot{\omega}_t = 0$. As proved above for surfaces under the assumption of the theorem, but which holds more generally for manifolds M of dimension $n > 1$, there exists an $\alpha \in \Omega^{n-1}(M, \mathbb{R})$ with $d\alpha = \omega_1 - \omega_0$. Define the time-dependent vector field X_t implicitly by $i_{X_t}\omega_t = -\alpha$. The flow ϕ_t generated by the field X_t satisfies $\phi_t^*\omega_t = \omega_0$. Indeed by Cartan's formula $\frac{d}{dt}\phi_t^*\omega_t = \phi_t^*(d\alpha + d(i_{X_t}\omega_t)) = 0$, so the family $\phi_t^*\omega_t$ is constant. $\qquad\square$

Above proof works also with a stronger hypothesis for symplectic forms.

Theorem 6.19 (J. Moser, Global Symplectic Rigidity along path). *Symplectic forms ω_0, ω_1 on a compact manifold M are isotopic if connected by a smooth path ω_t of symplectic forms, such that $[\omega_t] = [\omega_1] = [\omega_2]$ holds in $H_{dR}^2(M)$.* $\qquad\square$

6.6 Topology of the Two-Point Green Function

Let $S = (S, J)$ be a Riemann surface. The Green functions $h_{p,q}$, $p, q \in S$, $p \neq q$, are strongly related to the topology of the surface S. The function $h_{p,q}: S \to \mathbb{R} \cup \{\pm\infty\}$ is harmonic on $S \setminus \{p, q\}$. So, $h_{p,q}$ has no local maxima or minima. The function $h_{p,q}$ is locally the real part of a holomorphic function. Near p, q the function is the real part of a function $\frac{f}{z}$ with f holomorphic and z a local holomorphic coordinate at p or q. It follows that the critical points of $h_{p,q}$ are isolated and do not accumulate near p or q. Hence, the number of critical points is finite.

By choosing the points p, q in a general position we may assume that all critical points are non-degenerate and that the function $h_{p,q}$ separates the critical points. So for points p, q in general position the function $h_{p,q}$ is a Morse function. At a critical point c there exists a local centered holomorphic coordinate $z = x + iy$ with $h_{p,q} = \mathrm{Re}\,z^2 = x^2 - y^2$. So all critical points are saddle points of $h_{p,q}$.

Theorem 6.20. *Let $S = (S, J)$ be a Riemann surface and let p, q be distinct points in general position. The Green function $h_{p,q}$ is a Morse function. The number of critical points equals the number of critical values and is twice the genus g of S. Let $t_1 < t_2 < \cdots < t_{2g-1}$ be regular values that separate the critical values. The submanifold $h_{p,q}^{-1}(\{t_1, t_2, \ldots, t_{2g-1}\})$ is a union of embedded circles in S. The system of circles is a decomposition of the twice punctured surface $S \setminus \{p, q\}$ and also of the surface S into cylinders and pairs of pants. After removing for each cylinder one boundary curve, decompositions in pairs of pants remain.*

Proof. Let (p, q) be a generic pair of distinct points on a Riemann surface S of genus $g \geq 2$. Let $h_{p,q}$ have g' critical points and critical values. Connected components of the complement in S or in $S \setminus \{p, q\}$ of $h_{p,q}^{-1}(\{t_1, \ldots, t_{2g'-1}\})$ are cylinders or a pairs of pants. The number of pairs of pants that appear after having removed one boundary circle for each cylinder on $S \setminus \{p, q\}$ equals $2g' = \chi(S \setminus \{p, q\})$. On S there are two cylinders more and two pairs of pants less, so the final number of pairs of pants on S equals $2g' - 2 = \chi(S) = 2g - 2$. One concludes $g = g'$. $\qquad\square$

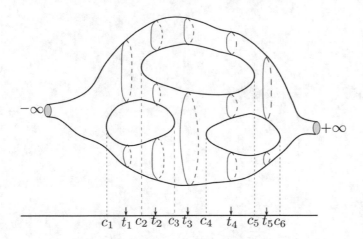

$c_1 \quad t_1 \; c_2 \; t_2 \quad c_3 \, t_3 \quad c_4 \qquad t_4 \qquad c_5 \, t_5 c_6$

Fig. 6.1 A genus 3 Green function with 4 cylinders and 6 pairs of pants.

Exercise 6.21. Let (S, J) be a compact connected Riemann surface. Call a pair of points (p, q), $p \neq q$, special if the Green function $h_{p,q} \colon S \setminus \{p, q\} \to \mathbb{R}$ is a Morse function that has no or only one critical value. Show that on the Riemann sphere all such pairs are special. Construct surfaces of higher genus that admit special pairs of points.

Chapter 7
Surfaces of Genus $g = 0$

7.1 The Uniformization Theorem, the Genus $g = 0$ Case

Theorem 7.1 (Uniformization Theorem for $g = 0$). *Let (S, J) be a J-surface of genus 0. The surface (S, J) is conformally isomorphic to the Riemann sphere $\mathbb{P}^1(\mathbb{C}) = \mathbb{C} \cup \{\infty\}$ and admits a conformally equivalent metric g_J of constant curvature $= +1$.*

Proof. Let (S^2, g_{S^2}) be the unit sphere in Euclidean 3-space with its induced metric g_{S^2}. Let $J_{g_{S^2}}$ be a J-field on S^2 that at each point $p \in S^2$ is a $\frac{\pi}{2}$-rotation. There exist two such J-fields that differ by sign. Let $J_{g_{S^2}}$ be one of them. The stereographic projection in $\mathbb{C} \times \mathbb{R}$ with center $(0, 2)$ maps the unit sphere $(S^2, J_{g_{S^2}})$ with center $(0, 1)$ conformally to the Riemann sphere $\mathbb{C} \cup \{\infty\}$.

Let λ be the closed 1-differential form of type $(1, 0)$ given by $\frac{dz}{z}$. The form is holomorphic on $\mathbb{C} \setminus \{0\} \subset \mathbb{P}^1(\mathbb{C})$ and with simple poles at 0 and ∞. For any homotopy class of a closed curve $\gamma \colon [0, 1] \to \mathbb{P}^1(\mathbb{C}) \setminus \{0, \infty\}$ the period $\int_\gamma \lambda$ is well defined. For the class c of the oriented unit circle $\int_c \lambda = 2\pi i$ holds.

Fix two points p, q on S and consider a, b on $\mathbb{P}^1(\mathbb{C}) = \mathbb{C} \cup \{\infty\}$, $a = 0$, $b = \infty$. By the local rigidity of J-fields we may choose a diffeomorphism $\phi \colon S \to \mathbb{P}^1(\mathbb{C})$ with $\phi(p) = a$, $\phi(q) = b$, such that the pull-back J-field $J_0 := \phi_*^* J_{\mathbb{P}^1(\mathbb{C})}$ on S coincides locally near p and q with the field J. Put $\alpha_0 := \phi^* \lambda$, which is a closed 1-differential form of J_0-type $(1, 0)$ on $(S \setminus \{p, q\}, J_0)$. Put $J_1 = J$ and interpolate between J_0 and J_1 by J_t, $t \in [0, 1]$, as before by pointwise hyperbolic geodesics $t \mapsto J_{t,p}$, $p \in S$. We want to construct a family of closed forms α_t of J_t-type $(1, 0)$ on $S \setminus \{p, q\}$ such that the period $\int_{\phi^* c} \alpha_t$ is independent of t. Hence we put $\alpha_t = \alpha_0 + dv_t$, where v_t is a family of functions on S. As in previous proofs we get for the family v_t of functions the differential equation

$$\dot{v}_t = -\Delta_{J_t}^{-1}(d((\alpha_0 + dv_t) \circ \dot{J}_t))$$

with initial condition $v_0 = 0$. Observe that the poles of α_0 at p, q do not harm the equation since $\dot{J}_t$ vanishes locally near p, q. This differential equation has a

© The Author(s), under exclusive license to Springer Nature Switzerland AG 2021
N. A'Campo, *Topological, Differential and Conformal Geometry of Surfaces*, Universitext,
https://doi.org/10.1007/978-3-030-89032-2_7

global solution on $[0, 1]$. The resulting functions v_t are smooth, hence the (J_1, m_i)-meromorphic forms $\alpha_t = \alpha_0 + dv_t \in \Omega_J^{1,0}(S)$ have poles only at p, q.

The residues at the poles p, q remain ± 1 since the period along ϕ^*c did not change. Hence both poles are simple. It follows that the forms α_t have no zero since pole or zero orders sum up to $-\chi(S) = -2 = \text{degr}((\alpha_t))$.

Let $e \in S$ be different from p, q. For $s \in S \setminus \{p, q\}$ let γ be a path from e to s and define

$$f_t(s) = e^{\int_\gamma \alpha_t} = e^{\int_e^s \alpha_t} \in \mathbb{C}.$$

The quantity $f_t(s)$ does not depend on the chosen path γ by Stokes' Theorem and since for the only period $2\pi i$ the identity $e^{2\pi i} = 1$ holds.

The map $z_J = f_1 \colon S \setminus \{p, q\} \to \mathbb{C}$ is (J_1, m_i)-holomorphic and extends to a holomorphic bijection $z_J \colon (S, J) \to \mathbb{C} \cup \{\infty\}$. This implies the first assertion of the theorem. The second assertion about the metric g_J is obtained by stereographic projection. $\qquad\square$

7.2 Strong J-Rigidity

Let J be a J-field on $\mathbb{R}^2$ that outside of a compact subset coincides with the $90°$-degree counter-clockwise rotation. The local rigidity provides for every point $p \in \mathbb{R}^2$ an open neighborhood U equipped with a (J, m_i)-holomorphic coordinate $z \colon (U, J) \to \mathbb{C}$.

Theorem 7.2 (Strong J-Rigidity). *Let J be a J-field on $\mathbb{R}^2$ that outside of a compact subset A coincides with the $90°$-degree counter-clockwise rotation. Then there exists a bijective (J, m_i)-holomorphic coordinate $z \colon (\mathbb{R}^2, J) \to \mathbb{C}$.*

Proof. Assume without restricting generality $0 \notin A$. Let J_t, $t \in [0, 1]$, interpolate between J_0, the $90°$-degree counter-clock-wise rotation, and the given field J. Use the previous Ansatz, $\alpha_t = \frac{dx + idy + (x+yi)dv_t}{x+yi}$, and solve the differential equation

$$\dot{v}_t = -\Delta_{J_t}^{-1}\left(d\left(\frac{dx + idy + (x + yi)dv_t}{x + yi} \circ J_t\right)\right)$$

with $v_0 = 0$ as initial condition. Also identify $(\mathbb{R}^2, J_0)$ and $(\mathbb{C}, m_i)$. Again it follows that α_1 is a (J, m_i)-meromorphic form with simple poles at 0 and ∞ and residues $\pm 2\pi i$. Hence, $dv_1 = \alpha_1 - \frac{dx+idy}{x+yi}$ is m_i-holomorphic on $(\mathbb{C} \cup \{\infty\}) \setminus A$. Then $z_J(p) = e^{\int_0^p \alpha_1}$ is a (J, m_i)-holomorphic coordinate $z_J \colon (\mathbb{R}^2, J) \to \mathbb{C}$. $\qquad\square$

7.3 Strong J-Rigidity and Volume Stretching

As before, let J be a field on $\mathbb{R}^2$ that coincides in the complement of a compact subset A with the field m_i of rotation by $90°$ degrees. How do the volume forms $dx \wedge dy$ and $\frac{i}{2}dz_J \wedge d\bar{z}_J$ on $\mathbb{R}^2$ compare? Think of $(\mathbb{R}^2 \setminus A, J)$ as $(\mathbb{C} \setminus A, m_i)$. Write

$e^{2u} dx \wedge dy = \frac{i}{2} dz_J \wedge d\bar{z}_J$. Also write dv_1 from the previous proof as

$$dv_1 = v(dx + idy) + w(dx - idy).$$

Remember, the function v is m_i-holomorphic and the function w vanishes on $(\mathbb{C} \cup \{\infty\}) \setminus A$. From

$$z_J(p) = \exp\left(\int_0^P \frac{dx + idy + (x + iy)(v(dx + idy) + w(dx - idy))}{x + iy}\right)$$

follows

$$dz_J = \frac{z_J}{x + iy}(dx + idy + (x + iy)(v(dx + idy) + w(dx - idy)))$$

and

$$\frac{i}{2} dz_J \wedge d\bar{z}_J =$$

$$\frac{z_J \bar{z}_J}{x^2 + y^2}(1 + x(v + \bar{v}) + y(iv + \bar{i}v) + (x^2 + y^2)(v\bar{v} - w\bar{w}))dx \wedge dy.$$

The smooth volume forms $\frac{i}{2} dz_J \wedge d\bar{z}_J$ and $dx \wedge dy$ are of equal orientation, so the proportionality factor is positive and can be written as e^{2u_J} for a smooth function $u_J : \mathbb{R}^2 \to \mathbb{R}$.

The proof of the above Uniformization Theorem, genus $g = 0$, has shown that the form dv has a holomorphic extension to $(\mathbb{C} \setminus A) \cup \{\infty\}$. Hence, the function u_J extends smoothly to $\mathbb{C} \cup \{\infty\}$.

The following vanishing is used in the next section.

Theorem 7.3. $\int_{\mathbb{R}^2} \Delta_J(u_J) = 0$.

Proof. From the Laplace Inversion Theorem it follows that

$$\int_{\mathbb{R}^2} \Delta_J(u_J) = \int_{\mathbb{C}\cup\{\infty\}} \Delta_J(u_J) = 0,$$

since $\mathbb{P}^1(\mathbb{C}) = \mathbb{C} \cup \{\infty\}$ is a compact J-surface. $\qquad\square$

Chapter 8
Surfaces with Riemannian Metric

8.1 Riemannian Curvature

The curvature tensor of a Riemannian metric g on a manifold M is a 2-differential form $R_g = R$ on M with values in $\text{End}(TM) = TM \otimes T^*M$. So at a point $p \in M$ the value $R_{g,p}$ associates to a pair (u, v) of tangent vectors in T_pM an endomorphism $R_{g,p}(u, v)$ of T_pM. The construction of R_g proceeds in two steps. In the first step one introduces a binary internal composition $\nabla = \nabla^g$ of vector fields. We denote by $\partial(M)$ the space of smooth vector fields on M. (Warning: do not forget the brackets, since ∂M commonly denotes the boundary of a manifold M.) The binary internal composition is a map

$$\nabla \colon \partial(M) \times \partial(M) \to \partial(M), \ (X, Y) \mapsto \nabla_X Y$$

which we would like to think of as the directional derivative of the field Y in the direction of the field X. This is a kind of wishful thinking. Why? The values of the fields Y_p, Y_q at points $p, q \in M$, $p \neq q$, belong to different spaces and do not fit into a three-term additive expansion that would define something like the directional derivative $d_X f$ of a function. An expansion $Y_{p+h} = Y_p + A(h) + R(h)$ is possible in a chart, but will depend on the chart!

The best we can do is to set up a list of axioms, say wishes, that makes $\nabla_X Y$ look like a directional derivative. An optimal list of wishes would be such that one can fulfill these wishes but only in a unique way.

Here is the list of axioms: Letters X, Y, Z represent vector fields, letters f, g functions and letters a, b numbers. *Smooth linearity* of an expression E means that smooth functions behave like scalars. Typically $E(fX) = fE(X)$.

1. The expression $\nabla_X Y$ is smoothly linear in the argument X and (scalarly) linear in the argument Y. This means

    $$\nabla_{fX_1 + gX_2} Y = \nabla_{fX_1} Y + \nabla_{gX_2} Y = f\nabla_{X_1} Y + g\nabla_{X_2} Y$$

    and

    $$\nabla_X (aY_1 + bY_2) = \nabla_X (aY_1) + \nabla_X (bY_2) = a\nabla_X Y_1 + b\nabla_X Y_2.$$

© The Author(s), under exclusive license to Springer Nature Switzerland AG 2021
N. A'Campo, *Topological, Differential and Conformal Geometry of Surfaces*, Universitext,
https://doi.org/10.1007/978-3-030-89032-2_8

2. Smooth linearity in the argument Y does not hold, but instead a Leibniz rule still holds. This means

$$\nabla_X(fY) = f\nabla_X Y + (\mathrm{d}_X f)\, Y.$$

3. The binary operation $\nabla_X Y$ is not commutative, but the following rule holds.

$$\nabla_X Y - \nabla_Y X = [X, Y]_{\mathrm{Lie}}.$$

4. The following chain rule of directional derivation $Zg(X, Y) = \mathrm{d}_Z g(X, Y)$ of the function $g(X, Y)$ holds, without any need for deriving the metric g,

$$\mathrm{d}_Z g(X, Y) = g(\nabla_Z X, Y) + g(X, \nabla_Z Y).$$

Axioms 1, 2 and 3 do not involve the metric g. An operation $\nabla_X Y$ satisfying axioms 1 and 2 is called a *covariant derivation of vector fields*. If axiom 3 is satisfied the covariant derivation is called *torsion-free*. If axiom 4 is satisfied the covariant derivation is called *isometric*.

These wishes are optimally fulfilled! This means that these wishes are fulfilled and that there is no room left for additional wishes.

Theorem 8.1 (Covariant Torsion Free Isometric Derivation of Vector Fields).
Every Riemannian manifold (M, g) admits a unique covariant derivation $\nabla^g = \nabla$ of vector fields that is isometric and torsion-free.

The covariant derivation ∇^g of vector fields of this theorem is called the *Levi-Cività connection*[1] or *covariant derivation* of the Riemannian metric.

Proof. For vector fields X, Y, Z the following Ansatz, due to J.-L. Koszul,[2] with 6 terms for the quantity $2g(\nabla_X Y, Z)$ defines implicitly an internal binary composition $(X, Y) \mapsto \nabla_X Y$ on $\partial(M)$.

$$\begin{aligned}
2g(\nabla_X Y, Z) = {} & Xg(Y, Z) + Yg(X, Z) - Zg(X, Y) \\
& + g([X, Y]_{\mathrm{Lie}}, Z) - g([Y, Z]_{\mathrm{Lie}}, X) - g([X, Z]_{\mathrm{Lie}}, Y).
\end{aligned}$$

Indeed, for X, Y given and for all vector fields Z the function $g(\nabla_X Y, Z)$ is known as the Koszul Ansatz. Knowing $g(\nabla_X Y, Z)$ for all fields Z determines the field $\nabla_X Y$, since for all points $p \in M$ the field value g_p is a scalar product.

By a systematic expansion of terms it is checked that this implicitly defined composition $(X, Y) \mapsto \nabla_X Y$ satisfies the axioms $1, 2, 3, 4$. This will prove the existence of a binary composition $(X, Y) \mapsto \nabla_X Y$ that satisfies the 4 axioms.

Let

$$\begin{aligned}
A_{\mathrm{Koszul}}(X, Y, Z) = {} & Xg(Y, Z) + Yg(X, Z) - Zg(X, Y) \\
& + g([X, Y]_{\mathrm{Lie}}, Z) - g([Y, Z]_{\mathrm{Lie}}, X) - g([X, Z]_{\mathrm{Lie}}, Y)
\end{aligned}$$

---

[1] Named after Italian mathematician and physicist Tulio Levi-Cività (1873–1941).

[2] Jean-Louis Koszul (1921–2018), French mathematician.

be a shorthand for the right-hand side of the Koszul Ansatz. The expression $A_{\text{Koszul}}(X, Y, Z)$ is scalarly linear in X, Y, so axiom 1 is partially checked. The smooth linearity of axiom 1 is checked by the following expansion

$$
\begin{aligned}
A_{\text{Koszul}}(fX, Y, Z) &= fXg(Y, Z) + Yg(fX, Z) - Zg(fX, Y) \\
&\quad + g([fX, Y]_{\text{Lie}}, Z) - g([Y, Z]_{\text{Lie}}, fX) - g([fX, Z]_{\text{Lie}}, Y) \\
&= fXg(Y, Z) + fYg(X, Z) + (\mathrm{d}_Y f)g(X, Z) - fZg(X, Y) \\
&\quad - (\mathrm{d}_Z f)g(X, Y) + fg([X, Y], Z) - (\mathrm{d}_Y f)g(X, Z) \\
&\quad - fg([Y, Z]_{\text{Lie}}, X) - fg([X, Z]_{\text{Lie}}, Y) + (\mathrm{d}_Z f)g(X, Y) \\
&= fA_{\text{Koszul}}(X, Y, Z)
\end{aligned}
$$

The Leibniz rule of axiom 2 is checked by the expansion

$$
\begin{aligned}
A_{\text{Koszul}}(X, fY, Z) &= Xg(fY, Z) + fYg(X, Z) - Zg(X, fY) \\
&\quad + g([X, fY]_{\text{Lie}}, Z) - g([fY, Z]_{\text{Lie}}, X) - g([X, Z]_{\text{Lie}}, fY) \\
&= Xg(fY, Z) + (\mathrm{d}_X f)g(Y, Z) + fYg(X, Z) - fZg(X, Y) \\
&\quad - (\mathrm{d}_Z f)g(X, Y) + fg([X, Y]_{\text{Lie}}, Z) \\
&\quad + (\mathrm{d}_X f)g([X, Y]_{\text{Lie}}, Z) - fg([Y, Z]_{\text{Lie}}, X) + \mathrm{d}_Z g(Y, X) \\
&\quad - fg([X, Z]_{\text{Lie}}, Y) \\
&= fA_{\text{Koszul}}(X, Y, Z) + 2(\mathrm{d}_X f)g(Y, Z).
\end{aligned}
$$

The commutativity defect of axiom 3 follows from

$$
\begin{aligned}
&A_{\text{Koszul}}(X, Y, Z) - A_{\text{Koszul}}(Y, X, Z) \\
&= Xg(Y, Z) + Yg(X, Z) - Zg(X, Y) + g([X, Y]_{\text{Lie}}, Z) - g([Y, Z]_{\text{Lie}}, X) \\
&\quad - g([X, Z]_{\text{Lie}}, Y) - Yg(X, Z) - Xg(Y, Z) + Zg(Y, X) - g([Y, X]_{\text{Lie}}, Z) \\
&\quad + g([X, Z]_{\text{Lie}}, Y) + g([Y, Z]_{\text{Lie}}, X) \\
&= 2g([X, Y]_{\text{Lie}}, Z).
\end{aligned}
$$

For axiom 4 expand and verify

$$
A_{\text{Koszul}}(Z, X, Y) + A_{\text{Koszul}}(Z, Y, X) = 2Zg(X, Y).
$$

The first 3 terms of Koszul's Ansatz, which has defined the composition $(X, Y) \mapsto \nabla_X Y$, can be expanded using axiom 4 as a sum of 6 terms

$$
\begin{aligned}
Xg(Y, Z) &+ Yg(X, Z) - Zg(X, Y) \\
&= (g(\nabla_X Y, Z) + g(Y, \nabla_X Z)) \\
&\quad + (g(\nabla_Y X, Z) + g(X, \nabla_Y Z)) \\
&\quad - (g(\nabla_Z X, Y) + g(X, \nabla_Z Y))
\end{aligned}
$$

and the last 3 terms using axiom 3 can be expanded as a sum of 6 terms

$$g([X,Y]_{\text{Lie}}, Z) - g([Y,Z]_{\text{Lie}}, X) - g([X,Z]_{\text{Lie}}, Y)$$
$$= (g(\nabla_X Y, Z) - g(\nabla_Y X, Z))$$
$$- (g(\nabla_Y Z, X) - g(\nabla_Z Y, X))$$
$$- (g(\nabla_X Z, Y) - g(\nabla_Z X, Y)).$$

Adding up yields that the Koszul Ansatz is a formula that follows from the axioms satisfied by $\nabla_X Y$. Hence, the uniqueness of the binary composition $(X, Y) \mapsto \nabla_X Y$ satisfying the axioms is proved. $\square$

An important example is the upper half plane

$$\mathbb{C}_+ = \{p \in \mathbb{R}^2 \mid y(p) > 0\} = \{p \in \mathbb{C} \mid \text{Im}(z(p)) > 0\}$$

with metric $g_{\mathbb{C}_+} = \frac{1}{y^2} g_{\text{Euclid}}$. Let E_1, E_2 be the constant vector fields with value $1, i \in \mathbb{C}$ respectively. The Koszul Ansatz simplifies from 6 to 3 terms since all Lie brackets $[E_i, E_j]_{\text{Lie}}$ vanish.

The Christoffel functions $\Gamma_{ij}^k \colon \mathbb{C}_+ \to \mathbb{R}$ allow us to express the vector fields $\nabla_{E_i} E_j$ as a linear combination of the fields E_k by

$$(\nabla_{E_i} E_j)_p = \sum_k \Gamma_{ij}^k(p) E_{k,p}, \ p \in \mathbb{C}_+$$

or, if using the Einstein summation convention, by

$$\nabla_{E_i} E_j = \Gamma_{ij}^k E_k.$$

The functions Γ_{ij}^k are commonly called *Christoffel symbols*.[3]
As an example

$$2g(\nabla_{E_1} E_1, E_1) = E_1 g(E_1, E_1) + E_1 g(E_1, E_1) - E_1 g(E_1, E_1) = 0 + 0 - 0,$$

$$2g(\nabla_{E_1} E_1, E_2) = E_1 g(E_1, E_2) + E_1 g(E_1, E_2) - E_2 g(E_1, E_1) = 0 + 0 - \frac{-2}{y^3},$$

so

$$\Gamma_{11}^1 = 0, \ \Gamma_{11}^2 = \frac{1}{y}, \ \nabla_{E_1} E_1 = \frac{1}{y} E_2.$$

By the same method of computation we have

$$\Gamma_{22}^1 = 0, \quad \Gamma_{22}^2 = -\frac{1}{y}, \quad \nabla_{E_2} E_2 = -\frac{1}{y} E_2,$$

$$\Gamma_{12}^1 = -\frac{1}{y}, \quad \Gamma_{12}^2 = 0, \quad \nabla_{E_1} E_2 = -\frac{1}{y} E_1.$$

Also by this computation or by axiom 2 we have

---

[3] Introduced by Elwin Bruno Christoffel (1829–1900), German mathematician, Professor in Zürich, Berlin and Strassburg.

$$\nabla_{E_2} E_1 = -\frac{1}{y} E_1.$$

The differential of a function can be used as a criterion for being *locally constant*. Given a smooth function f on an open subset U in Euclidean space $\mathbb{R}^n$, the criterion is: For every smooth parametrized curve $\gamma \colon [0, 1] \to U$ the equation

$$d_{\dot\gamma(t)} f(\gamma(t)) = (df)_{\gamma(t)}(\dot\gamma(t)) = 0, \ t \in [0, 1]$$

holds. Also the differential df can be used to reconstruct the function along the image of a curve γ by integration $f(\gamma(t)) = f(\gamma(0)) + \int_0^t (df)_{\gamma(s)}(\dot\gamma(s)) ds$.

Given a covariant derivative ∇ on vector fields $X \colon U \to \mathbb{R}^n$, one may ask for a similar application.

Let $\gamma \colon [0, 1] \to U$ be a smooth curve. A *vector field along the curve* γ is a map $X \colon t \in [0, 1] \mapsto X_{\gamma(t)} \in T_{\gamma(t)} U$. The above criterion for functions becomes, for vector fields along a curve, a definition of being "constant=parallel" along that curve. We say that the vector field X is *parallel along the curve* γ if

$$\nabla_{\dot\gamma(t)} X_{\gamma(t)} = 0$$

holds.

It is possible to reconstruct a vector field X that is parallel along a curve γ from its initial value $u = X_{\gamma(0)}$ by solving the smooth differential equation $\nabla_{\dot\gamma(t)} X_{\gamma(t)} = 0$ with initial condition $X_{\gamma(0)} u = u$. Moreover, given a curve γ and $u \in T_{\gamma(0)} U$, by solving this differential equation locally near $t = 0$ one constructs a parallel vector field $X \colon [0, T] \to TU$ along the restriction of γ to $[0, T]$ with initial value $X_{\gamma(0)} = u$.

For an isometric covariant differentiation ∇, global solutions along curves exist. This fact allows us to define the *parallel transport* $T_\gamma \colon T_p M \to T_q M$ of tangent vectors. More precisely, let $\gamma \colon [0, 1] \to M$ be a smooth curve on a Riemannian manifold (M, g) with $p = \gamma(0)$ and $q = \gamma(1)$. Let ∇ be the Levi-Cività covariant differentiation of (M, g). For $e \in T_p M$ let $X_{\gamma(t)}$ be the ∇-parallel vector field along the curve γ with $X_{\gamma(0)} = e$. Define $T_\gamma(e) = X_{\gamma(1)}$. The map T_γ is linear and moreover an isometry between the Euclidean spaces $(T_p M, g_p)$ and $(T_q M, g_q)$ since the connection ∇ is isometric.

As an example, see Fig. 8.1, consider the curve $\gamma \colon [-2\pi, 4\pi] \to \mathbb{C}_+$, $t \mapsto t + i$, and at $\gamma(-2\pi)$ the tangent vector $\frac{1}{4}\frac{\partial}{\partial x}$. Let $V \colon [-2\pi, 4\pi] \to T_{\mathbb{C}_+}$ be the vector field along γ, with $V_{-2\pi} = \frac{1}{4}\frac{\partial}{\partial x}$, that is parallel for the Levi-Cività connection of the hyperbolic metric. The vector field does not stay tangent to the curve γ but turns in the clockwise sense such that the oriented angle at $\gamma(t)$ between $\dot\gamma(t)$ and $V_{\gamma(t)}$ is $-t$. So along γ the field V_t makes 3 complete clockwise turns.

Let $U \subset M$ be a coordinate chart. For $p \in U$, $u, v \in T_p U = T_p M$ and $s \in [0, 1]$ let $P_s(p, u, v)$ be the parallelogram with vertices $p, p + su, p + su + sv, p + sv$. Parallel transport that begins at p and follows the oriented boundary of the parallelogram $T_{\partial P_s(p, u, v)} \colon T_p M \to T_p M$ defines an isometry $T_{p,s}(u, v)$ of $(T_p M, g_p)$. Clearly, $T_{p,0}(u, v) = \mathrm{Id}_{T_p M}$. The difference $T_{p,s}(u, v) - \mathrm{Id}_{T_p M} \in \mathrm{End}(T_p M)$ is of interest. As for the definition of the exterior derivative $d\omega$ of 1-differential forms, the Riemannian curvature is defined as the limit

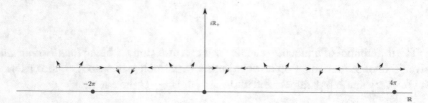

Fig. 8.1 Parallel transport.

$$R_p(u,v) = \lim_{s \to 0} \frac{1}{s^2}(T_{p,s}(u,v) - \mathrm{Id}_{T_pM}) \in \mathrm{End}(T_pM).$$

This is not really a definition since it might depend on the chart U. A better approach, but perhaps less clear at first inspection, is as follows.

The *Riemannian curvature R* of a Riemannian manifold (M, g) is a differential 2-form having as value a field of endomorphisms of the tangent spaces: $R \in \Omega^2(M, \mathrm{End}(TM))$. At a point $p \in M$, for vectors $u, v \in T_pM$ the value $R_p(u, v)$ maps T_pM linearly to T_pM. The Riemannian curvature R of (M, g) is expressed in terms of the Levi-Cività covariant differentiation ∇ as follows. For vector fields X, Y, Z,

$$R(X,Y) \colon \partial(M) \to \partial(M)$$

$$Z \mapsto R(X,Y)(Z) = (\nabla_X \circ \nabla_Y - \nabla_Y \circ \nabla_X - \nabla_{[X,Y]_{Lie}})(Z).$$

Theorem 8.2. *The expression*

$$R(X,Y)(Z) = (\nabla_X \circ \nabla_Y - \nabla_Y \circ \nabla_X - \nabla_{[X,Y]_{\mathrm{Lie}}})(Z)$$

is smoothly linear in all three arguments X, Y, Z. Hence the Riemannian curvature R becomes a tensor field.

Proof. The expression is anti-symmetric in X, Y, so it suffices to check smooth linearity in the arguments X and Z. We compute

$$
\begin{aligned}
R(fX,Y)(Z) &\\
&= (\nabla_{fX} \circ \nabla_Y - \nabla_Y \circ \nabla_{fX} - \nabla_{[fX,Y]_{\mathrm{Lie}}})(Z) \\
&= (f\nabla_X \circ \nabla_Y - f\nabla_Y \circ \nabla_X - \mathrm{d}_Y f \nabla_X - f\nabla_{[X,Y]_{\mathrm{Lie}}} + \mathrm{d}_Y f \nabla_X)(Z) \\
&= fR(X,Y)(Z)
\end{aligned}
$$

and

$$
\begin{aligned}
R(X,Y)(fZ) &= (\nabla_X \circ \nabla_Y - \nabla_Y \circ \nabla_X - \nabla_{[X,Y]_{\mathrm{Lie}}})(fZ) \\
&= \nabla_X(f\nabla_Y(Z) + \mathrm{d}_Y fZ) - \nabla_Y(f\nabla_X + \mathrm{d}_X fZ) \\
&\quad - f\nabla_{[X,Y]_{\mathrm{Lie}}}(Z) - \mathrm{d}_{[X,Y]_{\mathrm{Lie}}} fZ \\
&= (f\nabla_X \circ \nabla_Y - f\nabla_Y \circ \nabla_X - f\nabla_{[X,Y]_{\mathrm{Lie}}})(Z) \\
&\quad + ((\mathrm{d}_X \mathrm{d}_Y - \mathrm{d}_Y \mathrm{d}_X)f - \mathrm{d}_{[X,Y]_{\mathrm{Lie}}} f)Z \\
&= fR(X,Y)(Z).
\end{aligned}
$$

$\square$

If one wishes to think of the curvature tensor R on a Riemannian manifold (M, g) of $\mathrm{Dim}(M)$ locally in a coordinate chart $U \subset M$ as a square matrix of square matrices of dimension $\mathrm{Dim}(M)$, one needs as coefficients $\mathrm{Dim}^4(M)$ fields of numbers, or functions

$$R_{ijl}^k : U \to \mathbb{R}, \ 1 \le i, j, k, l \le \mathrm{Dim}(M),$$

$$p \in U \mapsto R_{ijl}^k(p) := g_p(R_p(e_{i,p}, e_{j,p})(e_{l,p}), e_{k,p}) \in \mathbb{R},$$

where the vector fields e_h, $h = 1, 2, \ldots, \mathrm{Dim}(M)$, evaluate at every point $p \in U$ to a basis of $T_p M$. In an overlap of charts $U_\alpha \cap U_\beta$, the transition of charts map $\phi_{\beta\alpha}$ acts at each point on these numbers through the matrix of its differential in a complicated way.

So, in principle for a locally framed surface one needs at every point $2^4 = 16$ numbers.

This situation simplifies drastically if working with an orthonormal basis (e_h). The following symmetries induce relations among the functions R_{ijl}^k.

Theorem 8.3. *The tensor field R is a 2-differential form, hence $R(X, Y) = -R(Y, X)$ or $R_{ijl}^k = -R_{jil}^k$. The values of $R(X, Y)$ are infinitesimal isometries of (TM, g), hence $R_{ijl}^k = -R_{ijk}^l$.*

For surfaces it follows that many of these values are 0 and that only 1 number (instead of 16) is enough to characterize the curvature R at a given point. It is traditionally the number $k_g(p) = R_{122}^1(p)$, provided that (e_1, e_2) is an orthonormal frame. As an extra simplification, the value $g_p(R_p(e_{1,p}, e_{2,p})(e_{2,p}), e_{1,p})$ is independent of the chosen orthonormal frame (e_1, e_2). The number $k_g(p) \in \mathbb{R}$ is called the *Gaussian curvature* at the point p.

Observe that $g_p(R_p(e_{1,p}, e_{2,p})(e_{2,p}), e_{1,p})^2 = \mathrm{Det}(R_p(e_{1,p}, e_{2,p}))$.

Important examples of Riemannian metrics on surfaces are metrics on open subsets U of $\mathbb{R}^2$ which are conformal to the standard Euclidean metric g_{Euclid}, hence given by $g_p^u(e_i, e_j) = e^{2u(p)}\delta_{ij}$, where u is a function on U. From the Koszul formula one gets the following beautiful formula for the curvature, due to Gauss:

Theorem 8.4 (Special Gauss Curvature Formula).

$$k_{g^u} = e^{-2u}\Delta(u) := e^{-2u}\left(-\frac{\partial^2}{\partial x^2} - \frac{\partial^2}{\partial y^2}\right)(u).$$

Proof. The constant vector fields $e_1 = (1, 0)$ and $e_2 = (0, 1)$ on $\mathbb{R}^2$ commute, are mutually g^u-perpendicular, but are of g^u-length e^u. The last three terms of the Koszul formula for $2g^u(\nabla_{e_i} e_j, e_k)$ vanish. One gets

$$g^u(\nabla_{e_1} e_1, e_1) = e^{2u} u_x, \quad g^u(\nabla_{e_1} e_1, e_2) = -e^{2u} u_y,$$

$$g^u(\nabla_{e_1} e_2, e_1) = e^{2u} u_y, \quad g^u(\nabla_{e_1} e_2, e_2) = e^{2u} u_x,$$

$$g^u(\nabla_{e_2} e_2, e_2) = e^{2u} u_y, \quad g^u(\nabla_{e_2} e_2, e_1) = -e^{2u} u_x,$$

$$g^u(\nabla_{e_2} e_1, e_1) = e^{2u} u_x, \quad g^u(\nabla_{e_2} e_1, e_2) = e^{2u} u_y$$

and

$$\nabla_{e_1} e_1 = u_x e_1 - u_y e_2, \quad \nabla_{e_1} e_2 = u_y e_1 + u_x e_2,$$
$$\nabla_{e_2} e_2 = -u_x e_1 + u_y e_2, \quad \nabla_{e_2} e_1 = u_y e_1 + u_x e_2.$$

It follows that

$$
\begin{aligned}
R(e_1, e_2)(e_2) &= (\nabla_{e_1} \circ \nabla_{e_2} - \nabla_{e_2} \circ \nabla_{e_1})(e_2) \\
&= \nabla_{e_1}(-u_x e_1 + u_y e_2) - \nabla_{e_2}(u_y e_1 + u_x e_2) \\
&= -u_{xx} e_1 - u_x(u_x e_1 - u_y e_2) + u_{yx} e_2 + u_y(u_y e_1 + u_x e_2) \\
&\quad - u_{yy} e_1 - u_y(u_y e_1 + u_x e_2) - u_{xy} e_2 - u_x(-u_x e_1 + u_y e_2) \\
&= (-u_{xx} - u_{yy}) e_1 = \Delta(u) e_1
\end{aligned}
$$

and

$$g^u(R(e_1, e_2)(e_2), e_1) = e^{2u} \Delta(u).$$

The vectors fields $f_1 = e^{-u} e_1$ and $f_2 = e^{-u} e_2$ evaluate at each point to an orthonormal frame. The Riemannian curvature R and the metric g^u are tensor fields, hence by smooth linearity,

$$
\begin{aligned}
k_{g^u} &= g^u(R(f_1, f_2)(f_2), f_1) \\
&= e^{-4u} g^u(R(e_1, e_2)(e_2), e_1) \\
&= e^{-4u} e^{2u} \Delta(u) = e^{-2u} \Delta(u). \qquad \square
\end{aligned}
$$

An important example is the upper half plane

$$\mathbb{C}_+ = \{p \in \mathbb{R}^2 \mid y(p) > 0\} = \{p \in \mathbb{C} \mid \mathrm{Im}(z(p)) > 0\}$$

with metric $g_{\mathbb{C}_+} = \frac{1}{y^2} g_{\mathrm{Euclid}}$, hence $u = \frac{1}{2}\log(\frac{1}{y^2})$ and $e^{-2u} = y^2$. The Gauss formula yields

$$k_{\mathbb{C}_+} = y^2 \left(-\frac{\partial^2}{\partial x^2} - \frac{\partial^2}{\partial y^2} \right) \left(\frac{1}{2}\log\left(\frac{1}{y^2}\right) \right) \equiv -1.$$

The push forward metric s on $\mathbb{R}^2$ by the stereographic projection with pole $(0, 0, 2)$ of the Euclidean metric of the unit sphere with center $(0, 0, 1)$ in $\mathbb{R}^3$ satisfies $s_p(e_i, e_j) = \frac{4}{4 + x(p)^2 + y(p)^2} \delta_{ij}$.

Exercise 8.5. Check that the Gaussian curvature $k_s \equiv +1$. Conclude that the Gaussian curvature of the unit sphere S^2 in Euclidean 3-space is +1.

Exercise 8.6. Let K be the surface $K = \{p \in \mathbb{R}^3 \mid x(p)^2 + y(p)^2 - z(p)^2 = 1\}$ with induced metric from the Euclidean space $\mathbb{R}^3$ with standard Euclidean metric. The surface has the shape of a hyperboloid that one observes in nuclear power plant cooling towers. The surface K is a surface of revolution with axis the z-coordinate axis of the ambient space. The surface projects radially onto an open portion T of the unit sphere S^2, $\pi\colon p \in K \mapsto q = p/\|p\|_{\mathrm{Euclidean}} \in T \subset S^2$. Through every point $q \in T$ pass two great circles C'_q and C''_q of the sphere S^2 that lie in the closure of T but not in T. Show that $\pi^{-1}(C'_q), \pi^{-1}(C'_q)$ are straight lines L'_p, L''_p on K that pass through $p = \pi^{-1}(q)$. This fact allows an easy, stable, cheap construction of the cooling towers. Unfortunately it is the only stable, cheap aspect of nuclear power generation. Is the map π conformal? Compute the Gaussian curvature of K.

Exercise 8.7. Let $a, b \colon \mathbb{R}^2 \to \mathbb{R}$ be a smooth functions.
Compute the Gaussian curvature k on $\mathbb{R}^2$ for the metric

$$g_p(u, v) = e^{b(p)}x(u)x(v) + e^{-b(p)}y(u)y(v).$$

Answer: $k = \frac{-1}{2}(e^b)_{yy} + \frac{-1}{2}(e^{-b})_{xx}$.
Compute the Gaussian curvature k for the metric

$$g_p(u, v) = x(u)x(v) - 2a(p)x(u)y(v) + (1 + a(p)^2)y(u)y(v).$$

Answer: $k = -aa_{xx} - a_x^2 + a_{xy}$.

The ratio of two volume forms is a positive function, so two smooth volume forms ω and ω_0 satisfy $\omega = e^{2u}\omega_0$ for a smooth function u. The following theorem of Gauss relates the Gaussian curvatures of the conformal metrics $g(J, \omega), g(J, e^{2u}\omega)$. Note that $g(J, e^{2u}\omega) = e^{2u}g(J, \omega)$.

Theorem 8.8 (Gauss Curvature Formula). *With the previous notations:*

$$k_{g(J,e^{2u}\omega)}e^{2u} = \Delta_{g(J,\omega)}(u) + k_{g(J,\omega)} = \Delta_{J,\omega}(u) + k_{g(J,\omega)}. \qquad \square$$

Remember the previous definition $\Delta_{g(J,\omega)} = \Delta_{J,\omega} = \Delta_J \%\omega$. The integrand of the Gauss–Bonnet Theorem, which we will present soon, is the 2-differential form $K_{g(J,\omega)} = k_{g(J,\omega)}\omega$. An equivalent formulation of the Gauss Curvature Formula, but more "additive", can be stated in terms of the Gaussian curvature 2-form $K_{g(J,\omega)}$.

Theorem 8.9 (Gauss Curvature Addition). *With the previous notations:*

$$K_{g(J,e^{2u}\omega)} = \Delta_J(u) + K_{g(J,\omega)}.$$

Proof (Gauss Curvature Formula). The statement of the theorem is local, so we can assume that we are working near a point p in a coordinate chart U of S. The local rigidity theorem of J-fields (eventually after shrinking U) provides a local complex coordinate $z \colon U \to \mathbb{C}$ that is (J, m_i)-holomorphic. Let $x, y \colon U \to \mathbb{R}$ be the real coordinates with $z = x + iy$. The proof proceeds in two steps. First assume that the restriction of ω to U is given by $\omega_0 = dx \wedge dy$. Let e_1, e_2 be the vector fields $\frac{\partial}{\partial x}, \frac{\partial}{\partial y}$. Observe $J(e_1) = e_2$, $J(e_2) = -e_1$. We have $g(J, e^{2u}\omega_0)(e_1, e_2) = 0$ and $g(J, e^{2u}\omega_0)(e_1, e_1) = g(J, e^{2u}\omega_0)(e_2, e_2) = e^{2u}$. These coordinates are so-called *isothermal* coordinates. In this case, from the Gauss Curvature Formula, it follows that

$$k_{g(J,e^{2u}\omega_0)}e^{2u} = \Delta_{J,\omega_0}(u) = \left(-\frac{d^2}{dx^2} - \frac{d^2}{dy^2}\right)(u)$$

since $k_{g(J,\omega_0)} = 0$.

In the next step assume that the restriction of ω to U is any volume form $\omega = e^{2v}\omega_0$. Now applying the previous step twice, and using the definition

$$\Delta_{J,e^{2v}\omega_0} := \Delta_J \%_0(e^{2v}\omega_0) = e^{-2v}\Delta_{J,\omega_0}$$

one gets

$$
\begin{aligned}
k_{g(J,e^{2u}\omega)}e^{2u} &= e^{-2v}k_{g(J,e^{2u+2v}\omega_0)}e^{2u+2v} \\
&= e^{-2v}\Delta_{J,\omega_0}(u+v) \\
&= e^{-2v}\Delta_{J,\omega_0}(u) + \Delta_{J,\omega_0}(v)e^{-2v} \\
&= \Delta_{J,\omega}(u) + k_{g(J,\omega)},
\end{aligned}
$$

proving the theorem.   □

How does the Gaussian curvature $k_{g(\omega,J)}$ or the Gaussian curvature form $K_{g(\omega,J)}$ vary with J, keeping ω fixed? We begin with a study in the case where $S = \mathbb{R}^2$ with standard basis (e_1, e_2), coordinates x, y and volume form $\omega = dx \wedge dy$. Let J be an oriented smooth J-field on S. Its matrix $M_p = \left(\begin{smallmatrix} h & * \\ k & -a \end{smallmatrix}\right)$ on the basis (e_1, e_2) at a point $p \in S$ satisfies $k > 0$ and can be written as

$$M_p = \begin{bmatrix} a(p) & -e^{-b(p)}(1 + a(p)^2) \\ e^{b(p)} & -a(p) \end{bmatrix},$$

where $a, b : S \to \mathbb{R}$ are smooth functions on S. The fixed point $\mathrm{Fix}(M_p)$ in the upper half plane $\mathbb{C}_+$ is given by

$$\mathrm{Fix}(M_p) = e^{-b(p)}(a(p) + i) \in \mathbb{C}_+.$$

Observe that the parametrized curves

$$b \in \mathbb{R} \mapsto e^b(a(p) + i) \in \mathbb{C}_+$$

have as image open Euclidean straight half lines $L_{a(p)}$ through $a(p) + i$ with $0 \in \mathbb{C}$ as extremity. Only the image of the curve L_0 with $a(p) = 0$ is a hyperbolic geodesic. The curves $L_{a(p)}$ are equidistant lines at distance $\mathrm{arcsinh}(|a(p)|)$ to the hyperbolic geodesic L_0. The parametrized curves

$$a \in \mathbb{R} \mapsto e^{-b}(a + 1) \in \mathbb{C}_+$$

are horocycles with base point at ∞ and perpendicular to L_0.

The matrix G_p for the metric $g(\omega, J_p)$ at p is

$$G_p = \begin{bmatrix} e^{b(p)} & -a(p) \\ -a(p) & e^{-b(p)}(1 + a(p)^2)) \end{bmatrix}.$$

Exercise 8.10. For two smooth functions $\alpha, \beta : \mathbb{R}^2 \to \mathbb{R}$ let $J(\alpha, \beta)$ on $\mathbb{R}^2$ be the J-field $J_p = J(\alpha(p), \beta(p))$ with

$$J(\alpha(p), \beta(p))(e_1) = \sinh(\alpha(p))e_1 + e^{\beta(p)\cosh(\alpha(p))}e_2$$

and let $g_p(u, v) = \omega_p(u, J_p(v))$ be the corresponding metric with volume form $\omega = dx \wedge dy$. Study the locus $\mathrm{Fix}(J(\alpha(p), \beta(p))$ in $\mathbb{C}_+$. Compute the Gaussian curvature k_g.

We denote by $J(a, b)$ the J-field given by functions a, b with

$$J(a(p), b(p))(e_1) = a(p)e_1 + e^{b(p)}(1 + a(p)^2)e_2.$$

The field $J(0, 0)$ is the field m_i on $\mathbb{C} = \mathbb{R}^2$, where $x(p)e_1 + y(p)e_2 \in \mathbb{R}^2$ is identified with $x(p) + y(p)i \in \mathbb{C}$. Observe that $\mathrm{Fix}(J(0, 0)) = i \in \mathbb{C}_+$.

The following is an *infinitesimal avant goût* of the Gauss–Bonnet Theorem.

Theorem 8.11 (Avant Goût: Gauss–Bonnet). *Let J be a J-field on $\mathbb{R}^2$. Assume $J(e_1) = e_2$ for $p \in \mathbb{R}^2$ with $|x(p)|, |y(p)| > 1$. Then*

$$\int_{\mathbb{R}^2} k_{g(dx \wedge dy, J)} dx \wedge dy = 0$$

holds.

Proof. Let $z_J : \mathbb{R}^2 \to \mathbb{C}$ be the global coordinate provided by the Strong J-Rigidity (Theorem 7.2) and let $\omega = \frac{i}{2}dz_J \wedge d\bar{z}_J$ be the corresponding volume form. The metric $g(J, \omega)$ has 0 Gaussian curvature, hence $\int_{\mathbb{R}^2} k_{g(J, \omega)}\omega = 0$. Let u_J be the function satisfying $\omega = e^{2u_J} dx \wedge dy$. With the Gauss Curvature Addition Theorem 8.9

$$k_{g(J, dx \wedge dy)} dx \wedge dy = \Delta_J(u_J) + k_{g(J, \omega)}\omega$$

it follows that

$$\int_{\mathbb{R}^2} k_{g(J, dx \wedge dy)} dx \wedge dy = \int_{\mathbb{R}^2} \Delta_J(u_J) + \int_{\mathbb{R}^2} k_{g(J, \omega)}\omega = 0 + 0 = 0. \qquad \square$$

Exercise 8.12. Construct for $c \in \mathbb{R}$ a Riemannian metric g on $\mathbb{R}^2$ such that the function $p \mapsto k_g(p)$ has compact support and $\int_{\mathbb{R}^2} k_g dx \wedge dy = c$ holds.

8.2 Topology of Surfaces and Curvature

If (S, g) is an oriented Riemannian surface, we have two ways to associate a volume form to the curvature tensor R_g.

The Gaussian curvature $k_{g,p}$ is the number $g_p(R_{g,p}(e_1, e_2)(e_2), e_1)$, where (e_1, e_2) is an orthonormal basis of (T_pS, g_p). A more intrinsic definition is $k_{g,p} = \mathrm{Pfaff}(R_{g,p})$. The endomorphism $R_{g,p}$ of (T_p, g_p) is anti-symmetric, so has an invariant $\mathrm{Pfaff}(R_{g,p})$.[4]

---

[4] Named after German mathematician Johann Friedrich Pfaff (1765–1825), teacher of Carl Friedrich Gauss (1777–1855).

Important for us is that the Pfaffian is a square-root of the determinant, so $\text{Pfaff}(R_{g,p})^2 = \text{Det}(R_{g,p})$ holds. The curvature tensor R_g evaluates at $p \in S$, for two tangent vectors $e, f \in T_p S$, to an infinitesimal isometry $R_p(e, f)$ of the oriented Euclidean plane $(T_p S, g_p)$. Infinitesimal isometries of oriented Euclidean planes can be thought of as angular speeds of rotations, hence as real numbers. So the Pfaffian $\text{Pfaff}(R)$ of the curvature tensor R is a 2-differential form on S. Let $\text{dVol}_g(S)$ be the volume form of the oriented Riemannian surface (S, g).

The first volume form $k_g \text{dVol}_g(S)$ uses the Gaussian curvature k_g, hence the Pfaffian of the R_g, the second one $k_g^2 \text{dVol}_g(S)$ uses the square k_g^2 of the Gaussian curvature, hence the determinant of R_g.

The celebrated Gauss–Bonnet Theorem[5] tells us that the integral of $k_g \text{dVol}_g(S)$ over a compact surface S does not depend upon the metric g but computes up to the factor 2π the Euler number $\chi(S)$.

Theorem 8.13 (Gauss–Bonnet Theorem). *Let S be a compact connected surface with Riemannian metric g and of genus p.*

$$\int_S \text{Pfaff}(R_g) = \int_S k_g \text{dVol}_g(S) = 2\pi\chi(S) = 2\pi(2 - 2p).$$

Proof. A Riemannian metric g on an oriented surface S is equal to a Riemannian metric $g(J, \omega)$, where ω is a 2-form and where J is a J-field J. Denote by $\text{GB}(S, J, \omega)$ the Gauss–Bonnet integral $\int_S k_{g(J,\omega)}\omega$.

The proof consists of three claims.

(i) For every genus p, there exists a Riemannian surface (S_p, g_0) of genus p with metric $g_0 = g(J_0, \omega_0)$ for which the Gauss–Bonnet identity $\text{GB}(S_P, J_0, \omega_0) = 2\pi\chi(S)$ holds.

(ii) The Gauss–Bonnet integral $\int_S k_{g(J,\omega)}\omega$ does not depend upon the volume form ω.

(iii) The Gauss–Bonnet integral $\int_S k_{g(J,\omega)}\omega$ does not depend upon the J-field J.

For genus $p = 0$ take the unit sphere in $\mathbb{R}^3$, for $p = 1$ the torus $\mathbb{R}^2/\mathbb{Z}^2$ with the quotient metric and for $p \geq 2$ the previously constructed hyperbolic surface S_p by identifying edges of a regular hyperbolic $4p$-gon with angles $\frac{2\pi}{4p}$. All three satisfy the Gauss–Bonnet identity, so the first claim is established.

Two volume forms ω_0, ω_1 are proportional by a factor e^{2u}. Hence, the second claim follows by the Gauss Curvature Addition Theorem 8.9 and the vanishing of the integral $\int_S \Delta_J(u)$.

Let $J_t, t \in [0, 1]$, be the pointwise geodesic path with end points J_0, J_1. The following establishes the third claim:

$$\frac{d}{dt}\int_S k_{g(J_t,\omega)}\omega = 0.$$

Fix $t \in [0, 1]$. Let (U_α, z_α) be a holomorphic finite z-atlas on S having as J-field the field J_t. Such an atlas exists since S is compact by the Local Rigidity Theorem for J-fields in dimension 2 (Theorem 5.30).

---

[5] Pierre Ossian Bonnet (1819–1892), French mathematician.

Let (ϕ_α) be a smooth partition of unity with Support$(\phi_\alpha) \subset U_\alpha$. Put $J_{\alpha,s} = J_{t+s\phi_\alpha}$. Observe that

$$\sum_\alpha \frac{d}{ds} J_{\alpha,s}{}_{|s=0} = \frac{d}{dt} J_t$$

and

$$\frac{d}{dt} \int_S k_{g(J_t,\omega)} \omega = \sum_\alpha \frac{d}{ds} \int_S k_{g(J_{\alpha,s},\omega)} \omega_{|s=0}.$$

All terms in the last sum vanish by the previous Avant Goût version of the Gauss–Bonnet Theorem 8.11. Indeed, let ω_α be a volume form on S that coincides on U_α with $\frac{i}{2} dz_\alpha \wedge d\bar{z}_\alpha$. Then by the second claim $\int_S k_{g(J_{\alpha,s},\omega)} \omega = \int_S k_{g(J_{\alpha,s},\omega_\alpha)} \omega_\alpha$. The integrand $k_{g(J_{\alpha,s},\omega_\alpha)} \omega_\alpha$ only varies with s at points in the support of ϕ_α, hence

$$\frac{d}{ds} \int_S k_{g(J_{\alpha,s},\omega_\alpha)} \omega_\alpha = \frac{d}{ds} \int_{U_\alpha} k_{g(J_{\alpha,s},\frac{i}{2}dz_\alpha \wedge d\bar{z}_\alpha)} \frac{i}{2} dz_\alpha \wedge d\bar{z}_\alpha.$$

The coordinate z_α maps the field $J_{\alpha,s}$ to a field $K(\alpha, s)$ on $z_\alpha(U_\alpha) \subset \mathbb{C}$ that differs from the field m_i on a compact subset $A_\alpha \subset U_\alpha$. The integral changes after extending $K(\alpha, s)$ by m_i to $\mathbb{C}$ to the integral

$$\int_{z_\alpha(U_\alpha)} k_{g(K(\alpha,s),dx \wedge dy)} dx \wedge dy = \int_\mathbb{C} k_{g(K(\alpha,s),dx \wedge dy)} dx \wedge dy \equiv 0,$$

which does not vary with the parameter s by the Avant Goût Theorem 8.11. $\qquad \square$

Exercise 8.14. Let g be a Riemannian metric on $\mathbb{R}^2$ such that $p \mapsto k_g(p)$ is non-negative. Moreover, assume that g is the Euclidean metric in the complement of a compact set. Prove that g is isotopic to the Euclidean metric by a compactly supported isotopy.

Exercise 8.15. Let g be a Riemannian metric on the hyperbolic plane $\mathbb{H}^2$. Moreover, assume that g is the hyperbolic metric in the complement of a compact set K. For a hyperbolic disk D of radius r that contains K compute the Gauss–Bonnet integral $\int_D k_g(p) d\text{Vol}_g(p)$.

8.3 Hyperbolic Length and Extremal Length

Let (S, J) be a compact connected J-surface of genus ≥ 2. Let γ_0 be a simply closed smooth essential curve on S and let Γ be the set of all curves on S that are homotopic to γ_0. Each volume form ω on S together with the complex structure J gives a Riemannian metric $g_{J,\omega}$ by $g_{J,\omega}(u, v) = \omega(u, Jv)$. The length of a tangent vector u at the point p is given by $\|u\| = \sqrt{\omega_p(u, J_p u)}$. The length $L_{J,\omega}(\gamma)$ of a smooth curve $\gamma \colon [0, 1] \to S$ on S is defined by the integral

$$L_{J,\omega}(\gamma) := \int_0^1 \|\dot{\gamma}(t)\| dt.$$

Of special interest is measuring lengths of simply closed curves on S or of homotopy classes γ_h of such curves. We denote by

$$l_{J,\omega}(\gamma) = \operatorname{Inf}_{\gamma \in \gamma_h} L_{J,\omega}(\gamma)$$

the infimum of the lengths of curves over the homotopy class γ_h of the curve γ.

There are essentially two ways of introducing a length for simple closed curves that only depends on the J-field. One way is to work with a volume form that is characterized by special properties. The most celebrated one is given by the Uniformization Theorem, that will be proved later. This theorem claims the existence of a unique volume form ω_J such that the metric g_{J,ω_J} has constant Gaussian curvature -1. In other words the Riemannian surface (S, g_{J,ω_J}) is a hyperbolic surface with charts in the hyperbolic plane and chart transition by hyperbolic isometries. The length

$$L_J(\gamma) = L_{J,\omega_J}(\gamma)$$

is called the hyperbolic length of the curve γ and $l_J(\gamma)$ the minimal hyperbolic length of γ.

A second way is to eliminate the volume form ω by imposing a maximal constraint value with respect to the volume form ω for the length. As before define $l_{J,\omega}(\gamma) = \operatorname{Inf}_{\gamma \in \gamma_h} L_{J,\omega}(\gamma)$. The quantity $l_{J,\omega}(\gamma)$ is of weight $\frac{1}{2}$ with respect to ω, i.e. $l_{J,a\omega}(\gamma) = a^{1/2} l_{J,\omega}(\gamma), a > 0$. The quantity $l_{J,\omega}(\gamma)^2$ is of weight 1, as is the quantity $\int_S \omega$. This allows the following definition

$$\lambda_J(\gamma) = \operatorname{Sup}_\omega \frac{l_{J,\omega}(\gamma)^2}{\int_S \omega}.$$

The quantity $\lambda_J(\gamma)$ is called the extremal length of the homotopy class of the curve γ.

The definition of the extremal length of a simply closed curve γ proceeds in two steps: first minimize $l_{J,\omega}(\gamma)$ over its homotopy class γ_h, next maximize over volume forms ω that are constrained by $\int_S \omega = 1$. To ensure that the maximum is achieved, allow instead of smooth volume forms ω positive Radon probability measures μ. In view of the Gauss–Bonnet Theorem the volume constraint $\int_S \omega = -2\pi\chi(S)$ would be more natural.

The following theorem follows rather directly from the definitions and estimates the extremal length from below in terms of the hyperbolic length.

Theorem 8.16 (Domination of hyperbolic length by extremal length). *Let S be a hyperbolic surface of genus $g \geq 2$. Let γ be a simply closed geodesic on S of length $l(\gamma)$ and extremal length $\lambda(\gamma)$. Then the inequality*

$$\lambda(\gamma) \geq \frac{l^2(\gamma)}{-2\pi\chi(S)}$$

holds.

Proof. Let J and ω_J be the J-field and volume form that underlie the hyperbolic metric on S. Recall from the Gauss–Bonnet Theorem: $\int_S \omega_J = -2\pi\chi(S)$. From the definition of the extremal length it follows that

$$\lambda(\gamma) = \lambda_J(\gamma) = \mathrm{Sup}_{\omega, \int_S \omega = -2\pi\chi(S)} \frac{l_{J,\omega}(\gamma)^2}{-2\pi\chi(S)}$$

$$\geq \frac{l_{J,\omega_J}(\gamma)^2}{-2\pi\chi(S)}$$

$$= \frac{l^2(\gamma)}{-2\pi\chi(S)}. \qquad \square$$

The standard ring A_r of radius $r > 1$ is the J-surface

$$A_r = \{p \in \mathbb{C} \mid 1/r < |p| < r\}$$

with J-field induced from the complex plane. A ring ρ in a hyperbolic surface (S, J, ω_J) is a conformal injective diffeomorphism $\rho \colon A_r \to S$. A maximal ring in S is a ring ρ with $\int_{A_r} \rho^*\omega_J = -2\pi\chi(S)$. A maximal ring is in the class of a simply closed curve γ if the image by ρ of the unit circle is homotopic to γ. Maximal rings of a given class γ exist by a classical exhaustion argument.

A maximal modulus ring in the class of a simply closed curve γ is a maximal ring $\rho \colon A_r \to S$ in that class, such that moreover its modulus $2\log(r)$ is maximal.

A more geometric understanding of the notion of extremal length is provided by:

Theorem 8.17. *Let γ be an essential simply closed curve on a hyperbolic surface S. Let $\rho \colon A_r \to S$ be a maximal modulus ring in the class of γ. Then $\lambda(\gamma) = \frac{1}{\pi}\log(r)$ holds.*

From hyperbolic trigonometry follows the so-called Collar Lemma.

Theorem 8.18 (Collar Lemma). *Let S be a complete hyperbolic surface. Let γ be a simply closed geodesic on S of length $l(\gamma)$. The open collar neighborhood of γ of width* $\mathrm{arcsinh}(\frac{1}{\sinh(l(\gamma)/2)})$ *is an embedded annulus in S.* $\qquad \square$

The Collar Lemma provides a non-maximal ring for the class of a simply closed geodesic. Explicit inversion of hyperbolic trigonometric functions leads to the Maskit inequality, which estimates the extremal length from above in terms of hyperbolic length [74, 77].

Theorem 8.19 (Maskit inequality). *Let S be a complete hyperbolic surface. Let γ be a simply closed geodesic on S of length $l(\gamma)$. Let $\lambda(\gamma)$ be the extremal length of γ in S with its underlying conformal structure. The inequality*

$$\lambda(\gamma) \leq \frac{1}{2}l(\gamma)e^{\frac{1}{2}l(\gamma)}$$

holds. $\qquad \square$

For the proofs of the Collar Lemma and the Maskit inequality, see the book of Peter Buser [18].

Chapter 9
Outline: Uniformization by Spectral Determinant

9.1 A Theorem of Mueller–Wendland and Osgood–Phillips–Sarnak

Let S be a compact, oriented surface of genus $g(S) \geq 2$. We denote by $\mathrm{Vol}^+(S) \subset \Omega^2(S, \mathbb{R})$ the space of positive volume forms on S with total volume $2\pi(2g(S) - 2)$. It is an open convex subset in the affine space $\{\omega \in \Omega^2(S) \mid \int_S \omega = -2\pi\chi(S)\}$. The quantity $Z(S, J, \omega)$ is a spectral invariant of the Laplacian $\Delta_{J,\omega}$ of the surface S with Riemannian metric $g(u, v) = \omega(u, Jv)$. It involves the eigenvalues and has the formal properties of a determinant.

By the law of Weyl and Karamata the nth-eigenvalue λ_n is multiplicatively uniformly close to $n\chi(S)$, so the Spectral Zeta function $s \mapsto \zeta(s, S, J, \omega) = \sum_{n \geq 1} \lambda_n^{-s}$ behaves with respect to convergence and holomorphy in s like the classical Riemann Zeta function $\zeta(s) = \sum_{n \geq 1} \frac{1}{n^s}$. We have $\zeta(s, S, J, \omega) \sim \frac{1}{\chi(S)^s} \zeta(s)$. In particular, $\zeta(s, S, J, \omega)$ has a holomorphic extension at $s = 0$. The quantity $Z(S, J, \omega)$, called the *regularized determinant* of $\Delta_{J,\omega}$, is defined as

$$Z(S, J, \omega) = \exp\left(-\frac{\mathrm{d}}{\mathrm{d}s}\zeta(0, S, J, \omega)\right).$$

Notice that for a linear map $L \colon \mathbb{R}^N \to \mathbb{R}^N$ with positive real eigenvalues the analogous method computes the determinant of L.

Here we give the precise definition of the quantity $Z(S, J, \omega)$ and state the Theorem of Mueller–Wendland [82] and Osgood–Phillips–Sarnak [86]:

Theorem 9.1. *Let S be a compact, oriented surface of genus $g(S) \geq 2$ with a J-field J. The function $\omega \in \mathrm{Vol}^+(S) \mapsto Z(S, J, \omega) \in \mathbb{R}$ is strictly concave. Its non-degenerate maximum is achieved at the form ω for which the Riemannian metric $g_{J,\omega}$ is of constant curvature.* □

© The Author(s), under exclusive license to Springer Nature Switzerland AG 2021
N. A'Campo, *Topological, Differential and Conformal Geometry of Surfaces*, Universitext,
https://doi.org/10.1007/978-3-030-89032-2_9

9.2 Uniformization by Spectral Determinant, $g \geq 0$

An important corollary is the Uniformization Theorem:

Theorem 9.2. *Let S be a compact connected complex surface of genus $g(S) \geq 0$. Then S admits an atlas with charts in the round two sphere S^2 if $g = 0$, in the Gaussian $\mathbb{C}$ if $g = 1$, and in the hyperbolic plane $\mathbb{H}^2$ if $g \geq 2$. Transitions are given by restrictions of oriented spherical, Euclidean or hyperbolic isometries depending on the genus.*

Proof. Let J be the J-field on S that underlies the given complex structure. Let $\omega_{\max} \in \text{Vol}(S)$ be the maximizer of $Z(S, J, \omega)$. The scalar product $g_{J, \omega_{\max}}$ is a Riemannian metric of curvature $+1$, 0 or -1. Hence, S has a spherical, Euclidean or hyperbolic atlas, as claimed. $\square$

In the next chapter we give an alternative proof of the above corollary, i.e. of the Uniformization Theorem. Our proof is largely inspired by the work [82] and [86]. Instead of using the functional $Z(S, J, \omega)$ we will use an energy functional $E(S, J, \omega)$ on the space of volume forms that measures the quadratic deviation from constant Gaussian curvature in the spirit of the least squares method, also due to Gauss.

9.3 Polyakov's String Dynamics

The extremal value $E(J)$ of $E(S, J, \omega)$ on the space of normalized volume forms, which is in fact a minimum, does not depend on the conformal structure J, see Section 10.1. So the functional $E(J)$ cannot be used as a Hamiltonian on the moduli spaces. The extremal value $Z(J)$, a maximum, of the functional $\omega \mapsto Z(S, J, \omega)$ depends on J. The function $J \mapsto Z(J)$ was used by Polyakov as a Hamiltonian on the (symplectic) moduli spaces in order to define string dynamics.

The energy $E(J)$ related to the Poincaré metric of constant curvature does not depend upon J. How about the energy $E_B(J)$ of the Bergman metric (see Section 19.1 for the definition)? Does $E_B(J)$ together with the symplectic structure on $\mathbb{T}_g$ lead to interesting dynamics on Teichmüller space?

It is striking that both quantities $Z(J)$ and $E(J)$ are extrema of expressions involving determinants. The first $Z(S, J, \omega)$ is an infinite product of the eigenvalues of the Laplacian, the second $E(S, J, \omega)$ the integral of the determinant of the Riemannian curvature.

The critical point ω_J of the mapping $\omega \mapsto Z(S, J, \omega)$ which mainly depends upon the spectrum of the Laplace operators of the metrics $g_{J, \omega}$ determines the hyperbolic metric g_{J, ω_J}.

The following seminal result of Heinz Huber[1] is a forerunner [53]: here, by the *length spectrum* of a Riemannian manifold we mean the set of lengths of the closed geodesics.

---

[1] Heinz Huber (1926–2000), Swiss mathematician, Professor in Basel.

Theorem 9.3 (Heinz Huber). *Let S_1, S_2 be two hyperbolic surfaces of genus $g \geq 2$ with Laplace operators having equal spectra. Then the length spectra of S_1, S_2 are equal.*

So, the spectrum of the Laplace operator at the critical point almost determines the hyperbolic metric.

Chapter 10
Uniformization by Energy

10.1 Energy and Curvature

As seen before, the volume form $k_g d\text{Vol}_g^+$ is related to the topology of the surface by the Gauss–Bonnet Theorem (Theorem 8.13). The second volume form $\text{Det}(R_g)d\text{Vol}_g^+$ is also related to the topology of compact, oriented surfaces of genus $p \geq 2$, but this time by an inequality instead of an equality.

Define $\text{Vol}_\chi(S)$ to be the space of smooth volume forms ω on S with $\omega(u, Ju) > 0$ and of total volume $-2\pi\chi(S) = 2\pi(2p - 2)$. The space $\text{Vol}_\chi(S)$ depends only on the orientation given by the J-field J. It is the space of positive volume forms on the oriented surface S having total volume equal to $2\pi(2p - 2)$. The space $\text{Vol}_\chi(S)$ is an open convex subset in the affine space of 2-forms ω normalized by $\int_S \omega = 2\pi(2p-2)$.

We define $E_J(\omega) := \int_S k_{g(J,\omega)}^2 \omega$ as the J-curvature energy of the volume form $\omega \in \text{Vol}_\chi(S)$. For our purposes it is remarkable that the functional $E_J : \text{Vol}_\chi(S) \to \mathbb{R}$ has the following main extremal property, like the functional $Z(S, J, \omega)$.

Theorem 10.1 (Extremal Energy Theorem, genus $\geq$ 2). *Let S be a J-surface of genus $g \geq 2$. For $\omega \in \text{Vol}_\chi(S)$ the inequality*

$$E_J(\omega) := \int_S k_{g(J,\omega)}^2 \omega \geq -2\pi\chi(S)$$

holds, with equality if and only if the Gaussian curvature $k_{g(J,\omega)}$ is constant of value -1. *For $\omega_0, \omega_1 \in \text{Vol}_\chi(S)$, assume $E_J(\omega_0) = E_J(\omega_1) = -2\pi\chi(S)$. Then $\omega_0 = \omega_1$.*

Proof. Let $\delta = (-1) - k_{g(J,\omega)}$ be the deviation of the Gaussian curvature from -1. The ω-mean value of $k_{g(J,\omega)}$ equals -1 by the Gauss–Bonnet Theorem 8.13 for $\omega \in \text{Vol}_\chi(S)$. Hence $\int_S \delta\omega = 0$. It follows that

$$\int_S k_{g(J,\omega)}^2 \omega = \int_S (-\delta - 1)^2 \omega = \int_S (1 + \delta^2)\omega \geq \int_S \omega = -2\pi\chi(S)$$

with equality if and only if $\int_S \delta^2 \omega = 0$, i.e. $\delta \equiv 0$ and $k_{g(J,\omega)} \equiv -1$.

© The Author(s), under exclusive license to Springer Nature Switzerland AG 2021
N. A'Campo, *Topological, Differential and Conformal Geometry of Surfaces*, Universitext,
https://doi.org/10.1007/978-3-030-89032-2_10

Let ω_0, ω_1 be volume forms with total volume $-2\pi\chi(S)$ and such that the Gaussian curvature of $g(J, \omega_0)$ and of $g(J, \omega_1)$ is constantly -1. Let u be the function with $\omega_1 = e^{2u}\omega_0$. From

$$-e^{2u} = k_{g(J, e^{2u}\omega_0)}e^{2u} = \Delta_{g(J,\omega_0)}(u) + k_{g(J,\omega_0)} = \Delta_{g(J,\omega_0)}(u) - 1$$

it follows that $\Delta_{g(J,\omega_0)}(u) = 1 - e^{2u}$. Remember that the Laplacian $\Delta_{g(J,\omega_0)}$ satisfies $\Delta_{g(J,\omega_0)}(u)(p) \geq 0$ at a maximum p of u and $\Delta_{g(J,\omega_0)}(u)(q) \leq 0$ at a minimum q of u. Hence from $\Delta_{g(J,\omega_0)}(u)(p) = 1 - e^{2u(p)} \geq 0$ follows $u(p) \leq 0$ and from $\Delta_{g(J,\omega_0)}(u)(q) = 1 - e^{2u(q)} \leq 0$ follows $u(q) \geq 0$. The function u has to be non-positive at each maximum of u and non-negative at each minimum of u. We conclude that $u \equiv 0$, and $\omega_0 = \omega_1$. $\qquad\square$

We did not prove that the energy functional E_J achieves its minimum. So the Uniformization Theorem does not follow yet. In the next section the existence of a minimum is proved by lifting minima above pointwise geodesics in the space $\mathbb{J}(TS)$ of J-fields on surfaces S of genus ≥ 2.

For J-surfaces S of genus 0 or 1 the following Extremal Energy Theorem holds. For a J-surface S of genus 0 let $\mathrm{Vol}_\chi(S)$ be the space of smooth J-oriented volume forms ω with $\int_{[S]} \omega = 2\pi\chi(S)$. On a J-surface of genus 1 let $\mathrm{Vol}_\chi(S)$ be the space of J-oriented smooth volume forms ω with $\int_{[S]} \omega = 1$.

Theorem 10.2 (Extremal Energy Theorem, genus 0 or 1). *Let S be a J-surface of genus 0 or 1. For $\omega \in \mathrm{Vol}_\chi(S)$ the energy inequality*

$$E_J(\omega) = \int_S k^2_{g(J,\omega)}\omega \geq 2\pi\chi(S)$$

holds. In both cases of genus 0 or 1 surfaces equality holds if and only if the Gaussian curvature is constant, constantly $+1$ for genus 0, constantly 0 for genus 1 surfaces.

The space $\{\omega \in \mathrm{Vol}_\chi(S) \mid E_J(S) = 2\pi\chi(S)\}$ of minimizers is reduced to one element for the case of genus 1 and to a 3-dimensional family parametrized by the hyperbolic 3-space $\mathbb{H}^3$ in the case of genus 0 surfaces.

Proof. The energy inequality is clear in the case of genus 1 since we have $k^2_{g(J,\omega)} \geq 0$ and $\chi(S) = 0$. Equality is achieved by the Euclidean metric on $S = \mathbb{C}/\Gamma$. This Euclidean metric g_0 is in fact the metric $g(J, \frac{i}{2}dz \wedge d\bar{z})$. Let $\omega_1 = \frac{ie^{2u}}{2}dz \wedge d\bar{z}$ be a second minimizer. The Γ-periodic function u on $\mathbb{C}$ will be harmonic, so constant. It follows that $e^{2u} = 1$ by the normalization of volume forms, so $u = 0$ and $\omega_1 = \frac{i}{2}dz \wedge d\bar{z}$.

The energy inequality in the case of genus 0 surfaces is obtained by the same method as before in the case of genus ≥ 2. The minimum 4π is realized by the oriented Euclidean unit sphere S^2 in $\mathbb{R}^3$. Let $\omega_0 \in \mathrm{Vol}_\chi(S)$ be the corresponding Euclidean volume form. Let $\omega_1 \in \mathrm{Vol}_\chi(S)$ be second minimizer. Let the function u be such that e^{2u} is the ratio ω_1/ω_0 of the two minimizer. From the Gauss Curvature Addition Theorem 8.9 follows

$$\Delta_{S^2}(u) = -1 + e^{2u} = 2u + \frac{(2u)^2}{2!} + \frac{(2u)^3}{3!} + \frac{(2u)^4}{4!} + \cdots$$

In the case of genus ≥ 1 the corresponding equation differs by a sign and it was possible to exclude non-trivial solutions by showing that maxima/minima of u have to be negative/positive. In contrast, in the case of genus 0 surfaces, geometry will help out and provide a 3-dimensional family of solutions. The remaining task will be to exclude further solutions.

Let (S, J, ω_0) be the Euclidean two-sphere with standard J-field J and normalized and J- oriented volume form ω_0. The group of automorphisms $\text{Aut}(S, J)$ of the Riemann surface $(S, J) = \mathbb{C} \cup \{\infty\}$ is the group of complex homographic substitutions isomorphic to $\text{PSL}(2, \mathbb{C})$. It follows that for each $A \in \text{Aut}(S, J)$, the pullback form $\omega_A = A^* \omega_0$ is a minimizer too, since the diffeomorphism A is an isometry from $(S, g(J, \omega_A))$ to (S, J, ω_0). Observe $\{A \in \text{Aut}(S, J) = \text{PSL}(2, \mathbb{C}) \mid A^* \omega_0 = \omega_0\} = SO(3) = PU(2)$. The space of minimizers that we obtain is parametrized by the homogeneous space $\text{PSL}(2, \mathbb{C})/PU(2)$. This space is of real dimension 3 and is a model of the hyperbolic 3-space $\mathbb{H}^3$.

Let ω_1 be a minimizer. The Riemannian surface $(S, g(J, \omega_1))$ is an oriented two-sphere with constant Gaussian curvature $+1$. From differential geometry we know that in case of simply connected, two-dimensional, complete Riemannian manifolds of constant Gaussian curvature the curvature determines the isometry type [10]. So $(S, g(J, \omega_1))$ and $(S, g(J, \omega_0))$ are isometric by an oriented isometry A. Clearly, $A \in \text{Aut}(S, J)$ and $\omega_1 = \omega_A$. $\qquad \square$

Remark. In his thesis [16] Walter Brägger introduced special partially defined metrics on the conformal sphere (S^2, J). Such a metric is a pair (U, g_U) where U is a non empty connected open subset in S^2, and g_U is a smooth Riemannian metric on U, such that the following properties hold:

1. The metric is in the conformal class of the restriction of J to U.
2. The Gaussian curvature k_{g_U} is constant, and equal to $-1, 0$ or $+1$.
3. Let G_U be the subgroup of the group of conformal automorphisms of (S^2, J) consisting of those automorphisms that globally fix U and that restrict on U to an isometry of (U, g_U). The required property is that G_U acts in an isotropically transitive way on U.

We call such a metric a *Brägger metric* on (S^2, J). Let $B(S^2, J)$ be the space of Brägger metrics up to scaling by a constant. Observe that the scaling class of (U, g_U) consists of one element if $k_{g_U} \neq 0$.

The space $B(S^2, J)$ is naturally the union of 3 subspaces $B_k(S^2, J)$, $k = -1, 0, +1$ according to the Gaussian curvature of g_U. As shown above, the space $B_{+1}(S^2, J)$ is a model of the hyperbolic 3-space $\mathbb{H}^3$, the subspace $B_0(S^2, J)$ is identified with the conformal sphere (S^2, J), and finally the space $B_{-1}(S^2, J)$ carries a Lorentzian metric. The space $B(S^2, J)$ has a very rich geometry with group of automorphisms isomorphic to the group $\text{PSL}(2, \mathbb{C})$. As a topological space $B(S^2, J)$ is homeomorphic to the real 3-dimensional vector space $\mathbb{R}^3$.

The open subsets U that underlie a Brägger metric (U, g_U) are of the following types: $U = S^2$ for $k = +1$, $U = S^2 \setminus \{p\}$, $p \in S^2$ for $k = 0$ and U is a disk bounded by a Moebius circle for $k = -1$.

10.2 The Uniformization Theorem, Case $g \geq 2$, By Energy

Theorem 10.3. Uniformization Theorem, compact surfaces $g \geq 2$. *Let* (S, J) *be a compact J-surface of genus* $g \geq 2$. *The surface* (S, J) *admits a unique conformally equivalent hyperbolic metric* g_J.

The metric g_J supplied by this theorem is called the Poincaré metric of the J- or z- surface (S, J). We recommend reading the work [30, 31] of the European family of Paul Henri de Saint-Gervais, living in France, on the Uniformization Theorem.

First we give a short outline of the following longer proof. We have already constructed an oriented hyperbolic surface S_g for every genus $g \geq 2$. Let (S_g, J_0) be the underlying J-surface. Let J be another oriented J-field on S_g. Let J_t be the family of J-fields that traces pointwise geodesic segments in $\mathbb{H}_J$ and with $J_1 = J$.

Let ω_0 be the hyperbolic volume form on (S_g, J_0). We look for a family ω_t of volume forms such that the metrics $g(J_t, \omega_t)$ all have constant curvature -1. The constraint of keeping the curvature of $g(J_t, \omega_t)$ equal to -1 on S for all times $t \in [0, 1]$ implicitly defines a differential equation for the family of forms ω_t. The family ω_t will appear as the solution of a differential equation

$$\frac{\mathrm{d}}{\mathrm{d}t} \omega_t = K(J_t, \dot{J}_t, \omega_t)$$

with initial condition at ω_0 for $t = 0$. Putting $\omega_t = e^{2u_t} \omega_0$, this differential equation transforms to a differential equation for the family of functions u_t on S with initial condition $u_0 = 0$. The proof will consist in showing that this differential equation has a global solution for $t \in [0, 1]$.

A *pair of pants decomposition* consists of a system of $3g - 3$ pairwise disjoint simply closed essential curves c_i, $1 \leq i \leq 3g - 3$, that cut the surface in $-\chi(S) = 2g - 2$ connected components.

A *pants marking* on S, or *marking* for short, is a pair (P, P') consisting of a pair of pants decomposition $P = \{c_i, 1 \leq i \leq 3g - 3\}$ and a system of smooth simple essential curves $P' = \{c_i', 1 \leq i \leq 3g - 3\}$ such that each curve c_i' is not homotopic to the curve c_i, but intersects transversally in two points or one point the curve c_i. Moreover, the curve c_i' is disjoint from the curves c_j, $j \neq i$. There can be two or one connected component adjacent to a curve c_i of a pair of pants decomposition and the curve c_i', except for its intersection with c_i, stays in the union of those components.

The following proof uses that a marking (P, P') on a hyperbolic surface (S, m) is isotopic to a unique marking (P_m, P_m') given by a system of curves $c_{i,m}, c_{i,m}'$ that are simply closed geodesics. Curves in (P, P') and (P_m, P_m') of equal label i are isotopic.

Proof. Let J_t, $t \in [0, 1]$, be the pointwise geodesic family of J-fields on $S = S_g$ connecting J_0 and J. Let ω_t be a family of smooth volume forms of total volume $-2\pi\chi(S)$ such that at $t = 0$ the metric $g(J_0, \omega_0)$ is hyperbolic with constant Gaussian curvature -1. Let k_{J_t, ω_t} be the Gaussian curvature function of the metric $g(J_t, \omega_t)$.

The $\frac{\mathrm{d}}{\mathrm{d}t}$-derivative of k_{J_t, ω_t} can be computed by the chain-rule and two partial derivatives, the first with respect to J_t, namely

$$\frac{\partial}{\partial J} k_{J_t,\omega_t} := \frac{d}{ds} k_{J_{t+s},\omega_t}|_{s=0}$$

and the second with respect to ω_t, namely

$$\frac{\partial}{\partial \omega} k_{J_t,\omega_t} := \frac{d}{ds} k_{J_t,\omega_{t+s}}|_{s=0}.$$

The family ω_t of volume forms can be written as $\omega_t = e^{2u_t}\omega_0$, where u_t is a family of smooth functions with $u_0 = 0$.

From the Gauss Curvature Formula (Theorem 8.8)

$$k_{J_t,\omega_t}\omega_t = (\Delta_{g(J_0,\omega_0)}(u_t) + k_{J_0,\omega_0})\omega_0$$

we can deduce the partial $\frac{d}{dt}$-derivative with respect to ω

$$\frac{\partial}{\partial \omega} k_{J_t,\omega_t} = (\Delta_{g(J_0,\omega_0)}(\dot{u}_t) + 2\dot{u}_t - 2\dot{u}_t \Delta_{g(J_0,\omega_0)}(u_t))e^{-2u_t}.$$

The family of functions k_{J_t,ω_t} are independent of t if the total $\frac{d}{dt}$-derivative of k_{J_t,ω_t} vanishes, hence if the family (u_t) of functions satisfies the smooth differential equation

$$(\Delta_{g(J_0,\omega_0)}(\dot{u}_t) + 2\dot{u}_t - 2\dot{u}_t \Delta_{g(J_0,\omega_0)}(u_t))e^{-2u_t} = -\frac{\partial}{\partial J} k_{J_t,\omega_t}$$

with initial value $u_0 = 0$.

Let $t \in [0,T[\mapsto u_t$ be a local solution. We have $k_{g(J_t,e^{2u_t}\omega_0)} = -1$ on $[0,T[$. If we choose $t \in [0,T[$ as the new time origin and put $u_{t+s} = u_t + v_s$ we can rewrite for $s \in [0,T-t[$ a differential equation for the family v_s as follows:

$$(\Delta_{g(J_{t+s},e^{2v_s}\omega_t)}(\dot{v}_s) + 2\dot{v}_s - 2\dot{v}_s\Delta_{g(J_t,\omega_t)}(v_s))e^{-2v_s} = -\frac{\partial}{\partial J} k_{J_{t+s},\omega_{t+s}}$$

with $v_0 = 0$ as initial condition. Observe that $\dot{v}_s = \dot{u}_{t+s}$ and at $s = 0$ we have $\Delta_{g(J_t,\omega_t)}(v_s) = 0$. This yields the equation

$$(\Delta_{g(J_t,\omega_t)} + 2\mathrm{Id})(\dot{u}_t) = -\frac{\partial}{\partial J} k_{J_t,\omega_t}.$$

The operator $(\Delta_{g(J_t,\omega_t)} + 2\mathrm{Id})$ is invertible on the space of smooth functions, so we obtain, as outlined above, before the proof, a differential equation for the family of functions u_t

$$\dot{u}_t = -(\Delta_{g(J_t,e^{2u_t}\omega_0)} + 2\mathrm{Id})^{-1}\left(\frac{\partial}{\partial J} k_{J_t,e^{2u_t}\omega_0}\right).$$

This equation, being Lipschitz, has local solutions $t \in [0,T[\mapsto u_t$ for some $T \in]0,1]$. In a first step we will show that the J-surface (S, J_T) admits a conformally equivalent hyperbolic metric. So, for the maximal such $T \in [0,1]$ one has $T = 1$.

Finally it follows that the surface $(S, J_1) = (S, J)$ admits a conformally equivalent hyperbolic metric.

The parametrized path $\{J_t \mid t \in [0, 1]\}$ in the space of J-structures on S with its natural uniform convergence topology of fields is compact. For every $t \in [0, 1]$ the J-surface (S, J_t) is a Riemann surface with a holomorphic atlas by the local rigidity of J-fields in real dimension two (Theorem 5.30). Hence we can use quantities depending on the associated conformal structure.

One such quantity is the extremal length $\lambda(\gamma)$ of essential simply closed curves γ on S. From the above compactness of $\{J_t \mid t \in [0, 1]\}$ we deduce for every simply closed curve γ the existence of a constant Λ_γ such that the extremal length of γ measured with J_t is bounded from above by Λ_γ.

By Theorem 3.47 (Degeneration of Hyperbolic $\perp$-Hexagons) and Theorem 8.16 (Domination of Hyperbolic Length by Extremal Length) it follows that for no essential simply closed curve does the g_t-length, $t \in [0, T[$, shrink to zero or its length approach ∞ as $t \to T$. We claim, see the argument below, that the J-surface (S, J_T) is hyperbolic. In fact, we will construct on the surface (S, J_T) a hyperbolic structure with underlying J-structure equal to J_T by producing explicitly an atlas with charts in the hyperbolic plane and transition maps by local hyperbolic isometries.

We can conclude that the differential equation for the family of functions u_t admits a solution for $t \in [0, 1]$.

The corresponding family of metrics $g(J_t, e^{u_t}\omega_0)$ is hence hyperbolic with curvature -1. In particular at $t = 1$ we have proved that the arbitrary J-surface (S, J) admits a conformally equivalent hyperbolic metric.

The argument that justifies the above claim is as follows. We choose for the hyperbolic surface (S, J_0, g_0) a marking (P, P') consisting of $2g - 2$ pants P_j having boundary curves $c_1, \ldots, c_{3g-3}$ and a system $c_1', \ldots, c_{3g-3}'$ of dual curves.

Assume for simplicity that the dual trivalent graph of the pairs of pant decomposition P has no short loops consisting of one edge.

For $t \in [0, T[$ let $c_{i,t}, c_{i,t}'$ be the closed geodesic representatives for c_i, c_i' in the surface (S, g_t).

We denote by $P_{1,t}, P_{2,t}, \ldots, P_{2g-2,t}, t \in [0, T[$, the connected components of the complement in S of the union of the curves $c_{i,t}$. From the assumption on the dual trivalent graph it follows that the closures $\bar{P}_{j,t}, j = 1, 2, \ldots, 2g - 2$, are closed sub-surfaces with 3 closed geodesics as boundary. We call the sub-surfaces $\bar{P}_{j,t}$ the geodesic pants of the marked hyperbolic oriented surface (S, g_t). The labeling of the curves $c_1, \ldots, c_{3g-3}$ induces an order and labeling by $l = 1, 2, 3$ on the set of boundary components $b_{j,t}^l$ of each geodesic pant $\bar{P}_{j,t}$. Moreover each boundary component of a pant $\bar{P}_{j,t}$ has a special point, namely the foot point of the shortest arc $a_{j,t}^l$ in $\bar{P}_{j,t}$ connecting $b_{j,t}^l$ with $b_{j,t}^{l++}$. The operation $l \mapsto l++$ is cyclic, $1 \to 2 \to 3 \to 1$. Each curve $c_{i,t}$ is the boundary curve of two geodesic pants P_t', P_t'', so each $c_{i,t}$ has two special points $a_{i,t}', a_{i,t}''$, where $a_{i,t}'$ resp. $a_{i,t}''$ is the foot point of the arc in the geodesic pants P_t' resp. P_t''. We map the geodesic $c_{i,t}$ with constant speed and preserving its induced orientation from P_t' to the oriented unit circle S^1 in $\mathbb{C}$, mapping $a_{i,t}'$ to 1. Let $\theta_{i,t} \in S^1$ be the image of $a_{i,t}''$. Note that the value of $\theta_{i,t}$ remains unchanged if P_t'' is used instead of P_t'.

By Theorem 8.16 (Domination of Hyperbolic Length by Extremal Length) there exists a constant $0 < M$ such that the lengths of the curves $c_{i,t}$, $c'_{i,t}$, $t \in [0, T[$, belong to the interval $[0, M]$. By Theorem 3.47 (Degeneration of Hyperbolic $\perp$-Hexagons) if a curve length of $c_{i,t}$, $c'_{i,t}$ approached 0 the dual companion would have length growing to infinity, which is excluded. So there exists a constant $0 < m < M$ such that the lengths of the curves $c_{i,t}$, $c'_{i,t}$, $t \in [0, T[$, belong to the interval $[m, M]$.

Let $0 \leq t_1 < t_2 < \cdots < t_n < \cdots < T$ be a sequence converging to T such that all limits $\lim_{k \to \infty} \text{length}(c_{i,t_k}) \in [m, M]$ and $\lim_{k \to \infty} \theta_{i,t_k} \in S^1$ exist. We denote by C_i, Θ_i the corresponding limits. The numbers C_i define $2g - 2$ hyperbolic pants Q_j, which can be glued along boundary components in a unique way to a hyperbolic surface Σ, using the dual trivalent graph Γ_P and the angles Θ_i. Let $(\Sigma, J_\Sigma, g_\Sigma)$ the triple of a surface, a J-field, and a Riemannian metric. The lengths of the arcs a^l_{j,t_k} also converge with limits $A^l_j > 0$.

The union of the curves c_{i,t_k} and the arcs a^l_{j,t_k} build an embedded graph h_{t_k} in the hyperbolic surface (S, g_{t_k}). The edges of h_{t_k} are geodesic segments and the closures of the connected components of $S \setminus h_{t_k}$ are isometric to right-angled isometric hexagons H'_{j,t_k}, H''_{j,t_k} in the hyperbolic plane with $\bar{P}_{j,t_k} = H'_{j,t_k} \cup H''_{j,t_k}$. The hyperbolic surface (S, J_{t_k}) has the pant decomposition Q_{j,t_k}, each pant being the union of two isometric hexagons H'_{j,t_k}, H''_{j,t_k}.

Let K'_{j,t_k}, K''_{j,t_k} be the open regular neighborhoods of H'_{j,t_k}, H''_{j,t_k} in the hyperbolic surface (S, J_{t_k}), obtained by ϵ-thickening each hexagon for a sufficient small $\epsilon > 0$. Let $\phi'_{j,t_k} : K'_{j,t_k} \to \mathbb{C}_+$ and $\phi''_{j,t_k} : K''_{j,t_k} \to \mathbb{C}_+$ be the isometric embeddings that map the first foot point a'^1_{j,t_k} to $i \in \mathbb{C}_+$ and the arc u^1_{j,t_k} to $[i, +i\infty]$.

The family of maps ϕ'_{j,t_k}, ϕ''_{j,t_k}, together with their domains is an atlas for the surface (S, J_{t_k}) with holomorphic and hyperbolic transitions of charts in $\mathbb{C}_+$.

The previous ϵ can be chosen independently of the index k, for instance $\epsilon = \text{Min}\{m/3, \text{Length}(a^l_{j,t_k})/3\}$. The trigonometry of right-angled hexagons gives an a priori estimate $\epsilon(n, M)$ from below for the quantity $\text{Length}(a^l_{j,t_k})$ depending on the bounds n, M:

$$\epsilon(n, M) = \text{Min}_{a,b,c \in [m/2, M/2]} \frac{\cosh(a) + \cosh(b)\cosh(c)}{\sinh(b)\sinh(c)}.$$

So a better uniform choice valid a priori for $t \in [0, 1]$ is $\epsilon = \text{Min}\{m/3, \epsilon(n, M)/3\}$.

It is important to observe that the transition mappings of the atlas are hyperbolic isometries that depend continuously upon the lengths of the curves c_{i,t_k} or arcs a_{i,t_k} and the angles $\theta_{i,t_k} \in S^1$. Moreover, the transition mappings converge as $t_k \to T$. This atlas is compatible with both structures on $(S, J_{t_k}) = (S, g_{t_k})$, the structure of a J-surface and also the structure of a hyperbolic surface. The limit for $t_k \to T$ as a J-surface is (S, J_T). The limit as a hyperbolic surface is the surface Σ, its underlying J-structure J_Σ is the J-field J_T. This finishes the proof of the above claim and completes the proof of existence of the metric g_J.

The uniqueness of the metric g_J follows from Theorem 10.1 (Extremal Energy Theorem, genus ≥ 2).

This finishes the proof of the Uniformization Theorem for compact J- or z-surfaces of genus $g \geq 2$. $\square$

As a first consequence one obtains the existence and uniqueness of a normalized volume form of minimal energy.

Theorem 10.4. *Let* (S, J) *be a surface of genus* $g \geq 2$. *The Energy* E_J *achieves the minimal value* $-2\pi\chi(S)$ *at a unique volume form* $\omega_{\min}$ *normalized by* $\int_S \omega_{\min} = -2\pi\chi(S)$. $\qquad\square$

Theorem 10.5. *Let* (S, J) *be a surface of genus* $g \geq 2$. *The Hessian of the Energy* E_J *is non-degenerate at its minimum* $\omega_{\min}$.

Proof. Using the Gauss Curvature Formula (Theorem 8.8), the Hessian is

$$\omega \mapsto \int_S (\Delta_{J,\omega_{\min}}(u))^2 \omega_{\min},$$

where u is a function with $\omega = e^{2u}\omega_{\min}$ and $\int_S e^{2u}\omega_{\min} = -2\pi\chi(S)$. $\qquad\square$

The previous theorem implies:

Theorem 10.6. *Let* S *be a surface of genus* $g \geq 2$. *The unique minimum* $\omega_{\min}$ *of the Energy* E_J *depends smoothly on the field* $J \in \mathbb{J}(TS)$. $\qquad\square$

Exercise 10.7. Let P be a pant decomposition of a compact hyperbolic surface S given by $3g(S) - 3$ closed geodesics all of different lengths $< \log(2 + \sqrt{3})$. Show that the group of isometries of S has no more than 2 elements. Show that a conformal bijection of S equals the identity.

Exercise 10.8. On a hyperbolic surface S of genus g construct a covering by $4g - 4$ non-overlapping right angled hyperbolic hexagons. Show that the number $M(L)$ of closed geodesics on S of length $< L$ is finite.

A theorem of Heinz Huber gives the asymptotic $M(L) \sim \frac{e^L}{L}$ for $L \to \infty$ [53], [54].

A theorem of Maryam Mirzakhani gives the asymptotic $N(L) \sim L^{6g-6}$ for the number $N(L)$ of simply closed geodesics of length $< L$ [78]. [1]

10.3 The Uniformization Theorem, Case $g = 1$

Theorem 10.9 (Uniformization Theorem, $g = 1$). *Let* (S, J) *be a* J-*surface of genus* 1. *The surface* (S, J) *admits a Euclidean metric* g_J *in the conformal class of* J.

Proof. Let $0 \neq \alpha \in \mathrm{Holo}(S, J)$ be a closed 1-differential form in the 1-dimensional complex vector space $\mathrm{Holo}(S, J)$. Locally at a point $p \in S$ the form can be written as $\alpha = f \mathrm{d}z$ with f being holomorphic. If $\alpha_p = 0$ at some $p \in S$, the order of vanishing $\mathrm{order}_p(f)$ of the function f at p is independent of the chosen local J-holomorphic

---

[1] Maryam Mirzakhani (1977–2017), Iranian mathematician, Professor at Stanford.

coordinate z and is defined as the order of vanishing $\text{order}_p(\alpha)$ of the form α. Observe that $\sum_{p \in S} \text{order}_p(\alpha) = -\chi(S) = 0$. It follows that $\alpha_p \neq 0$, $p \in S$.

Put

$$\omega := \frac{i}{2}\alpha \wedge \bar{\alpha}.$$

Observe that $\alpha \circ J = i\alpha$ and $\bar{\alpha} \circ J = -i\alpha$. The 2-form is of J-type $(1,1)$ and is real-valued. Moreover, at $p \in S$ for $u \in T_p(S)$ we have

$$\omega_p(u, J_p(u)) = \frac{i}{2}\det\begin{vmatrix} \alpha_p(u) & i\alpha_p(u) \\ \bar{\alpha}_p(u) & -i\bar{\alpha}_p(u) \end{vmatrix} = |\alpha_p(u)|^2.$$

Since $\alpha_p \neq 0$ and $\omega_p(u, J_p(u)) > 0$, $u \neq 0$, the 2-differential form ω is a volume form on S that moreover agrees with the orientation given by the field J. This form ω together with the field J defines a Riemannian metric g_J by $g_{J,p}(u, v) = \omega_p(u, J_p(v))$.

The form α being locally exact, on an open neighborhood U of $p \in S$ there exists a function $z \colon U \to \mathbb{C}$ with $dz = \alpha$. The function z is (J, m_i)-holomorphic.

One checks that the map $z \colon U \to \mathbb{C}$ is an isometry from (U, g_J) to $(z(U), g_{\mathbb{C}})$, where $g_{\mathbb{C}}$ is the standard Euclidean metric $g_{\mathbb{C}}(u, v) = \text{Re}(u\bar{v})$ on $\mathbb{C}$. Since the Euclidean metric $g_{\mathbb{C}}$ is conformally equivalent with the multiplication m_i on $\mathbb{C}$ and since the map z is (J, m_i)-holomorphic, the metric g_J is locally Euclidean and conformally equivalent with the field J. $\qquad\square$

Let (S, J) be a genus 1 surface equipped with a J-field J with the Euclidean metric g_J and $p \in S$ a point. Let $e_p \in T_p(S)$ be a tangent vector with $g_{J,p}(e_p, e_p) = 1$.

The tangent space $T_p(S)$ is equipped with the endomorphism $J_p \colon T_p(S) \to T_p(S)$ and the anti-symmetric bilinear form $\omega_p \colon T_p(S) \times T_p(S) \to \mathbb{R}$ with the scalar product $g_{J,p}(u, v) = \omega_p(u, J_p(v))$. There exists a unique $\mathbb{R}$-linear map $I_p \colon \mathbb{C} \to T_p(S)$ with $I_p(i) = e_p$ and $I_p \circ m_i = J_p \circ I_p$.

For $u \in T_p(S)$ let the curve $\gamma_u \colon [0,1] \to S$ be the geodesic with respect to the metric g_J with $\gamma_u(0) = p$ and $\dot{\gamma}(0) = u$. We define the map $E_p \colon T_p(S) \to S$ by $E_p(u) = \gamma_u(1)$.

Theorem 10.10. *Let (S, J) a genus 1 surface with J-field J. Let p, I_p, E_p be as before. The composed map $E_p \circ I_p \colon \mathbb{C} \to S$ is (m_i, J)-holomorphic and factors over $\mathbb{C}/\Gamma$ by the quotient map $\mathbb{C} \to \mathbb{C}/\Gamma$ and an (m_i, J)-holomorphic bijection $\mathbb{C}/\Gamma \to (S, J)$.* $\qquad\square$

As a corollary we have the previously stated theorem.

Theorem 10.11. *Let Σ be a Riemann surface of genus 1. Then Σ is bi-holomorphic to $\mathbb{C}/\Gamma_\tau, \tau \in \mathbb{C}_+$, for a suitable lattice Γ_τ in $\mathbb{C}$.* $\qquad\square$

10.4 Comments About Uniformization, $g = 0, 1$ or $g \geq 2$

Given a J-surface (S, J) one can ask for Riemannian metrics of constant curvature for which the (one of the) $\frac{\pi}{2}$-rotations in the tangent spaces coincide with the given field J. In short, one says that the metric is conformally equivalent to the given J-field. The Uniformization Theorems provide the existence of such metrics. Differences depending upon the genus appear concerning uniqueness and the value of the curvature.

In the case of genus $g \geq 2$ two such metrics g_1 and g_2 differ by a numerical factor $\lambda > 0$. One has $g_2 = \lambda g_1$. There are two ways of normalizing: one by prescribing the volume of the Riemannian manifold (S, g_J), one by prescribing the curvature. Prescribing the volume to $-2\pi\chi(S)$ or the curvature to -1 are equivalent normalizations by the Gauss–Bonnet Theorem 8.13.

In the case of genus $g = 1$, two metrics that are conformally equivalent to the given J-field differ by a numerical factor. Only one normalization works, that of prescribing the volume, say to 1.

In the case of genus $g = 0$, two metrics that are conformally equivalent to the given J-field differ pointwise by a factor that can vary. One has $g_2 = fg_1$, where f is a real-valued positive function. The functions that appear are the so-called Poisson kernels. Of interest is the space M_J of all metrics g_J of constant curvature $+1$ that are conformally equivalent to the given J-field. It turns out that the space M_J has a very rich geometry, it is a model of the hyperbolic 3-dimensional space $\mathbb{H}^3$.

We have (almost) seen that the group $PSL(2, \mathbb{R})$ of projective automorphisms of the projective real line $\mathbb{P}^1(\mathbb{R})$ is also the group of oriented isometries of the hyperbolic plane. Remember that a projective automorphism of $\mathbb{P}^1(\mathbb{R})$ preserves the cross-ratio and the distance in $\mathbb{H}$ was defined by using a cross-ratio.

The above model of the hyperbolic three space explains why the group $PGL(2, \mathbb{C})$ of projective automorphisms of the complex projective line $\mathbb{P}^1(\mathbb{C}) = \mathbb{C} \cup \{\infty\}$ is also the group of oriented isometries of $\mathbb{H}^3$.

10.5 Consequences of the Uniformization Theorem for Surfaces of Genus ≥ 2

Let (S, J) be a Riemann surface of genus ≥ 2. The Uniformization Theorem provides a unique volume form ω_J such that firstly $g_J(X, Y) = \omega_J(X, JY)$ is a Riemannian metric and secondly that the curvature of the metric is constant -1.

As a first immediate consequence we have

Theorem 10.12. *Let (S, J) be a Riemann surface of genus ≥ 2. Let $\phi \colon S \to S$ be a diffeomorphism that fixes the J-field J. Then ϕ preserves the volume form ω_J and the Riemannian metric g_J.* $\square$

Metrical properties of the metric g_J are in fact also conformal properties of (S, J). Recall that the Riemannian metric g_J induces on S a metric $\mathrm{Dist}_J(p, q) = \mathrm{Inf}_\gamma \mathrm{Length}_{g_J}(\gamma)$, where the infimum is taken over all smooth paths connecting

p, q. An important metrical property is the *injectivity radius* $\rho(S, g_J)$. Denote by $B^S(p, r)$ the closed Dist_J-ball in (S, g_J) with center $p \in S$ and radius r. For all $p \in S$ there exist $\rho_p > 0$ such that $B^S(p, \rho_p)$ is not homeomorphic to a closed ball $B^{\mathbb{H}}(q, \rho_p)$ in the hyperbolic plane, but the open balls $b^S(p, \rho_p)$ and $b^{\mathbb{H}}(q, \rho_p)$ are homeomorphic. The injectivity radius $\rho(S, J) = \rho(S, g_J)$ is defined by

$$\rho(S, g_J) := \text{Inf}_{p \in S} \, \rho_p.$$

Exercise 10.13. For (S, J) compact of genus ≥ 2 the quantity $\rho(S, J) = \rho(S, g_J)$ is strictly positive. For $p, q \in S$ with $\text{Dist}_J(p, q) < \frac{1}{2}\rho(S, J)$ there exists a unique smooth curve $\sigma_{p,q} : [0, \text{Dist}_J(p, q)] \to S$ of length $\text{Dist}_J(p, q)$ connecting p, q. Moreover, the curve $\sigma_{p,q}$ depends continuously upon p, q. The image of $\sigma_{p,q}$ is the segment $[p, q]$ parametrized by arc length.

A first consequence that uses a homotopy-theoretic assumption is

Theorem 10.14. *Let (S, J) be a Riemann surface of genus ≥ 2. Let g_J be the Riemannian metric of constant curvature -1 given by the Uniformization Theorem. Let $\gamma : S^1 \to S$ be a cyclicly continuously parametrized curve that is not homotopic to a constant map. Then there exists a curve $\gamma_{\min} : S^1 \to S$, unique up to rigid reparametrization, that is homotopic to γ and minimizes length in its homotopy class.*

Proof. Consider $S^1 \subset \mathbb{C}$ with its standard metric. Use the uniform continuity of γ to choose $\delta > 0$ such that $\text{Dist}(\gamma(t), \gamma(s)) < \frac{1}{2}\rho(S, J)$, $|t - s| < \delta$, holds. Choose N regularly spaced points $t_k = e^{\frac{2\pi k}{N}} \in S^1$, $k = 1, 2, \ldots, N$, with $\frac{1}{N} < \delta$. Let γ_0 be the piecewise geodesic consisting of the unique segments that connect $\gamma(t_k)$ with $\gamma(t_{k+1})$ and $\gamma(t_N)$ with $\gamma(t_1)$. The curves γ and γ_0 are homotopic by $H(t, s) = \sigma_{\gamma(t), \gamma_0(t)}(s)$. The curve $\gamma_0 : S^1 \to S$ is Lipschitz continuous with Lipschitz constant $N\rho(S, J)$. Now replace γ_0 by a piecewise geodesic curve γ_1 where the pieces connect the midpoints of the segments in the curve γ_0 by segments. The total length decreases, and γ_1 is still Lipschitz continuous with constant $N\rho(S, J)$. Replace γ_1 by the same procedure with γ_2, etc., ... This sequence converges to a curve γ_∞, homotopic to γ, and that admits a geodesic reparametrization.

Let $\gamma', \gamma'' : S^1 \to S$ be two geodesics in the homotopy class of γ. Choose a parametrization of constant speed such that the periodic function $f : t \in \mathbb{R} \mapsto \text{Dist}_S(\gamma'(e^{2\pi t}), \gamma''(e^{2\pi t}))$ achieves at $t = 0$ its minimal value. The function f is convex, hence constant, since it is periodic. If $f(0) \neq 0$, the function will be strictly convex near 0. If follows that $f = 0$, showing uniqueness up to cyclic reparametrization. $\qquad\square$

Exercise 10.15. Fill in the details of the previous proof. Hint: Study Herbert Busemann's work [17].

A shorter way of expressing the content of this theorem is by saying: A bijective holomorphic map $\phi : (S, J) \to (S, J)$ is an isometry of (S, g_J).

The next consequence with a homotopy-theoretic assumption is

Theorem 10.16. *Let (S, J) be a Riemann surface of genus ≥ 2. Let $\phi \colon S \to S$ be a diffeomorphism that fixes the J-field J and that is homotopic to the identity map of S. Then ϕ is the identity map of S.*

Proof. Since ϕ fixes the field J, the map ϕ is an isometry of the hyperbolic metric μ for which J is the $90°$-degree rotation. Let a, b be two oriented μ-geodesics that are simply closed curves on S with one point of intersection. The map ϕ fixes both oriented geodesics a, b, hence also the intersection point of a and b. Now it follows that the isometry ϕ is the identity of S. $\square$

A more qualitative reformulation is

Theorem 10.17. *Let (S, J) be a compact Riemann surface of genus ≥ 2. Then there exists an $\epsilon > 0$ such that every diffeomorphism fixing the J-field J that has uniform distance less than ϵ from the identity is in fact equal to the identity.*

Proof. Two points $p, q \in S$ are in cut locus position if there exist two different geodesics minimizing length and connecting p with q. Let C be the set of pairs on S that are in cut locus position for the hyperbolic metric g_J. Let ρ_S be the infimum of all g_J-distances realized by pairs in C. Every open ball $b(p, \rho_S)$ in (S, g_J) is isometric with an open ball of radius ρ_S in the hyperbolic plane and there exists $p \in S$ such that the closed ball $B(p, \rho_S)$ is not isometric to a closed ball in the hyperbolic plane. Observe $\rho_S > 0$. Observe that a continuous map $f \colon S \to S$ such that for all $p \in S$ the g_J-distance $g_J(p, f(p)) < \rho_S$ is homotopic to the identity. Using the previous theorem, the theorem follows for every $0 < \epsilon < \rho_S$. $\square$

Theorem 10.18. *Let (S, J) be a compact Riemann surface of genus ≥ 2. The group $\mathrm{Aut}(S, J)$ of diffeomorphisms fixing the J-field is finite.*

Proof. The group $\mathrm{Aut}(S, J)$ is in fact the group of isometries with uniform topology of the compact metric space (S, g_J), so is compact. By the above theorem the topology is also discrete. $\square$

The above proof gives no hint about $\#\mathrm{Aut}(S, J)$. From hyperbolic geometry follows the bound $\#\mathrm{Aut}(S, J) \leq 84(g - 1)$ obtained by A. Hurwitz[2] [55]. See the book [71].

A reformulation is

Theorem 10.19 ((Almost) Free Action Theorem). *Let (S, J) be a compact Riemann surface of genus ≥ 2. The group $\mathrm{Diff}_0(S)$ of diffeomorphisms that are homotopic to the identity map of S acts freely on the space $\mathbb{J}(TS)$ of J-fields on S. The group $\mathrm{Diff}^+(S)$ of orientation-preserving diffeomorphisms acts with finite stabilizers on the space $\mathbb{J}(TS)$.*

The next theorem is about the orbits of the $\mathrm{Diff}_0(S)$ action on $\mathbb{J}(TS)$.

Theorem 10.20. *The $\mathrm{Diff}_0(S)$ orbit $\mathrm{Orb}(J) = \{\phi_*^*J \mid \phi \in \mathrm{Diff}_0(S)\}$ of $J \in \mathbb{J}(TS)$ is a sub-manifold of $\mathbb{J}(TS)$ on which J° induces a complex structure and $\omega^{\circ, \mu}$ induces a symplectic structure.*

---

[2] Adolf Hurwitz (1859–1919), German mathematician.

Proof. Let ϕ_t, $t \in [0, 1]$, be an isotopy generated by the non-autonomous smooth vector field X_t. Let $J_t = \phi_{t*}J$ be the corresponding curve on the orbit of $J \in \mathbb{J}(TS)$. The initial speed vector $H = \dot{J}_0 \in T_J \mathbb{J}(TS)$ at J is tangent to $\mathrm{Orb}(J)$ and every vector in $T_J \mathrm{Orb}(J)$ can be obtained this way. Let ψ_t be the isotopy generated by the vector field $Y_t = J(X_t)$. Let K be the initial speed vector of the curve $\psi_{t*}J$.

For a vector field Z on the surface S denote by L_Z the field of endomorphisms of the space of smooth vector fields $\Gamma^\infty(TS)$ given by $s \in \Gamma^\infty(TS) \mapsto [Z, s]_{\mathrm{Lie}}$. Observe $H = L_{X_0} \circ J - J \circ L_{X_0}$ and $K = L_{Y_0} \circ J - J \circ L_{Y_0}$.

Claim: $J^\circ(H) = K$. From the claim it follows that the tangent space $T\mathrm{Orb}(J)$ is J° invariant and that J° induces a complex structure on $\mathrm{Orb}(J)$.

The claim is local on S and locally we can work on $U \subset S$ with z-coordinates $z = x + yi$ for which the field J is constant $J = \left(\begin{smallmatrix} 0 & -1 \\ 1 & 0 \end{smallmatrix} \right)$.

Let $X = X_0$ be the vector field $X = a\frac{\partial}{\partial x} + b\frac{\partial}{\partial y}$. Then $L_X(\frac{\partial}{\partial x}) = [L_X, \frac{\partial}{\partial x}]_{\mathrm{Lie}} = -a_x\frac{\partial}{\partial x} - b_x\frac{\partial}{\partial y}$ and $L_X(\frac{\partial}{\partial y}) = [L_X, \frac{\partial}{\partial y}]_{\mathrm{Lie}} = -a_y\frac{\partial}{\partial x} - b_y\frac{\partial}{\partial y}$, so $L_X = \left(\begin{smallmatrix} -a_x & -a_y \\ -b_x & -b_y \end{smallmatrix} \right)$. Similarly $Y = Y_0 = J(X) = -b\frac{\partial}{\partial x} + a\frac{\partial}{\partial y}$, $L_Y = \left(\begin{smallmatrix} b_x & b_y \\ -a_x & -a_y \end{smallmatrix} \right)$. Finally

$$H = \begin{pmatrix} -b_x-a_y & -b_y+a_x \\ -b_y+a_x & b_x+a_y \end{pmatrix}, \quad K = \begin{pmatrix} -a_x+b_y & -a_y-b_x \\ -a_y-b_x & a_x-b_y \end{pmatrix}$$

and the claim $J^\circ(H) = J \circ H = K$ holds on U.

The form ω° restricts to a closed 2-form on $\mathrm{Orb}(J)$. Again $\omega^\circ(H, J^\circ H) \geq 0$ and > 0 if $H \neq 0$, $H \in T_J \mathrm{Orb}(J)$, hence the restriction of ω° is non-degenerate too. $\square$

10.6 The "Turn" Map $\tau: \mathbb{M}(S) \to \mathbb{J}(TS)$

For a surface S let $\mathbb{M}(S)$ be the space of smooth Riemannian metrics on S. For an oriented surface the map $\tau: \mathbb{M}(S) \to \mathbb{J}(TS)$ attaches to a Riemannian metric $m \in \mathbb{M}(S)$ the J-field $\tau(m) \in \mathbb{J}(TS)$ that at $p \in S$ is the oriented $\frac{\pi}{2}$-rotation in $(T_p(S), m_p)$. If m is given in an oriented (x, y) coordinate chart by the field $\left(\begin{smallmatrix} m_{11} & m_{12} \\ m_{21} & m_{22} \end{smallmatrix} \right)$ of symmetric 2×2 matrices, then $\tau(m)$ is given by the field $f \left(\begin{smallmatrix} m_{12} & -m_{22} \\ m_{11} & -m_{21} \end{smallmatrix} \right)$ with $f = \frac{1}{\sqrt{m_{11}m_{22}-m_{12}m_{21}}}$. So, the matrix of $\tau(m)$ is obtained by turning the symmetric matrix of m, changing signs in the second column, and scaling by the positive factor f making a trace-less matrix of determinant 1. If the coordinates (x, y) are chosen such that $dx \wedge dy$ represents the Riemannian m-volume, no scaling is needed.

The Uniformization Theorem for surfaces S of genus $g \geq 2$ can be reformulated as: the restriction τ_{-1} of τ to the space $\mathbb{M}_{-1}(S)$ is a diffeomorphism between the Fréchet manifolds $\mathbb{M}_{-1}(S)$ and $\mathbb{J}(TS)$. Moreover, the map τ_{-1} is $\mathrm{Diff}_0(S)$-equivariant.

The real Fréchet manifold $\mathbb{M}(S)$ has been studied by Arthur Fischer, Antony Tromba and Sumio Yamada. The position of $\mathbb{M}_{-1}(S)$ in $\mathbb{M}(S)$ is not geodesic but curved, which was proved by Sumio Yamada by computing the extrinsic curvature.

In the present treatment we focus on the space $\mathbb{J}(TS)$. Pulling back by τ_{-1} we rediscover results of Fischer, Tromba and Yamada [38, 121, 129, 130].

New will be a natural complex Fréchet manifold structure J° together with a non-degenerate 2-differential form $\omega^{\circ,\mu}$ on $\mathbb{J}(TS)$. The $\mathrm{Diff}_0(S)$-orbits are a foliation by J°-submanifolds and the $\omega^{\circ,\mu}$-orthogonal distribution to the $\mathrm{Diff}_0(S)$-orbits is integrable too. Its leaves are J°-submanifolds and copies of the Teichmüller space $\mathbb{T}_g$.

Chapter 11
Families of Spaces

11.1 What Do Locally Trivial, Trivial and Constant Mean?

Let U be a connected open set in $\mathbb{C}$. Let X be a holomorphic manifold of complex dimension 2 and let $\pi \colon X \to U$ be a surjective holomorphic map. For $t \in U$ denote by X_t the set $\{p \in X \mid \pi(p) = t\}$. This set is called the fiber above t. We want to view these data as a varying object X_t parametrized by $t \in U$.

In a first step we will introduce extra data and will make extra assumptions that have the following consequences:

  (i) for each $t \in U$, the fiber X_t is not just a set but a compact surface,
 (ii) each smooth path $\gamma \colon [0,1] \to U$, provides a diffeomorphism $\phi_\gamma \colon X_{\gamma(0)} \to X_{\gamma(1)}$.

At this point we can conclude that the varying object is a compact smooth surface and that its type up to diffeomorphism is constant.

In order to do so, the following properties are required. First we work in the setting of smooth manifolds. Let $\pi \colon X \to U$ be a smooth map from a connected differentiable manifold X to an open set U in a real vector space V.

(a) The map π is a *submersion*, meaning that the rank of the differential

$$(\mathrm{D}\pi)_p \colon T_p X \to V, \quad p \in X,$$

   is equal to the dimension of V.
(b) The map π is *proper*, meaning that the pre-image $\pi^{-1}(K)$ is compact for every compact subspace K in U. The map π is surjective.

It follows from the Implicit Function Theorem that the fibers X_t, $t \in U$, are smooth sub-manifolds in X.

The Ehresmann *connection forms* provide important tools for the study of submersions. An Ehresmann connection form ω for a submersion π is a 1-differential form on X with values endomorphisms of TX that satisfies the following property: at $p \in X$, the endomorphism $\omega_p \colon T_p X \to T_p X$ is a projection from $T_p X$ to the tangent space $T_p(X_{\pi(p)})$ of the fiber $X_{\pi(p)}$ through p.

© The Author(s), under exclusive license to Springer Nature Switzerland AG 2021
N. A'Campo, *Topological, Differential and Conformal Geometry of Surfaces*, Universitext,
https://doi.org/10.1007/978-3-030-89032-2_11

The kernel $\text{Ker}(\omega_p)$ of ω_p is a linear subspace in $T_p X$ that is supplementary to the subspace $T_p(X_{\pi(p)})$. The differential $(D\pi)_p$ restricts to an isomorphism from $\text{Ker}(\omega_p)$ to $V = T_{\pi(p)} U$. The difference of two connection forms $\alpha = \omega_1 - \omega_2$ for the submersion π is a form that annihilates the kernel of the differential $D\pi$. Locally connection forms exist by the Implicit Function Theorem and globally by using partitions of unity.

An Ehresmann connection form ω will allow us to compare fibers $X_a, X_b, a, b \in U$ of a proper submersion π. More precisely, let $\gamma : [0, 1] \to U$ be a smooth path from a to b. A lift of γ with initial value $q \in X_a$ is a path $\Gamma_q : [0, 1] \to X$ such that: $\Gamma_q(0) = q$, $\pi \circ \Gamma_q(t) = \gamma(t)$, $\omega(\dot{\Gamma}_q(t)) = 0$, $t \in [0, 1]$.

Since the map π is proper by assumption, the differential equation $\omega(\dot{\Gamma}_q(t)) = 0$ for Γ_q has a unique smooth global solution. In fact, the value $\Gamma_q(1) \in X_b$ depends smoothly on the initial condition $q \in X_a$, hence, for every smooth path γ from a to b the map $\phi_\gamma : X_a \to X_b$ defined by $q \in X_a \mapsto \Gamma_q(1) \in X_b$ is a diffeomorphism between the fibers X_a and X_b. The inverse map $(\phi_\gamma)^{-1}$ is constructed by the liftings of the path $t \in [0, 1] \mapsto \phi(1 - t)$ with initial values in X_b.

Examples of maps π satisfying the above assumptions are *projection maps* $p_U^F : F \times U \to U$, where F is a compact smooth manifold. The tangent space $T_p(F \times U)$, $p = (f, u)$, splits as a direct sum $T_p(F \times U) = T_f F \oplus T_u U$ and the connection form is the corresponding direct sum projection.

Are all maps $\pi : X \to U$ as above projection maps? More precisely, does there exist an $a \in U$ and a diffeomorphism $\psi : X \to X_a \times U$ such that $\pi = p_U^{X_a} \circ \psi$ holds? The answer is negative in general, for example the map $z \mapsto z^2$ from $X = \mathbb{C} \setminus \{0\}$ to $U = \mathbb{C} \setminus \{0\}$ is not a projection map. Indeed, $X_1 = \{-1, +1\}$ and the path $t \in [0, 1] \mapsto \cos(2\pi t) + i \sin(2\pi t)$ from $a = 1$ to $b = 1$ does not lift to the identity of X_1 for any connection form. This is easy to check: the fiber is of dimension 0, so there is only one connection form.

In the above example the map π is not a projection map, but for every $t \in U$, there exists an open neighborhood $U' \subset U$ of t such that by restricting the domain to $Y = \pi^{-1}(U')$ the resulting map $\pi' : Y \to U'$ is a projection map.

This is a general fact for proper *submersions*, i.e. for proper differential maps $\pi : M \to N$ between smooth manifolds M, N such that at every point $p \in M$ the differential $(D\pi)_p : T_p M \to T_{\pi(p)} N$ is surjective. A *trivial fibration of smooth manifolds* is a map $\pi : M \to N$ such that for a point $t \in N$ the preimage $\pi^{-1}(t)$ is a submanifold in M and there exists a diffeomorphism $\phi : M \to N \times \pi^{-1}(t)$ with $\pi = P \circ \phi$, where $P : N \times \pi^{-1}(t) \to N$ is the projection map. A *fibration of smooth manifolds* is a map $\pi : M \to N$ such that for every point $t \in N$ there exists an open neighborhood U of t in N such that the restriction $\pi^U : \pi^{-1}(U) \to U$ of π above U is a trivial fibration of smooth manifolds. The following very useful general fact is due to Charles Ehresmann.[1]

Theorem 11.1 (Ehresmann Fibration Criterion). *A proper submersion* $\pi : M \to N$ *is a fibration of smooth manifolds.*

---

[1] Charles Ehresmann (1905–1979), French mathematician, Professor in Strasbourg, Paris and Amiens.

Proof. Use the Implicit Function Theorem along a fiber $F = \pi^{-1}(t)$, $t \in N$, in order to have locally trivial fibrations $\pi_\alpha \colon U_\alpha \to \pi_\alpha(U_\alpha)$, where π_α are restrictions of π. By properness, only finitely many U_α cover the fiber F. Let U' be a star-shaped open neighborhood of $t \in N$ with $U' \subset \cap_\alpha \pi(U_\alpha)$. The restriction π' of π above the open neighborhood U' of $t \in N$ is a trivial fibration of smooth manifolds. Indeed, the connection forms ω_α on each U_α can be combined with a partition of unity to a connection form ω on $\pi^{-1}(U')$. Since U' was chosen to be star-shaped, one can use paths $\gamma_s = [t, s]$, $s \in U'$, which depend smoothly upon s. The resulting path liftings yield a smooth local trivialization of π above U'. □

Let **SmoothMan** be the set of isomorphy classes of smooth compact manifolds. The datum of a proper submersion $\pi \colon M \to N$ gives a map $\Pi \colon N \to$ **SmoothMan** by $t \in N \mapsto [\phi^{-1}(t)] \in$ **SmoothMan**. The proof of the Ehresmann Fibration Criterion shows

Theorem 11.2. *The induced map* $\Pi \colon N \to$ **SmoothMan** *by a proper submersion* $\pi \colon M \to N$ *is locally constant, in fact constant on each path connected component of N.* □

It follows that the appropriate manifold structure on **SmoothMan** is the structure of a manifold of dimension 0. One can consider **SmoothMan** as the vertices of a graph by adding edges between manifolds of equal dimension that are cobordant. So one gains more structure, even more if one colors the edges by the minimal complexity of the manifold that realizes the cobordism. The decision when to put an edge becomes algorithmic due to the work of René Thom[2] [118].

11.2 The Legendre Family

The following family of Riemann surfaces introduced by Legendre provides an illustration of the above.

Put $N = \mathbb{C}^{**} = \mathbb{C} \setminus \{0, 1\}$ and define the real 4-dimensional manifold M by

$$M = \{(x : y : z, t) \in \mathbb{P}^2(\mathbb{C}) \times N \mid y^2 z - x(x - z)(x - tz) = 0\}.$$

Let $\pi \colon M \to N$ be the map induced by the projection $(x : y : z, t) \mapsto t$. The affine part of M given by $z = 1$ is defined by

$$M' = \{(x, y, t) \in \mathbb{C}^2 \times \mathbb{C}^{**} \mid y^2 = x(x - 1)(x - t)\}.$$

Observe that M, N are smooth manifolds and that the map π is a proper submersion. The typical fibers are Riemann surfaces of genus 1. The map π is a fibration of smooth manifolds by the Ehresmann Fibration Criterion (Theorem 11.1). The fibration is not trivial. Indeed, the inverse image by π of a small circle around $0 \in \mathbb{C}^{**}$ is not homeomorphic to the 3-dimensional torus. The induced map $\Pi \colon \mathbb{C}^{**} \to$ **SmoothMan** is constant.

---

[2] René Thom (1923–2002), French mathematician, Professor in Grenoble, Strasbourg, and member of the Institut des Hautes Études Scientifique in Bures-sur-Yvette.

The manifolds M, N are in fact holomorphic manifolds and the map π is holomorphic. Does the Ehresmann Fibration Criterion hold here in the holomorphic setting? More precisely, is the smooth fibration π locally trivial in the holomorphic setting? For $t \in N$, does there exist an open neighborhood U of t in N together with a smooth holomorphic diffeomorphism $\phi_U : \pi^{-1}(U) \to U \times \pi^{-1}(t)$ satisfying on $\pi^{-1}(U)$ the factorization $\pi = P \circ \phi_U$. Again P is the projection from $U \times \pi^{-1}(t)$ to U. Claim: no such holomorphic local trivializations exist for these manifolds.

The justification of the claim uses a big part of the material that we have developed so far.

The fibers $M_t = \pi^{-1}(t)$ are Riemann surfaces of genus 1. Let $\omega_t \neq 0$ be a holomorphic differential on M_t. Since the space $\mathrm{Hol}(M_t)$ of holomorphic differentials has dimension 1 by Riemann's Existence Theorem 12.4, the choice of ω_t is possible and unique up to a scalar factor. Remember that the form ω_t is closed. For $t \in N$ let $[0, 1]$ and $[1, t]$ be the straight paths in $\mathbb{C}$. The paths intersect conveniently only in t if $t \notin \mathbb{R}$ or $t \in \mathbb{R}$, $t > 1$, which we assume. Let $h_t : M_t' \to \mathbb{C}$ be the restriction of the coordinate map $(x, y, t) \in M_t' \mapsto x \in \mathbb{C}$. The map h_t is 2 to 1 except above $0, 1, t$. The inverse images $a_t = h_t^{-1}([0, 1])$ and $b_t = h_t^{-1}([1, t])$ are simply closed smooth curves on M_t that intersect transversally at one point. We parametrize the curves a_t and b_t such that the frame $(\dot{a}_t, \dot{b}_t)$ of speed vectors of a_t and b_t at their intersection point agree with the orientation of the Riemann surface M_t. We have two choices up to simultaneously switching the orientations. Both the periods $\int_{a_t} \omega_t$ and $\int_{b_t} \omega_t$ depend upon the choices, but the ratio of the periods $R(t) = \int_{a_t} \omega_t / \int_{b_t} \omega_t$ does not depend upon the possible choices.

The ratio $R(t)$ is a (little, partial) measurement attached to the Riemann surface M_t and the system of curves a_t, b_t on it and hence in Latin it deserves to be called a *modulus* for these data. Moreover, by the Ehresmann Fibration Criterion and Stokes' Theorem, the quantity R_t would be locally constant if the opposite of the above claim held. In the present case the forms ω_t can be explicitly given in terms of the coordinates x, y:

$$\omega_t = \frac{dx}{y} = \frac{dx}{\pm\sqrt{x(x-1)(x-t)}}.$$

The modulus $R(t)$ evaluates to

$$R(t) = \int_{[0,1]} \frac{dx}{\sqrt{x(x-1)(x-t)}} \bigg/ \int_{[1,t]} \frac{dx}{\sqrt{x(x-1)(x-t)}}.$$

A numerical computation with SAGE shows

$$R(2) = 0.99999\sqrt{-1}, \quad R(3) = 0.85458\sqrt{-1}.$$

The map $R : t \in \mathbb{C} \setminus \mathbb{R}_{\leq 1} \mapsto R(t) \in \mathbb{C}$ is holomorphic, and by the computation not constant. It follows that R is nowhere locally constant. This proves that the Ehresmann Fibration Criterion does not hold in the holomorphic setting for the proper holomorphic submersion π. Even more follows:

Theorem 11.3. *The Legendre map*[3] $\pi : M \to N$ *is a proper holomorphic submersion. For no non-empty open connected subset U' in $N = \mathbb{C}^{**}$ is the restriction π' of π to $\pi^{-1}(U')$ a trivial fibration in the holomorphic setting.* $\square$

The so-called elliptic integrals $\int_a^b \frac{dx}{\sqrt{P(x)}}$, where $P(x)$ is a polynomial with distinct roots of degree 3 or 4, have a very long history.

The substitution of $-x + 2$ for x yields

$$\int_{[1,2]} \frac{dx}{\sqrt{x(x-1)(x-2)}} = -\sqrt{-1} \int_{[0,1]} \frac{dx}{\sqrt{x(x-1)(x-2)}}$$

and proves $R(2) = \sqrt{-1}$. Let i be the symbol for $\sqrt{-1}$, so $R(2) = i$.

The Riemann surface $\mathbb{C}/\mathbb{Z} \oplus \mathbb{Z}i$ has the same ratio of periods, if computed with the form dz and paths $a = \mathbb{R}i/\mathbb{Z}i$, $b = \mathbb{R}/\mathbb{Z}$. It follows that the Riemann surfaces $\mathbb{C}/\mathbb{Z} \oplus \mathbb{Z}i$ and $M_2 = \{x : y : z \in \mathbb{P}^2(\mathbb{C}) \mid y^2z - x(x-z)(x-2z) = 0\}$ are isomorphic.

Let **RiemannSurf** be the set of isomorphism classes of Riemann surfaces. The Riemann surfaces M_{1+i} and $M_{\frac{1-i}{2}}$ are isomorphic, for instance by the substitution of $(y, \frac{x}{1+i})$ for (y, x), and the moduli $R(1 + i)$ and $R(\frac{1-i}{2})$ differ by 1. It can be verified by computation that they differ. Why by 1?

By definition

$$R(1+i) = \int_{a_{1+i}} \omega_{1+i} \bigg/ \int_{b_{1+i}} \omega_{1+i}$$

$$R\left(\frac{1-i}{2}\right) = \int_{a_{\frac{1-i}{2}}} \omega_{\frac{1-i}{2}} \bigg/ \int_{b_{\frac{1-i}{2}}} \omega_{\frac{1-i}{2}} .$$

The above substitution transforms the integrals over b-type cycles into one another, so

$$\int_{b_{1+i}} \omega_{1+i} = \int_{b_{\frac{1-i}{2}}} \omega_{\frac{1-i}{2}} .$$

As for the integrals over a-type cycles one has

$$\int_{a_{\frac{1-i}{2}}} \omega_{\frac{1-i}{2}} = \int_{a_{1+i}+b_{1+i}} \omega_{1+i} .$$

It follows that $R(\frac{1-i}{2}) = R(1 + i) + 1$, as claimed.

In conclusion, the moduli can differ, but the corresponding Riemann surfaces are isomorphic. It clearly follows that the choice of the integration cycles a, b matters.

Let S be a compact oriented surface of genus 1. Let $p \in S$ be a point. A *marking* $\mu = (a, b)$ on S is a pair (a, b) of oriented simply closed smooth curves on S, up to isotopy, that meet only at p and intersect transversally at the point p, such that the oriented vectors $u \in T_p a$ and $v \in T_p b$ form an oriented frame for $T_p S$. Given a marking $\mu = (a, b)$ one can make 4 new markings. Namely, by cutting the surface S along the curves a and b one obtains a combinatorial square that has two diagonals

---

[3] Named after French mathematician Adrien-Marie Legendre (1752–1833).

d and e. On S such a diagonal is a non-oriented simply closed curve that intersects transversally both curves a and b only at p. The four new markings are (a, d), (a, e), (d, b), (e, b). In such a new marking the curve a or b keeps its orientation and the other curve d or e will be oriented such that the new pair becomes a marking. By repeating the choice of a new marking, starting from a given marking, one obtains any other marking.

Let J be a J-field on S compatible with the orientation of S. Let ω_J be a holomorphic form of J-type $(1, 0)$. As before we define a modulus $R(J, a, b) = \int_b \omega_J / \int_a \omega_J$. Changing the marking changes the modulus as follows:

Theorem 11.4. *The modulus* $z = R(J, a, b)$ *of an oriented J-field and marking* (a, b) *belongs to the upper half plane* $\mathbb{C}_+$. *Replacing the marking* (a, b) *by* (a, d) *or* (a, e) *changes the modulus* z *to* $z \pm 1$. *Replacing* (a, b) *with* (d, b) *or* (e, b) *changes the modulus* z *to* $\frac{z}{\pm z + 1}$. □

If one thinks of the torus S as the quotient $S = \mathbb{C}/\mathbb{Z} \oplus \mathbb{Z}i$, each marking μ based at $p = 0 + \mathbb{Z} \oplus \mathbb{Z}i$ of S is represented by a pair $(u_1 + u_2i, v_1 + v_2i)$ of vectors in the lattice $\mathbb{Z} \oplus \mathbb{Z}i$ satisfying $\left(\begin{smallmatrix} u_1 & v_1 \\ u_2 & v_2 \end{smallmatrix} \right) \in \mathrm{SL}(2, \mathbb{Z})$. The corresponding oriented simply closed curves on S are the images of the oriented segments $[0, u]$, $[0, v] \subset \mathbb{C}$.

Changing the marking corresponds to multiplying the matrix $\left(\begin{smallmatrix} u_1 & v_1 \\ u_2 & v_2 \end{smallmatrix} \right)$ from the right by $A \in \mathrm{SL}(2, \mathbb{Z})$. The modulus changes by the *modular action* of A on $\mathbb{C}_+$.

The action of $\mathrm{SL}(2, \mathbb{Z})$ on the set of markings is simply transitive. The action of $\mathrm{SL}(2, \mathbb{Z})$ on the space $\mathbb{C}_+$ of moduli factors through a faithful discrete action of $\mathrm{PSL}(2, \mathbb{Z})$. Replacing a marking (a, b) by replacing one entry by a diagonal of the complementary square as above corresponds to acting by $A = \left(\begin{smallmatrix} 1 & \pm 1 \\ 0 & 1 \end{smallmatrix} \right)$ or by $A = \left(\begin{smallmatrix} 1 & 0 \\ \pm 1 & 1 \end{smallmatrix} \right)$.

Remember that the Riemann surfaces $S = \mathbb{C}/\mathbb{Z} \oplus \mathbb{Z}\omega$ and $S = \mathbb{C}/\mathbb{Z} \oplus \mathbb{Z}\omega'$, $\omega, \omega' \in \mathbb{C}_+$, are isomorphic if and only if the moduli $\omega, \omega' \in \mathbb{C}_+$ are in the same $\mathrm{PSL}(2, \mathbb{Z})$-orbit. As a consequence:

Theorem 11.5. *The map* $\Pi \colon \mathbb{C}^{**} = N \to$ **RiemannSurf** *induced by* $\pi \colon M \to N$ *is not constant. More precisely, the map* Π *is locally finite to one.* □

Chapter 12
Functions on Riemann Surfaces

12.1 Meromorphic Functions on Riemann Surfaces

Let S be a compact connected Riemann surface. We have seen that holomorphic functions on S are constant functions. This means that the ring of holomorphic functions on S is the field $\mathbb{C}$. Clearly, it is impossible to get knowledge about the surface S by studying its holomorphic functions. If one wishes to study the Riemann surface S through a ring of functions one needs to consider more functions than just the holomorphic ones (see Section 12.5 (Why functions?)).

Let M be a Riemann surface. A map $f : M \to \mathbb{C} \cup \{\infty\}$ is called a meromorphic function if for every point $p \in M$ there exists a local holomorphic coordinate $z : U \to \mathbb{C}$ on an open neighborhood of p in M with $z(p) = 0$ and $f(q) \neq \infty, q \in U$, for all $q \neq p$, such that for some integer k the map $q \in U \setminus \{p\} \mapsto z(q)^{-k} f(q) \in \mathbb{C}$ is holomorphic and bounded. The least such $k \in \mathbb{Z}$ is called the *order* $\mathrm{order}_p(f)$ at p.

Locally there exist meromorphic functions on Riemann surfaces. For instance, let U be a connected chart of a Riemann surface and let $f : U \to \mathbb{C}$ be a holomorphic function, $f \neq 0$ but 0 at some point in U, then the ratio $(f + 1)/f$ defines a non-constant meromorphic function on U. More generally, ratios of holomorphic functions $\frac{f}{g}$, $g \neq 0$, define meromorphic functions.

It is not immediately clear that on a compact Riemann surface of genus $g \geq 0$ non-constant meromorphic functions exist globally. However, this will follow for the case of genus $g \geq 2$ from the following section where we construct linearly independent holomorphic 1-differential forms of type $(1, 0)$.

12.2 *J*-Harmonic 1-Differential Forms on *J*-Surfaces

Let α be a 1-differential form on a J-surface S. The form α is locally the differential of a function, i.e. for every point $p \in S$ there exists an open neighborhood U of p and a function $f : U \to \mathbb{R}$ with $\alpha = df$, if and only if $d\alpha = 0$ holds.

© The Author(s), under exclusive license to Springer Nature Switzerland AG 2021
N. A'Campo, *Topological, Differential and Conformal Geometry of Surfaces*, Universitext,
https://doi.org/10.1007/978-3-030-89032-2_12

Remember a 1-form α is called closed if $d\alpha = 0$ holds, and is called exact if $\alpha = df$ for some function f. The de Rham cohomology group, the quotient space $H^1_{dR}(S, \mathbf{R})$ of closed forms modulo exact forms, is a real vector space of dimension $2g$.

A J-harmonic 1-form is a form such that both forms α and $\alpha \circ J$ are closed.

The following fundamental result is due to Riemann and was generalized by Hodge [52] and de Rham [29] to Riemannian manifolds and all cohomology groups.

Theorem 12.1 (J-Harmonic characterization of genus). *Let S be a compact J-surface. Let $\mathrm{Harm}_{\mathbb{R}}(S, J)$ be the space of J-harmonic 1-forms. The map $[\cdot]\colon \alpha \in \mathrm{Harm}_{\mathbb{R}}(S, J) \mapsto [\alpha] \in H^1_{dR}(S, \mathbb{R})$ is an isomorphism of real vector spaces.*

Proof. Injectivity: Assume $0 = [\alpha] \in H^1_{dR}(S, \mathbb{R})$ with $\alpha \in \mathrm{Harm}_{\mathbb{R}}(S, J)$. Then α is exact, i.e. $\alpha = df$. Moreover, $\Delta_J(f) = d_J\alpha = 0$. So the function f is harmonic, hence locally constant, and $df = \alpha = 0$ follows. More precisely, let ω be any volume form on S. Then $\Delta_{J,\omega}(f) = \Delta_J(f)\colon \omega = 0$. So $f\colon S \to \mathbb{R}$ is harmonic for the Riemannian metric $g(J, \omega)$ and hence locally constant.

Surjectivity: Let the closed 1-form β represent a class $c \in H^1_{dR}(S, \mathbb{R})$. There exists a function $f\colon S \to \mathbb{R}$ with $\Delta_J(f) = d(\beta \circ J)$. Put $\alpha = \beta - df$. Then α is J-harmonic, since $d\alpha = 0$ and $d_J\alpha = d_J\beta - d_J(df) = 0$. Clearly $[\alpha] = [\beta] = c$. $\square$

It follows that the cohomology group

$$H^1_{dR}(S, \mathbb{C}) = \{\alpha \in \Omega^1(S, \mathbb{C}) \mid d\alpha = 0\}/\{df \mid f \in \Omega^1(S, \mathbb{C})\}$$

is a complex vector space of dimension $2g$.

We denote by $\mathrm{Harm}_{\mathbb{C}}(S, J)$ the space of harmonic complex valued 1-differential forms, i.e. forms $\alpha \in \Omega^1(S, \mathbb{C})$ satisfying $d\alpha = 0$ and $d(\alpha \circ J) = 0$. The map $\alpha \in \mathrm{Harm}_{\mathbb{C}}(S, J) \mapsto [\alpha] \in H^1_{dR}(S, \mathbb{C})$ is an isomorphism of complex vector spaces.

12.3 Riemann's Theorem About the Sub-Space $\mathrm{Hol}(S, J)$ of Closed Forms in $\Omega_J^{1,0}(S, \mathbb{C})$

The aim of this section is the theorem of Riemann about holomorphic differentials. Elements α in $\Omega_J^{1,0}(S, \mathbb{C}) \subset \Omega^1(S, \mathbb{C})$ are complex valued 1-differential forms on $S = (S, J)$ such that at $p \in S$ the $\mathbb{R}$-linear map $\alpha_p\colon T_p(S) \to \mathbb{C}$ is J_p-m_i linear, so the equations $\alpha_p \circ J_p(u) = i\alpha_p(u)$, $u \in T_p(S)$, or $\alpha \circ J = i\alpha$ hold. For a closed form $\alpha \in \Omega_J^{1,0}(S, \mathbb{C})$, i.e. $\alpha \in \mathrm{Hol}(S, J)$, we conclude from

$$d(\alpha \circ J) = d(i\alpha) = id\alpha = 0$$

that the form $\alpha \circ J$ is closed too.

Let $(U_a, z_a), a \in A$, be a $\mathbb{C}$-valued system of coordinate charts with holomorphic transition function on S. Such a system exists by the J-rigidity theorem. In such a local coordinate z, a form α in $\Omega_J^{1,0}(S, \mathbb{C})$ can be written as $\alpha = f dz$ where f is a locally smooth function on S. For $\alpha \in \mathrm{Holo}(S, J)$ we deduce from $d\alpha = d(f dz) =$

$d''f \wedge dz = 0$, that $d''f = 0$ holds, which says that the coefficient function f is holomorphic. The converse holds too: a 1-form α which is locally of the form $f dz$ with f holomorphic is contained in $\mathrm{Holo}(S, J)$.

Elements $\alpha = f dz \in \mathrm{Hol}(S, J) = \mathrm{Hol}(S, z)$ are closed 1-forms of J-type $(1, 0)$ and are traditionally called *holomorphic differentials* on the Riemann surface (S, z) and even more traditionally *differentials of the first kind*.

Switching between J- and z-structures, we conclude from the above that the following two versions of Riemann's theorem imply each other. The proof of the theorem (in the J-Version) will be given.

Theorem 12.2 (Complex Analytic Characterization of Genus).

z-Version: Let (S, z) be a connected oriented compact surface with holomorphic structure z. The complex dimension of the space $\mathrm{Hol}(S, z)$ of holomorphic differential forms of type $(1, 0)$ is independent of the holomorphic structure z and is equal to the genus $g(S)$ of the surface S.

J-Version: Let (S, J) be a connected oriented compact surface with J-structure J. The complex dimension of the space $\mathrm{Hol}(S, J)$ is independent of the structure J and is equal to the genus $g(S)$ of the surface S.

Proof (J-version). The $\mathbb{R}$-linear maps $\alpha \in \mathrm{Harm}_{\mathbb{R}}(S, J) \mapsto \alpha - i\alpha \circ J \in \mathrm{Hol}(S, J)$ and $\alpha \in \mathrm{Harm}_{\mathbb{R}}(S, J) \mapsto \alpha + i\alpha \circ J \in \mathrm{Hol}(S, -J)$ are injective. It follows that $\mathrm{Dim}_{\mathbb{C}}(\mathrm{Hol}(S, J)) = \mathrm{Dim}_{\mathbb{C}}(\mathrm{Hol}(S, -J)) \geq g$. From the direct sum decomposition $\mathrm{Hol}(S, J) \oplus \mathrm{Hol}(S, -J) = \mathrm{Harm}_{\mathbb{C}}(S, J)$ follows the opposite inequality. $\qquad\square$

12.4 Explicit Basis of $\mathrm{Hol}(S, J)$ for the Hyperelliptic Surface Defined By $y^2 = -x^{2g+1} + 1$

The next theorem shows for every genus $g > 0$ the existence of a surface S with holomorphic structure and a g-dimensional space of holomorphic differentials of type $(1, 0)$. See for instance the book [10] of H.F. Baker[1] where for the space of holomorphic differentials of type $(1, 0)$ on hyperelliptic Riemann surfaces an explicit basis is constructed.

In the following we construct such a basis for the hyperelliptic Riemann surface S_g defined by $y^2 = -x^{2g+1} + 1$.

Let $A_g \subset \mathbb{C}^2$ be the Riemann surface defined by the equation $y^2 = -x^{2g+1} + 1$. This equation defines a Riemann surface by the Implicit Function Theorem. Indeed, for $p \in \mathbb{C}^2$ with $y(p)^2 + x(p)^{2g+1} - 1 = 0$ the differential at p of $f = y^2 + x^{2g+1} - 1$ does not vanish.

The restriction of the coordinate function x to A_g maps A_g to $\mathbb{C}$ with preimages consisting of two elements, except for the $(2g + 1)^{\text{th}}$ roots of unity whose preimages have only one element. Let $A_g^* \subset A_g$ be the subset with $y \neq 0$. On A_g^* we define holomorphic differential 1-forms of type $(1, 0)$ by $\alpha_i = \frac{x^{i-1}}{y} dx, i = 1, 2, \ldots$. Since

---

[1] Henry Frederick Baker (1866–1956), British mathematician.

$2y\,dy = -(2g+1)x^{2g}\,dx$ on A_g we see that these forms have holomorphic extensions to A_g.

Let S_g^{sing} be defined as the projective curve in the complex projective plane $\mathbb{P}^2(\mathbb{C})$ with homogeneous coordinates $x : y : z$ defined by the homogeneous equation $y^2 z^{2g-1} + x^{2g+1} - z^{2g+1} = 0$. The manifold $\mathbb{P}^2(\mathbb{C})$ is covered by three affine charts $U_0 = \{x \neq 0\}$, $U_1 = \{y \neq 0\}$, $U_2 = \{z \neq 0\}$. On U_0 the ratios $(\frac{y}{x}, \frac{z}{x})$ are coordinates that map U_0 to $\mathbb{C}^2$. Analogously, the ratios $(\frac{x}{y}, \frac{z}{y})$ map as coordinates U_1 to $\mathbb{C}^2$ and the ratios $(\frac{x}{z}, \frac{y}{z})$ map as coordinates U_2 to $\mathbb{C}^2$.

In the chart U_2, putting $z = 1$, the equation for $S_g^{\text{sing}} \cap U_0$ will be $y^2 + x^{2g+1} - 1 = 0$. The Implicit Function Theorem shows, as seen above, that $U_2 \cap S_g^{\text{sing}} = A_g$ is a smooth manifold with holomorphic atlas. In the chart U_1 the equation, putting $y = 1$, is $z^{2g-1} + x^{2g+1} - z^{2g+1} = 0$. The Implicit Function Theorem does not apply at the point $(0 : 1 : 0) \in S_g^{\text{sing}} \subset \mathbb{P}^2(\mathbb{C})$ if $g > 1$. The point p with homogeneous coordinates $0 : 1 : 0$ is a so-called singular point of the curve S_g^{sing}.

Define for $s \in \mathbb{C}$, $|s| < \frac{1}{2}$, the holomorphic function $f(s) = s(1 - s^2)^{\frac{1}{2g-1}}$. Here $(1 - s^2)^{\frac{1}{2g-1}}$ denotes the branch with value 1 at $s = 0$. The functions $v = f(z)$ and x are at $p \in U_1$ centered local coordinate functions such that the local equation for $U_1 \cap S_g^{\text{sing}}$ is the equation $v^{2g-1} + x^{2g+1} = 0$. It follows that the map $\phi: D_\epsilon^* \to S_g^{\text{sing}}$, $0 < \epsilon << 1$, $t \mapsto \phi(t)$, with $v(\phi(t)) = t^{2g+1}$, $x(\phi(t)) = -t^{2g-1}$ is a holomorphic parametrization of a punctured neighborhood of $p \in S_g^{\text{sing}}$. Remember $A_g = S_g^{\text{sing}} \setminus \{p\}$. Now, define the Riemann surface S_g to be the result of gluing along D_ϵ^* by ϕ the disk D_ϵ to A_g. This gluing adds a point to A_g, which we denote by $\infty_{S_g} \in S_g$.

The restriction of the coordinate function x to A_g extends to a holomorphic map $h: S_g \to \mathbb{C} \cup \{\infty\}$.

Theorem 12.3. *Let g be an integer > 0. The Riemann surface S_g has genus g. The 1-forms $\frac{x^{i-1}}{y}dx$, $i = 1, 2, \ldots, g$, extend to holomorphic forms α_i on S_g. The system α_i, $i = 1, 2, \ldots, g$, is a basis of the $\mathbb{C}$-vector space $\text{Hol}(S_g)$.*

Proof. The Euler characteristic $\chi(S_g)$ equals twice the Euler characteristic $\chi(\mathbb{C} \cup \{\infty\})$ minus the number of fibers of $h: S_g \to \mathbb{C} \cup \{\infty\}$ having only one element. It follows that $\chi(S_g) = 2 \cdot 2 - (2g + 2) = 2 - 2g$ and that the genus of S_g equals g.

The forms α_i, $i = 1, 2, \ldots, g$, are of type $(1, 0)$ and holomorphic in A_g. The forms α_i, $i = 1, 2, \ldots, g$, have a common zero at $(0, \pm 1)$ of order $i - 1$ and a zero or pole at ∞_{S_g}. The support of the divisor (α_i) is included in $\{(0, 1), (0, -1), \infty_{S_g}\}$ and $\deg(\alpha_i) = -\chi(S_g)$ holds. The order of α_i, $i = 1, 2, \ldots, g$, at ∞_{S_g} is $-\chi(S_g) - 2(i-1) = 2(g-1) - 2(i-1) \geq 0$. Hence the forms α_i, $i = 1, 2, \ldots, g$, are holomorphic on S_g.

Let $u = e^{\frac{2}{2g+1}\pi i}$ be the first $(2g + 1)^{\text{th}}$ root of unity. Let $I_j = [u^{2j-1}, u^{2j}]$ be the real segment in $\mathbb{C}$ with endpoints $u^{2j-1}, u^{2j}, j = 1, 2, \ldots, g$. The inverse image of I_j by the map h consists of two intervals with common endpoints, so $c_j = h^{-1}(I_j)$ is a simply closed curve c_j on S_g. The automorphism $T: S_g \to S_g$ mapping $p \in S_g$

with coordinates $(x(p), y(p), z(p))$ to the point $T(p)$ with coordinates $(x(T(p)) = u^2 x(p), y(T(p)) = y(p), z(T(p)) = u^2 z(p)$ permutes in cyclic order the simply closed curves c_j. We orient the curves c_j in a T-invariant way and get integration cycles, again denoted by c_j.

The path integrals $p_{ij} = \int_{c_j} \alpha_i$, also called periods, satisfy

$$
p_{11} = \int_{c_1} \alpha_1 = 2 \int_0^1 \frac{(u^2 - u)\,ds}{\sqrt{((s-1)u - su^2)^{2g+1} + 1}}
$$

$$
= 2(u^2 - u) \int_0^1 \frac{ds}{\sqrt{((s-1) - su)^{2g+1} + 1}} \neq 0
$$

and

$$
p_{ij} = u^{2(j-1)(i-1)} p_{11}.
$$

Let P be the $g \times g$-matrix with entries $p_{i,j}$. The period matrix P equals $p_{1,1} V_g$, where V_g is the Vandermonde–Cauchy matrix[2] with second column the powers u^{2i-2}, $i = 1, 2, \ldots, g$. Hence $\det(P) \neq 0$, since the entries

$$
V_g[i, 2] = u^{2i-2} = e^{\frac{4i-4}{2g+1} \pi \sqrt{-1}}, \quad i = 1, 2, \ldots, g
$$

in the second column of V_g differ pairwise. We conclude that the forms α_i, $i = 1, 2, \ldots, g$, are $\mathbb{C}$-linearly independent and that the integration cycles c_j are $\mathbb{C}$-linearly independent in the homology $H_1(S_g, \mathbb{C})$.

The real and imaginary parts of the 1-forms α_i, $i = 1, 2, \ldots, g$, build a system of $2g$ $\mathbb{R}$-linearly independent closed real 1-forms on S_g. Since $H^1_{\mathrm{dR}}(S_g, \mathbb{R})$ has real dimension $2g$, the classes of the real and imaginary parts of the 1-forms α_i, $i = 1, 2, \ldots, g$, build an $\mathbb{R}$-basis for $H^1_{\mathrm{dR}}(S_g, \mathbb{R})$. It follows that the system of forms α_i, $i = 1, 2, \ldots, g$, is a $\mathbb{C}$-basis for $\mathrm{Hol}(S_g)$. $\qquad\square$

Theorem 12.4 (Riemann's Existence Theorem). *On every compact Riemann surface S exist non-constant meromorphic functions.*

Proof. In the case $g = 0$ the mapping $z \colon S \to \mathbb{C} \cup \{\infty\}$, see above, is meromorphic and non-constant. If $g > 1$ we can construct meromorphic functions using Riemann's Theorem 12.2 on holomorphic 1-forms of type $(1, 0)$ by putting $f = \frac{\alpha_1}{\alpha_2}$, where α_1, α_2 are linearly independent holomorphic 1-forms of type $(1, 0)$. If $g = 1$ this method does not work, since the space of holomorphic 1-forms of type $(1, 0)$ is of dimension one.

We have seen that a genus one J-surface (S, J) is bi-holomorphic to a quotient $\mathbb{C}/\Gamma_\tau$. Now following Weierstrass[3] we can directly write down a Γ_τ-periodic meromorphic function on $\mathbb{C}$ by

$$
\wp(z) = \frac{1}{z^2} + \sum_{(n,m) \neq (0,0)} \frac{1}{(z + n + m\tau)^2} - \frac{1}{(n + m\tau)^2}.
$$

---

[2] Named after French musician, chemist and mathematician Alexandre-Théophile Vandermonde (1735–1796) and French mathematician Augustin-Louis Cauchy (1789–1857).

[3] Karl Theodor Wilhelm Weierstrass (1815–1897), German mathematician.

The corresponding function on the quotient $\mathbb{C}/\Gamma_\tau$ is meromorphic with one pole of order two at $0 + \Gamma_\tau \in \mathbb{C}/\Gamma_\tau$.   □

Exercise 12.5. Prove that the above expression for $\wp(z)$ defines a meromorphic function on the Riemann surface $\mathbb{C}/\Gamma_\tau$.

In fact, a meromorphic function $f : S \to \mathbb{C} \cup \{\infty\}$ on a compact Riemann surface defines a holomorphic mapping $f : S \to \mathbb{P}^1(\mathbb{C})$. For $p \in S$, $q \in \mathbb{P}^1(\mathbb{C})$, with $f(p) = q$ let z be a local coordinate on S centered at p and let w be a local coordinate on $\mathbb{P}^1(\mathbb{C})$ centered at q. Locally near p, the quantity $w(f(p'))$ admits a power series expansion $w(f(p')) = a_k z(p')^k + a_{k+1} z(p')^{k+1} + \cdots$ with $a_k \neq 0$, $k \geq 0$. The integer k that appears in this expansion is called the local multiplicity $m_p(f)$ of f at p. We denote the reduced local multiplicity $m_p(f) - 1$ by $m_p^*(f)$.

Recall that the complex projective line $\mathbb{P}^1(\mathbb{C})$, the Riemann sphere $\mathbb{C} \cup \{\infty\}$, and the oriented sphere S^2 in $\mathbb{R}^3$ are diffeomorphic by orientation-preserving diffeomorphisms. Mappings $f : S \to S^2$ underlying a meromorphic mapping $f : S \to \mathbb{C} \cup \{\infty\} = S^2$ or a holomorphic mapping $f : S \to \mathbb{P}^1(\mathbb{C}) = S^2$ are special smooth mappings.

For a smooth mapping $f : M \to N$ between connected, oriented, compact differentiable manifolds of equal dimension n the Brouwer degree $\mathrm{degr}_{\mathrm{Br}}(f)$ is the integer $\int_{[M]} f^* \omega$, where $\omega \in \Omega^n(N)$ is a volume n-differential form on N with $\int_{[N]} \omega = 1$.

Theorem 12.6 (Topology of meromorphic mappings). *Let $f : S \to S^2$ underlie a holomorphic non-constant mapping $f : S \to \mathbb{P}^1(\mathbb{C}) = S^2$. Let S^{2*} be the set of regular values of f, and put $S^* = f^{-1}(S^{2*})$. Then the following properties hold:*

 (i) *The map f has only finitely many critical points $c_1, \ldots, c_k$ and critical values $v_1, \ldots, v_h$, $h \leq k$.*
 (ii) *The Brouwer degree satisfies $\mathrm{degr}_{\mathrm{Br}}(f) \geq 1$.*
 (iii) *For $q \in S^{2*}$ the set $f^{-1}(q)$ is finite with $\mathrm{degr}_{\mathrm{Br}}(f)$ elements.*
 (iv) *For $q \in S^2$ the identity $\sum_{p \in f^{-1}(q)} m_p(f) = \mathrm{degr}_{\mathrm{Br}}(f)$ holds.*
 (v) *The difference $\mathrm{degr}_{\mathrm{Br}}(f)\chi(S^2) - \chi(S)$ equals $\sum_{p \in S} m_p^*(f)$.*
 (vi) $\mathrm{degr}_{\mathrm{Br}}(f) = \frac{1}{2}(\chi(S) + \sum_{p \in S} m_p^*(f))$
 (vii) *The restriction $f^* : S^* \to S^{2*}$ of f to S^* is a covering map with $\mathrm{degr}_{\mathrm{Br}}(f)$ sheets.*   □

The existence of non-constant meromorphic functions on S and the inequality $\mathrm{degr}_{\mathrm{Br}} \geq 1$ makes it possible to define the *gonality* $\mathrm{gon}(S)$ by

$$\mathrm{gon}(S) = \mathrm{Minimum}_f \ \mathrm{degr}_{\mathrm{Br}}(f),$$

where the minimum is taken over all meromorphic non-constant functions f.

Among compact Riemann surfaces S, the lowest possible value 1 for the gonality is only realized by the projective line $\mathbb{P}^1(\mathbb{C})$. Indeed, let $f : S \to \mathbb{P}^1(\mathbb{C})$ be meromorphic of Brouwer degree 1. From the above theorem, one concludes $m_p^*(f) = 0$, $p \in S$, and hence $\chi(S) = 2$ and $\mathrm{genus}(S) = 0$. It follows by the Riemann Uniformization Theorem that S is bi-holomorphic to $\mathbb{P}^1(\mathbb{C})$.

Theorem 12.7. *The gonality* gon(S) *equals* 2 *for all compact Riemann surfaces of genus* 1 *or* 2.

Proof. The gonality gon(S) of a compact Riemann surface of genus $g \geq 1$ is at least 2. The Weierstrass meromorphic function $\wp(z)$ on a Riemann surface of genus 1 is of Brouwer degree 2, hence gon(S) = 2 holds for all compact Riemann surfaces of genus 1.

Again, the gonality of a compact Riemann surface S of genus 2 is at least 2. Choose two linearly independent holomorphic 1-forms ω_1, ω_2 on S. The Brouwer degree of the meromorphic function $f = \frac{\omega_1}{\omega_2}$ does not exceed the number of zeros, counted with multiplicities, of ω_1, hence does not exceed $2 = -\chi(S)$. It follows that gon(S) = 2. $\square$

Given a compact connected Riemann surface S, instead of minimizing the degree one can also minimize the number of critical values of $f \colon S \to \mathbb{P}^1(\mathbb{C})$ for $f \in K(S)$. The corresponding complexity CV(S) is very interesting. Clearly:

Exercise 12.8. CV(S) ≤ 2 if and only if S is bi-holomorphic to $\mathbb{P}^1(\mathbb{C})$.

A celebrated theorem of Belyi[4] [11, 12] treats the case CV(S) ≤ 3 and shows that the condition CV(S) ≤ 3 has an important number-theoretic meaning.

Theorem 12.9 (Belyi). *A compact connected Riemann surface S is of complexity* CV(S) ≤ 3 *if and only if S is bi-holomorphic to a projective complex curve $C \subset \mathbb{P}^N(\mathbb{C})$ that is the zero set of polynomials with coefficients in a number field.*

Belyi's Theorem opens a link between combinatorics of planar graphs and Galois theory. The key word for this link is *Dessin d'Enfant* introduced by Grothendieck in his *Esquisse d'un Programme* [45]. See the books [40, 67] for a proof and much more.

Belyi's Theorem suggests a more refined complexity BC(S) for compact connected Riemann surfaces that are defined over number fields: define BC(S) to be the minimal degree of a Belyi map $f \colon S \to \mathbb{P}^1(\mathbb{C})$. This complexity gives an enumeration of compact connected Riemann surfaces defined over number fields since the number of surfaces S up to bi-holomorphic equivalence of BC complexity less than k is finite.

12.5 Why Functions?

If one wishes to study an object, a space, or the future, one tries to make measurements. For instance, if one wishes to make predictions for the future, firstly one tries to obtain a list of things, events, that can happen, and secondly one tries to estimate the probability that such an event will happen. Examples of such measurements are the measurement of the temperature of the Earth today and also, mostly indirectly by studying sediments, old ice, chronical writings, paintings, etc. in the past, hoping to

---

[4] Gennadii Vladimirovich Belyi (1951–2001), Russian-Ukrainian mathematician.

arrive at predictions concerning the future. In probability theory one introduces the space Ω of events, and studies the ring of numerical stochastic variables $X \colon \Omega \to \mathbb{R}$ and their density distributions and expectation values, hoping to obtain estimates concerning events of particular interest. In many situations funding is needed for measurements.

In topology, no funding is needed! If one wishes to study a topological space X, measurements on X are for free, namely these are all the functions $f \colon X \to \mathbb{R}$, and of course the continuous functions are of particular interest. The absence of constraints from funding allows to consider all continuous functions on the space X. The question is finally: how much information concerning the space X is carried by the ring $C^0(X, \mathbb{R})$ of continuous functions on X with its addition and multiplication? A seminal result is the following theorem of I.M. Gelfand.[5]

Theorem 12.10. *Let X be a compact metric space with metric d. Then the ring $C^0(X, \mathbb{R})$ determines the underlying topological space X in a constructive way.*

Proof. A point $p \in X$ determines the ideal $I_p = \{f \in C^0(X, \mathbb{R}) \mid f(p) = 0\}$ in the ring $C^0(X, \mathbb{R})$. The ideal I_p is maximal among all proper ideals $I \neq C^0(X, \mathbb{R})$, i.e. ideals I not containing the constant function 1. Indeed, let g be a function with $g \notin I_p$. The ideal $J = (I_p, g)$ equals the whole ring $C^0(X, \mathbb{R})$ since $e = D_p + g^2 \in J$ and $1/e \in C^0(X, \mathbb{R})$ so $1 \in J$, where D_p is the continuous function $D_p(q) = d(p, q)$.

Conversely, for every maximal proper ideal I in $C^0(X, \mathbb{R})$ there exists a point $p \in X$ with $I = I_p$. Indeed, assuming $I \neq I_p$ for all $p \in X$ will yield a contradiction. Observe that no inclusion $I \subset I_p$ or $I_p \subset I$ holds by the maximality of the ideals I and I_p. It would follow that for every $p \in X$ there exists a function $f_p \in I$ and $f_p \notin I_p$. Let U_p be the open subset in X defined by $U_p = \{q \in X \mid f_p(q) \neq 0\}$. Clearly, $\cup_{p \in X} U_p = X$. The space X being compact, there exist finitely many points $p_1, \ldots, p_k$ with $\cup_i U_{p_i} = X$. Putting $e = f_{p_1}^2 + \cdots + f_{p_k}^2 > 0$, observe that $e \in I, 1/e \in C^0(X, \mathbb{R})$, so $1 \in I$ contradicting the assumption that the ideal I is proper. Hence $I = I_p$ for some $p \in X$.

It follows that the set X can be recovered as the set of all proper maximal ideals in the ring $C^0(X, \mathbb{R})$.

For every maximal ideal I in $C^0(X, \mathbb{R})$ the quotient ring $C^0(X, \mathbb{R})/I$ is isomorphic to the field $\mathbb{R}$ of real numbers. The field of real numbers has only the identity as automorphism, so we may canonically identify $C^0(X, \mathbb{R})/I = \mathbb{R}$. This identification allows us to evaluate $f \in C^0(X, \mathbb{R})$ at a maximal ideal I by defining $f(I)$ to be the class of f modulo I in $C^0(X, \mathbb{R})/I = \mathbb{R}$. So elements of the ring $C^0(X, \mathbb{R})$ are real-valued functions on the set $\mathrm{Spec}(C^0(X, \mathbb{R}))$ of maximal ideals in $C^0(X, \mathbb{R})$. The topology on $\mathrm{Spec}(C^0(X, \mathbb{R}))$ is the coarsest such that all f in $C^0(X, \mathbb{R})$ become continuous functions $f \colon \mathrm{Spec}(C^0(X, \mathbb{R})) \to \mathbb{R}$. $\qquad\square$

The above proof uses the multiplication of functions but does not use the topology on $C^0(X, \mathbb{R})$ given by the maximum norm

$$\|f\|_{\mathrm{Max}} = \mathrm{Sup}_{p \in X} |f(p)|.$$

[5] Israel Moiseevich Gelfand (1913–2009), Russian mathematician.

The topological vector space $C^0(X, \mathbb{R})$ is a Banach space. A *topological vector space* (E, T) is a vector space E together with a topology T such that the addition $+: E \times E \to E$ and multiplication $\cdot: k \times E \to E$ are continuous, where the field k of scalars is $\mathbb{R}$ or $\mathbb{C}$. A *Banach space* is a topological vector space (E, T) such that the topology can be given by a norm $\| \cdot \|_E$ on the vector space E and such that every Cauchy sequence in E for the norm $\| \cdot \|_E$ converges in E.

Exercise 12.11. A metric space (E, d) is *complete* if every Cauchy sequence in E converges in E. Give an example of a topological space (E, T) such that the topology T underlies a metric d_1 and a metric d_2 such that (E, d_1) is complete and (E, d_2) isn't.

Exercise 12.12. Let $\| \cdot \|_1, \| \cdot \|_2$ be norms on a vector space E. Assume that the underlying topologies coincide. Show that if $(E, \| \cdot \|_1)$ is complete, then $(E, \| \cdot \|_2)$ is also complete.

A natural question is: Does the Banach space $C^0(X, \mathbb{R})$ determine the topology of X?

The answer is negative. Let X be the tree with 5 vertices of which 4 are terminal. The space X looks like the letter X. Let Y be the tree with 4 vertices of which 3 are terminal. The space Y looks like the letter Y. The topological spaces X and Y are not homeomorphic, but the Banach spaces $C^0(X, \mathbb{R})$ and $C^0(Y, \mathbb{R})$ are isomorphic as topological vector spaces.

More generally, let $\chi(X)$ denote the Euler characteristic:

Theorem 12.13 (Euler characteristic determines Banach space). *The Banach spaces $C^0(X, \mathbb{R})$ and $C^0(Y, \mathbb{R})$ of continuous functions on compact connected graphs X, Y, both having at least one edge, are isomorphic topological vector spaces if $\chi(X) = \chi(Y)$.*

The proof depends on a central theorem in topology and in the theory of Banach spaces. We explain this central theorem first. The following is our second example of a sharp compromise.

Theorem 12.14 (Banach–Schauder Isomorphism Theorem). *Let (E, T) be a Banach space with topology T. Let T' be a finer topology and T'' a coarser topology than the topology T, such that both (E, T') and (E, T'') are Banach spaces. Then the three topologies T', T, T'' coincide.*

The proof relies on the Baire property of complete metric spaces.

A subset A of a topological space X is called *meagre* if the interior of the closure of A in X is empty. This means no non-empty open subset in X is contained in the closure of A. Equivalently, a subset A is meagre in X if the interior of its complement in X is dense in X.

Exercise 12.15. The $1/3$-Cantor set C in $\mathbb{R}$ is meagre. The set C is obtained by the following process: at step 1, remove from $[0, 1]$ the $1/3$-middle open interval $]1/3, 2/3[$, at step 2, remove from the two remaining closed intervals $[0, 1/3]$ and $[2/3, 1]$ their $1/3$-middle open intervals. Continue in this way, at step $n \in \mathbb{N}$, remove the union U_n of the open middle intervals from the remaining 2^{n-1} closed intervals.

Finally, the set $C = [0, 1] \setminus \bigcup_n U_n$ is the Cantor set. The complement of C in $[0, 1]$ is open and dense in $[0, 1]$.

A topological space X has the *Baire property* if any countable union of meagre subsets of X is a meagre subset of X. Equivalently, the space X has the Baire property if any intersection of a countable family of open dense subsets of X is a dense subset of X.

The topological space $\mathbb{Q}$ of the rational numbers clearly does not have the Baire property, but as a consequence of the following seminal theorem of Baire[6] the space of real numbers $\mathbb{R}$ has the Baire property.

Theorem 12.16 (Baire). *A countable union of meagre subsets of a complete metric space is a meagre subset.*

Proof. Let $(U_n)_{n \in \mathbb{N}}$ be a countable family of open and dense subsets of a complete metric space X. We will show that for every point $p \in X$ and every open ball $b(p, 2r)$, $r > 0$, the intersection $b(p, 2r) \cap \bigcap_{n \in \mathbb{N}} U_n$ is non-empty. This proves that $\bigcap_{n \in \mathbb{N}} U_n$ is dense in X.

The open set U_1 is dense in X, so the intersection $b(p, r) \cap U_1$ is open and non-empty. So there exists a point $q_1 \in b(p, r) \cap U_1$ and radius $0 < r_1 < r/2$ such that the closed ball $B(q_1, r_1)$ is included in $b(p, r) \cap U_1$. The set U_2 is open in X, so the intersection $b(q_1, r_1) \cap U_2$ is open and non-empty, hence the existence of q_2 and $0 < r_2 < r/4$ with $B(q_2, r_2) \subset b(q_1, r_1) \cap U_2$. Inductively, one shows the existence of a sequence of points q_n and radii $0 < r_n < \frac{r}{2^n}$ with $B(q_n, r_n) \subset b(p, r) \cap \bigcap_{k=1}^{n} U_k$. The sequence $(q_n)_{n \in \mathbb{N}}$ is a Cauchy sequence with limit $q \in X$ by the completeness of the metric space X. The limit point q belongs to $\bigcap_{k=1}^{\infty} U_k$ by construction and to the closed ball $B(p, r)$ with center p and radius r since the distance $d(q_n, p)$ from q_n to p is estimated by $d(q_n, p) \leq r/2 + r/4 + \cdots r/2^n \leq r$. Hence $q \in b(p, 2r) \cap \bigcap_{n=1}^{\infty} U_n$. $\square$

Proof (Banach–Schauder Isomorphism Theorem). Let $\phi : (E, T) \to (E, T'')$ and $\psi : (E, T') \to (E, T)$ be the continuous, bijective linear maps, between Banach spaces, both given by $\phi(u) = \psi(u) = u$, $u \in E$. The inverse maps ϕ^{-1} and ψ^{-1} are clearly linear. The Banach–Schauder Theorem claims in fact that these inverse maps are moreover continuous or equivalently that the maps ϕ and ψ are open. We show here that the map ϕ^{-1} is continuous.

In order to do so we denote by X the Banach space (E, T) and by Y the Banach space (E, T''). Let $\| \cdot \|_X$ and $\| \cdot \|_Y$ be norms on X and on Y.

Let b^X be the unit ball in X. We have $\bigcup_{n \in \mathbb{N}} n b^X = X$ since a vector $u \in X$ belongs to $n b^X$ for $n > \|u\|_X$. The map ϕ is surjective, so the Banach space Y equals the countable union $\phi(\bigcup_{n \in \mathbb{N}} n b^X) = Y$. It follows by Baire's Theorem that for some $n \in \mathbb{N}$, $n \neq 0$, the set $\phi(n b^X)$ is not meagre in Y. Since the map $M_n : Y \to Y$ defined by multiplication with n is a homeomorphism that maps $\phi(b^X)$ to $\phi(n b^X)$, the set $\phi(b^X)$ is not meagre in Y. Hence, there exists $q \in Y$ and $r > 0$ such that the open ball $b^Y(q, r)$ of Y is contained in the closure $\overline{\phi(b^X)}$ in Y of $\phi(b^X)$. So also $b^Y(-q, r) \subset \overline{\phi(b^X)}$ since $-\overline{\phi(b^X)} = \overline{\phi(b^X)}$. This implies the inclusion

---

[6] René Baire (1874–1932), French mathematician, was Professor in Dijon.

$$b^Y(0, 2r) = b^Y(q, r) + b^Y(-q, r) \subset \overline{\phi(b^X)} + \overline{\phi(b^X)} = \overline{\phi(2b^X)}.$$

Up to this point we have used stretching by integral factors, that ϕ is surjective, and that the topology of the Banach space Y is the topology of a complete metric space.

Moreover, by using the homogeneity of norms and the linearity of ϕ we obtain for every $t > 0$ the inclusion

$$b^Y(0, 2rt) \subset \overline{\phi(b^X(0, 2t))}.$$

This means: $\forall q \in b^Y(0, 2rt)$ and $\forall \epsilon > 0$ there exists a $p \in b^X(0, 2t)$ such that $\|\phi(p) - q\|_Y < \epsilon$ holds.

The next claim is the inclusion $b^Y(0, r) \subset \phi(2b^X)$. Indeed, for $q \in b^Y(0, r)$ construct a sequence $(p_k) \in X$ inductively for $k = 1, 2, \ldots$. Take $t_1 = 1/2$ and $\epsilon_1 = r/2$, so choose $p_1 \in b^X(0, 1)$ with $\|q - \phi(p_1)\|_Y < \epsilon_1 = r/2$. Next, take $t_2 = t_1/2$, $\epsilon_2 = \epsilon_1/2$ and choose $p_2 \in b^X(0, 1/2)$ with $\|(q - \phi(p_1)) - \phi(p_2)\|_Y < \epsilon_2$. Continue by taking $t_{k+1} = t_k/2$, $\epsilon_{k+1} = \epsilon_k/2$ and choosing $p_{k+1} \in b^X(0, 2t_{k+1})$ with

$$\|(q - \phi(p_1) \cdots - \phi(p_k) - \phi(p_{k+1})\|_Y < \epsilon_{k+1}.$$

The series $\sum_k p_k$ converges in the complete space X with limit $p \in b^X(2)$. From linearity and continuity of ϕ one concludes $q = \phi(p)$, so the claim $b^Y(0, r) \subset \phi(b^X(0, 2))$ follows.

This implies the inclusion $\phi^{-1}(b^Y(0, r)) \subset b^X(0, 2)$, hence the estimate

$$\|\phi^{-1}(q)\|_X \le 2/r \, \|q\|_Y,$$

showing the continuity of the inverse map ϕ^{-1}. $\qquad \square$

In the above proof the injectivity of the mapping ϕ was not used. The above proof also yields:

Theorem 12.17 (Banach–Schauder Open Mapping Theorem). *Let X, Y be Banach spaces and let $\phi \colon X \to Y$ be a continuous surjective linear map. Then the map ϕ is open.*

Now we are ready to prove that the Euler characteristic determines the Banach space of continuous functions for compact connected graphs that have at least one edge.

Proof (Euler characteristic determines Banach space). First we prove that the Banach spaces of functions on the tree-like letters X, Y are isomorphic as topological vector spaces. We think of the graph X as four copies I_1, I_2, I_3, I_4 of the interval $[0, 1]$ that have the point 0 in common.

Let $C_0^0([0, 1])$ be the Banach space of continuous functions on $[0, 1]$ that vanish at the point 0.

Let I_{123} be the subtree of X consisting of the union of $I_1 \cup I_2 \cup I_3$. Define the map ϕ by

$$\phi \colon C^0(X) \to C^0(I_{123}) \oplus C_0^0([0, 1]), \quad f \mapsto (f_{123}, g),$$

where f_{123} is the restriction of f to I_{123} and where $g \in C_0^0([0,1])$ is given by $g(t) = -f(0) + f_4(t)$. Here f_i denotes the restriction of f to I_i. Clearly, the map ϕ is linear, continuous and bijective. So the Banach spaces $C^0(X)$ and $C^0(I_{123}) \oplus C_0^0([0,1])$ are isomorphic by the Banach–Schauder Theorem.

We think of the graph Y as the tree $I_{123} \cup I_4$. The union is such that the endpoint 1 of I_3 is identified with the endpoint 0 of I_4. The map

$$\psi: C^0(Y) \to C^0(I_{123}) \oplus C_0^0([0,1]), \, f \mapsto (f_{123}, g),$$

where as before f_{123} is the restriction of f and where $g \in C_0^0([0,1])$ is given by $g(t) = -f_3(1) + f_4(t)$, is linear, continuous and bijective. So the Banach spaces $C^0(Y)$ and $C^0(I_{123}) \oplus C_0^0([0,1])$ are isomorphic, and consequently the Banach spaces $C^0(X)$ and $C^0(Y)$ are isomorphic.

More generally, by the same method, one can reduce the number of terminal edges and prove that the Banach spaces of continuous functions on compact trees with at least two vertices are isomorphic topological vector spaces. More precisely, the Banach space of continuous functions on a compact tree is isomorphic to $C^0([0,1])$ or to $\mathbb{R}$.

Let $C_{0,1}^0[0,1]$ be the Banach space of continuous functions on $[0,1]$ that vanish at 0 and at 1.

Let G be a connected compact graph having at least one edge. By subdividing that edge we may assume that G has at least two vertices. We think of G as a finite union $I_1 \cup I_2 \cup \cdots \cup I_k$ of copies of $[0,1]$. The numbering is such that the union $I_{12...h}$ of the first h edges is a maximal subtree in G.

Define the map ϕ by

$$\phi: C^0(G) \to C^0(I_{12\cdots h}) \oplus \bigoplus_{h < i \le k} C_{0,1}^0[0,1], \, f \mapsto (f_{1,2,\dots,h}), g),$$

where $f_{12...h}$ is the restriction of f to the tree $I_{12...h}$ and where the i-component $g_i \in C_{0,1}^0[0,1]$ of g depends on the restriction f_i of f to I_i as follows:

$$g_i(t) = -(1-t)f_i(0) - tf_i(1) + f_i(t).$$

The map ϕ is linear, continuous and bijective. So by the Banach–Schauder Theorem and the above statement concerning trees, it follows that up to isomorphism of topological vector spaces the Banach space $C^0(G)$ depends only on $k - h = -\chi(G) - 1$. $\qquad\square$

The above together with ignorance and optimism induced the author to ask how much geometry of a compact metric space X is encoded in its Banach space $C^0(X, \mathbb{R})$.

Questions: Does $\chi(X) \ne \chi(Y)$ for connected compact graphs X, Y imply that the Banach spaces $C^0(X, \mathbb{R})$ and $C^0(Y, \mathbb{R})$ are not isomorphic as topological vector spaces? Does the Banach space of continuous functions on a connected compact graph determine the Euler characteristic of the graph? What about the same question for compact connected oriented surfaces?

I thank Nicolas Monod for his help concerning this question. He has informed me about the following result of Milyutin, see [6, 76]:

Theorem 12.18 (Milyutin, Thesis, Moscow State University, 1952). *Let X be an uncountable, metrizable, compact space. Then the Banach spaces $C^0(X, \mathbb{R})$ and $C^0([0, 1], \mathbb{R})$ are isomorphic.*

It follows that the answers to the above questions are negative. Moreover, Nicolas Monod observes that the isomorphisms between the Banach spaces of graphs with equal Euler characteristic are explicitly constructed and do not really depend upon the Banach–Schauder Theorem. So, little optimistic hope for discovering in the Banach space $C^0(X, \mathbb{R})$ the geometry of the topological space X remains, if one takes as input the Banach space $C^0(X, \mathbb{R})$ up to *explicitly constructed* isomorphisms. The problem is to define what *explicitly constructed* means.

12.6 The Field $K(S)$ of Meromorphic Functions

Let $f, g : S \to \mathbb{C} \cup \{\infty\}$ be meromorphic functions on a Riemann surface S. Their sum $f + g$ and their product fg are again meromorphic functions. The definition of the sum and product of meromorphic functions needs some care.

At $p \in S$ the value $(f + g)(p)$ is defined as $f(p) + g(p)$, if $f(p) \neq \infty \neq g(p)$. If $f(p) = \infty$, $g(p) \neq \infty$ or $g(p) = \infty$, $f(p) \neq \infty$ define $(f + g)(p) = \infty$. If $f(p) = \infty = g(p)$ choose an open neighborhood U of p in S such that the restrictions of f, g to U only have a pole at p. Let z be a coordinate function on U that is centered at p. Let k be an integer such that the functions $z^k f, z^k g$ are holomorphic on U. Define $z^{-k}(z^k f + z^k g)$ to be the restriction of $f + g$ to U. A similar construction defines the product fg as a meromorphic function on S. More explicitly, define $z^{-2k}((z^k f)(z^k g))$ to be the restriction of fg to U. Also, if a meromorphic function h is $\neq 0$, a meromorphic function $\frac{1}{h}$ can be defined as its multiplicative inverse. More explicitly, there exist $k \in \mathbb{Z}$ and $p \in U' \subset U$ such that $z^k h$ is holomorphic and nowhere 0 on the open neighborhood U' of p. Define $z^k(1/(z^k h))$ to be the restriction of $1/h$ to U'.

It follows that the set $K(S)$ of all meromorphic functions on S, together with the above sum and product operations, is a field.

The field $K(\mathbb{C} \cup \{\infty\})$ is the field of rational functions $\mathbb{C}(z)$, i.e. the field of fractions $\frac{P(z)}{Q(z)}$ where $P(Z), 0 \neq Q(Z) \in \mathbb{C}[Z]$ are polynomials in some indeterminate Z, and z is the linear coordinate function $z : \mathbb{C} \to \mathbb{C}$ with $z(1) = 1$. The group of automorphisms of the field $\mathbb{C}(z)$ is the group of homographic substitutions $z \mapsto \frac{az+b}{cz+d}$ where the matrix $\left(\begin{smallmatrix} a & b \\ c & d \end{smallmatrix} \right)$ belongs to $\mathrm{PGL}(2, \mathbb{C})$.

The holomorphic functions on S are the constant complex functions, and form a subfield of $K(S)$ canonically isomorphic to the field of complex numbers $\mathbb{C}$. Hence the field $K(S)$ of meromorphic functions is moreover a $\mathbb{C}$-algebra. We will prove that the $\mathbb{C}$-algebra $K(S)$ of meromorphic functions on a compact Riemann surface S determines the surface S. This means that the Riemann surface S can be reconstructed from the field $K(S)$ and its subfield of constant functions. A first step in the proof of this statement is the following possibility of separating pairs of points on S by values of meromorphic functions.

Theorem 12.19 (Meromorphic separation of pairs of points). *Let S be a compact, connected Riemann surface. Let p, q be distinct points on S. Then there exists a meromorphic function $g: S \to \mathbb{C} \cup \{\infty\}$ that takes distinct finite values $g(p) = 1$, $g(q) = 0$ at the points p and q.*

Proof. Let r be a third point on S. Let $\omega_{p,q}$, $\omega_{p,r}$ be meromorphic differentials with poles only at p, q and p, r respectively and residues $+1$ at p. The meromorphic function $f = \frac{\omega_{p,q}}{\omega_{p,r}}$ takes values $f(p) = 1$ and $f(q) = \infty$. The meromorphic function $g = g_{p,q} = \frac{1}{f}$ separates the points p, q with the finite values $g(p) = 1$, $g(q) = 0$. □

Theorem 12.20 (Meromorphic separation of finite sets of points). *Let F be a finite set of points on the compact Riemann surface S. There exists a meromorphic function that separates the points of F pairwise.*

Proof. Let $K_F(S)$ be the vector space of the restrictions to F of all the meromorphic functions with only finite values at F. For each pair (p, q) in F, $p \neq q$, the vector space $K_{p,q}$ of functions $f \in K_F(S)$ with $f(p) = f(q)$ is at least of codimension 1 in $K_F(S)$ by the previous theorem. Any function $g \in K(S)$ whose restriction $f \in K_F(S)$ does not lie in the finite union $\bigcup_{p,q \in F, p \neq q} K_{p,q}$ separates the points in F by distinct finite values.   □

Theorem 12.21. *Let S be a compact Riemann surface. Let $f: S \to \mathbb{C} \cup \{\infty\}$ be a non-constant meromorphic function of Brouwer degree n. The field $K(S)$ is an algebraic extension of degree n of its subfield $\mathbb{C}(f)$ generated by f. The substitution $z \mapsto f$ induces an isomorphism of fields $\mathbb{C}(z) \to \mathbb{C}(f)$.*

Proof. Given $g \in K(S)$, let $U_g \subset \mathbb{C}$ be the set of all the points p such that $f^{-1}(p)$ consists of n points and such that g on $f^{-1}(p)$ does not take the value ∞. The complement of the set U_g is finite. For $p \in U_g$ let $P(X, p) \in \mathbb{C}[X]$ be the monic polynomial of degree $\deg(f)$ having the values of g on $f^{-1}(p)$ as simple roots. The coefficients $c_i(p)$ define holomorphic functions on U_g. Moreover, the functions c_i take every value only finitely many times, in fact equal or less than n^2 times, hence the functions c_i have a meromorphic extension to $\mathbb{C} \cup \{\infty\}$. After writing $c_i(p) = \frac{P_i(z(p))}{Q_i(z(p))}$ and putting

$$P(X, Z) = X^n + \frac{P_1(Z)}{Q_1(Z)} X^{n-1} + \cdots + \frac{P_n(Z)}{Q_n(Z)} \in \mathbb{C}(Z)[X]$$

for $q \in S$ the equation

$$g(q)^n + \frac{P_1(f(q))}{Q_1(f(q))} g(q)^{n-1} + \cdots + \frac{P_n(f(q))}{Q_n(f(q))} = 0$$

holds. The equation is polynomial in g of degree n with coefficients in $\mathbb{C}(f)$. It follows that $K(S)$ is an extension of degree at most n. Choosing for $g \in K(S)$ a function that separates the n points of $f^{-1}(p)$, $p \in U_g$, with finite values the above equation has n distinct roots $g(q)$, $q \in f^{-1}(p)$, so g is not the root of a polynomial in $\mathbb{C}(f)[X]$ of degree less than n. It follows that $K(S)$ is an extension of degree n. □

Theorem 12.22. *Let $f, g \in K(S)$ be meromorphic functions. Assume that f is not constant of Brouwer degree n and that g separates the points of a fiber $f^{-1}(p)$ consisting of n points. Then the fields $K(S)$ and $\mathbb{C}(f, g)$ generated by f and g coincide.*

Proof. The function g separates the points in the fiber $f^{-1}(p)$. Replacing g by $g = \frac{g}{g-a}$, $a \in \mathbb{C}$, a not occurring as a value of g along $f^{-1}(p)$, the function g will separate the points of $f^{-1}(p)$ by finite values. Using the Vandermonde determinant, there exist uniquely defined constants $C_1, C_2, \ldots, C_n$ such that for a given $h \in K(S)$ and all $q \in f^{-1}(p)$ the identity

$$h(q) = C_1 g(q) + C_2 g(q)^2 + \cdots + C_n g(q)^n$$

holds. Let $U_{f,g}$ be the maximal open neighborhood of p in $\mathbb{C}$ such that for all $r \in U_{f,g}$ the function g separates the points of the fiber $f^{-1}(r)$ with n finite values. The complement of $U_{f,g}$ consists of finitely many points. Using the holomorphic Implicit Function Theorem, there exist holomorphic functions $c_i(r)$, $r \in U_{f,g}$, with $c_i(p) = C_i$ such that on $f^{-1}(U_{f,g})$ the identity

$$h(q) = c_1(f(q))g(q) + c_2(f(q))g(q)^2 + \cdots + c_n(f(q))g(q)^n$$

holds. They are meromorphic functions on $\mathbb{C}$ since they are finite to one on $U_{f,g}$. So, $h \in \mathbb{C}(f)[g]$, which is in fact even more than claimed.   □

Theorem 12.23. *Let S be a compact Riemann surface. There exist three meromorphic functions $f, g, h \in K(S)$ such that the product map $f \times g \times h \colon S \to \mathbb{P}^1(\mathbb{C})^3$ is a holomorphic embedding.*

Proof. Let $f \in K(S)$ be non-constant with Brouwer degree n, let $g \in K(S)$ separate the n points of a fiber $f^{-1}(p)$. Let E be the complement of $g^{-1}(U_{f,g})$, where $U_{f,g}$ as in the previous proof. The subset $E \subset S$ is finite. Let $h \in K(S)$ separate the points of E such that moreover no point of E is critical for h. The product map $f \times g \colon g^{-1}(U_{f,g}) \to \mathbb{P}^1(\mathbb{C})^2$ is a holomorphic embedding. The product map $f \times g \times h \colon S \to \mathbb{P}^1(\mathbb{C})^3$ is injective since h separates the points of E and is moreover a holomorphic embedding since h is not critical at points of E.   □

12.7 Reconstruction of the Riemann Surface S From $K(S)$ and its Subfield $K_0(S)$

Let S be a compact, connected Riemann surface. Denote by $K_0(S)$ the subfield of holomorphic functions in the field of meromorphic functions $K(S)$ on S. In a first step the set S will be identified with the set $\Sigma = \Sigma(K(S))$ of all proper subrings O that are in sandwich position

$$K_0(S) \subset O \subset K(S)$$

and satisfy the *maximal subring* property

$$f \in O \text{ or } 1/f \in O, \ f \in K(S).$$

At this stage, rings that satisfy the maximal subring condition are called *m-subrings*.

The subring $K_0(S)$ does not satisfy the maximal subring property, since all $f \in K(S) \setminus K_0(S)$ take both values 0 and ∞, so $1/f$ has a pole, hence $1/f \notin K_0(S)$. So m-subrings are in strict sandwich position.

For $p \in S$ let O_p be the subring of all meromorphic functions on S that are holomorphic on a neighborhood of p. Clearly $O_p \in \Sigma$ and moreover the map $p \in S \mapsto O_p \in \Sigma$ is injective. Indeed, for two distinct points p, q put $f_{p,q} = \frac{1}{g_{p,q}}$ where $g_{p,q}$ separates the points p, q with the values $g(p) = 1$, $g(q) = 0$ and observes $f_{p,q} \in O_p$ and $f_{p,q} \notin O_q$.

The m-subrings O have many special properties.

Theorem 12.24. *Let O be an m-subring in $K(S)$. Then the following hold:*

(1) *The ring O is a local ring. The set $m(O) = \{f \in O \mid 1/f \notin O\}$ is the unique maximal ideal in O.*

(2) *For $f \in K(S)$, $f \notin O$ it follows that $1/f \in m(O)$.*

(3) *A chain $(f_1, f_2, \ldots, f_n)$ of elements $f_i \in m(O)$ with $f_1 \neq 0$ and $f_i \in f_{i+1}O$, $i = 1, 2, \ldots, n-1$ satisfies $n \leq \text{degree}(f_1)$.*

(4) *The ideal $m(O)$ is a principal ideal.*

(5) *Let u be a generator of the ideal $m(O)$. The chain of ideals in O*

$$(u) \supset (u^2) \supset \cdots \supset (u^k) \supset \cdots$$

is not stationary.

(6) *$\bigcap_{n \in \mathbf{N}}(u^n) = \{0\}$.*

(7) *For $f \in K(S)$ there exists a $k \in \mathbf{N}$ with $u^k f \in O$.*

(8) *For $f \in K(S) \setminus O$ there exists a unique $k > 0$ with $u^k f \in O^*$.*

(9) *The m-subring O is a maximal proper subring in $K(S)$.*

(10) *The inclusion $K_0(S) \subset O$ induces the equality $K_0(S) = O/m(O)$ of fields.*

(11) *For every function $f \in O^*$ there exists a unique number $\lambda \in \mathbb{C}$ such that $f - \lambda \in m(O) = (u)$ holds.*

(12) *The zero set $\{q \in S \mid u(q) = 0\}$ of the function u is a one point set $\{p\}$ and $O = O_p$ holds.*

Proof. (1) We first show that $m(O)$ is an ideal. For $r \in O, g \in m(O)$ it follows that $(rg)^{-1} \notin O$ since otherwise $(rg)^{-1}r = 1/g \in O$, contradicting $g \in m(O)$. For $f, g \in m(O)$, $f, g \neq 0$, by the maximal subring property $f/g \in O$ or $g/f \in O$ holds. Assume $f/g \in O$. Then $f + g = (1 + f/g)g \in m(O)$.

Observe that the set $m(O)$ is the complement in O of the set of invertible elements O^*. It follows that $m(O)$ is the unique maximal ideal, since $m(O)$ is indeed an ideal.

(2) For $f \in K(S) \setminus O$ by the maximal subring property $1/f \in O$ holds and $1/f \in m(O)$ by the definition of $m(O)$.

(3) $K(S)$ is a $\mathbb{C}(f_1)$-vector space of dimension $n = \text{degree}(f_1)$. The claim follows, since elements of a chain $(f_1, f_2, \ldots, f_m)$ with $f_i \in m(O)$ and $f_1 \neq 0$ and

$f_i \in f_{i+1}O$, $i = 1, 2, \ldots, m-1$ are $\mathbb{C}(f_1)$-linearly independent. Indeed, assume the contrary. Then there exists a linear relation

$$a_1 f_1 + a_2 f_2 + \cdots + a_m f_m = 0, a_i \in \mathbb{C}(f_1).$$

After clearing denominators we may assume $a_i \in \mathbb{C}[f_1]$. After dividing the relation by a power of f_1 we achieve that not all a_i are polynomials with vanishing constant term. Let j be the maximal index such that a_j has a non-vanishing constant term. Coefficients a_i with $i > j$ can be written as $a_i = f_1 b_i$, $b_i \in \mathbb{C}[f_1]$ and the coefficient a_j as $a_j = c + f_1 b_j$, $b_j \in \mathbb{C}[f_1]$, $c \in \mathbb{C}$, $c \neq 0$. The above linear relation implies

$$-c = \sum_{i<j} \frac{f_i}{f_j} a_i + f_1 b_j + \sum_{i>j} \frac{f_1}{f_j} f_i b_i.$$

From the hypothesis $f_{i+1} \in f_i m(O)$, $i = 1, 2, \ldots, n-1$, follows $\frac{f_i}{f_j} \in m(O), i < j$. Hence, $-c \in m(O)$ in contradiction with $c \in \mathbb{C}$, $c \neq 0$.

(4) Assume that the ideal $m(O)$ is not principal. Choose $f_1 \in m(O)$. Choose for $i = 1, 3, \ldots, n = \text{degree}(f_1)$ elements $f_{i+1} \in m(O)$, $f_{i+1} \notin f_i O$. From $f_{i+1} \notin f_i O$ follows $f_{i+1}/f_i \notin O$ hence $f_i/f_{i+1} \in m(O)$ by the maximal property and the definition of $m(O)$. So, $f_i \in f_{i+1}O, i = 1, 2, \ldots, n = \text{degree}(f_1)$, contradicting the previous chain length bound.

(5) Indeed, $u^{k-1} \notin (u^k)$.

(6) Assume $0 \neq g \in \cap_{k \in \mathbb{N}} u^k O$. Put $n = \text{degree}(g)$ and $f_i = \frac{g}{u^i}$, $i = 0, 1, 2, \ldots, n$. The chain $(f_0 = g, f_1, f_2, \ldots, f_n)$ contradicts the above chain length bound.

(7) For $f \in K(S)$ assume $u^k f \notin O, k \in \mathbb{N}$. Put $f_k = (u^k f)^{-1} \in m(O)$. Observe that $f_k \in f_{k+1}, k = 0, 1, 2 \ldots$ contradicting above chain length bound.

(8) For $f \notin O$ take minimal $k \in \mathbb{N}$ with $u^k f \in O$. Clearly, $k > 0$ and $u^k f \in O^*$.

(9) For $f \notin O$ let R be the subring generated by O and f. For $k > 0$ minimal, as above, with $u^k f \in O^*$, the subring R is also generated by O and u^{-k}. For $g \in K(S)$ with $u^h g \in O, h \in \mathbb{N}$ choose $l \in \mathbb{N}$ with $lk \geq h$. Observe that $g = u^{-lk} u^{lk-h} u^h g \in R$, since $u^{-lk} \in R$ and $u^{lk-h}, u^h g \in O$. It follows that $R = K(S)$.

(10) Observe that $K_0(S) \cap m(O) \subset O^* \cap m(O) = \{0\}$. So the inclusion $K_0(S) \subset O$ induces an inclusion of fields $K_0(S) \subset O/m(O)$. Remember $K_0(S)$ is the field $\mathbb{C}$ of constant functions. Every $f \in O$, $f \neq 0$, is algebraic over the subfield $K_0(u)$, so satisfies a relation

$$a_d f^d + a_{d-1} f^{d-1} + \cdots + a_0 = 0$$

with $d \geq 1$, $a_i \in K_0(u)$. After clearing denominators and possibly after dividing by a power of u the above relation has coefficients $a_i \in K_0[u]$ such that not all constant terms of the polynomials $a_i \in K_0[u]$ vanish. Reducing modulo $m(O) = (u)$ a nontrivial polynomial relation with coefficients in $K_0(S)$ for the class of f modulo $m(O)$ follows. So $f + m(O) \in K_0(S)$ since $K_0(S)$ is algebraically closed.

(11) From $f - \lambda \in (u)$ and $f - \lambda' \in (u)$ follows $\lambda - \lambda' \in (u)$, hence $\lambda - \lambda' \notin O^*$ and $\lambda = \lambda'$, proving uniqueness. Existence follows from the equality $K_0(S) = O/m(O)$.

(12) Assume $u(q) = u(q') = 0, q, q' \in S, q \neq q'$. Consider $f \in O$ with distinct finite values at q, q'. For no $\lambda \in K_0(S)$ do both equalities $f(q) - \lambda = 0$

and $f(q') - \lambda = 0$ hold, so for no $\lambda \in K_0(S)$ does $f - \lambda \in (u) = m(O)$ hold, in contradiction with the above. □

The Ostrowski Reconstruction Theorem[7] can now be stated and proved.

Theorem 12.25 (Ostrowski Reconstruction Theorem). *Let $O \in \Sigma(K(S))$. Let $u \in m(O)$ be a generator of the ideal $m(O)$. Then the zero set of u consists of one point $p \in S$ and $O = O_p$ holds. The map $p \in S \mapsto O_p \in \Sigma$ is a bijection. Elements $f \in K(S)$ define mappings $\check{f}: \Sigma \to K_0(S) \cup \{\infty\} = \mathbb{C} \cup \{\infty\}, O \in \Sigma \mapsto \check{f}(O) = \infty$ if $f \notin O$, else $O \in \Sigma \mapsto \check{f}(O) = \lambda \in K_0(S), f - \lambda \in (u)$. The maps $\check{f}$ are continuous for the coarsest topology T_Σ on Σ. The topological space (Σ, T_Σ) is a surface and carries the unique structure of a Riemann surface for which the maps $\check{f}$, $f \in K(S)$, are holomorphic. A system of generators u_p, $p \in \Sigma = S$, of the ideals $m(O_p)$ provides a system of holomorphic local coordinates. The map $p \in S \mapsto O_p \in \Sigma(K(S))$ is an isomorphism of Riemann Surfaces.*

Proof. Clear! □

J. Kürschák[8] created the theory of valuations, see his papers [64, 65] published in 1912 and 1913. Ostrowski showed in his paper [87], published in 1913, that the valuations of the field of rational numbers correspond to prime numbers except for the usual absolute value.

In the publication [88], Ostrowski proved the above Reconstruction Theorem and showed moreover that it is also a Construction Theorem: Let K be a field of any characteristic and let L be a field extension of transcendence degree one. Then the set of maximal subrings in sandwich position between the algebraic closure $\bar{K}$ of K in L and L carries the structure of a Riemann Surface with coordinates having values in the Riemann sphere $\bar{K} \cup \{\infty\}$. He called this Riemann surface the "Absolute Riemann Surface" of the field extension L/K.

The language used by Ostrowski came from valuation theory. A (discrete) valuation on L is a surjective group homomorphism $v: L^* \to \mathbb{Z}$ satisfying $v(f + g) \geq \mathrm{Min}(v(f), v(g))$. The maximal subring O of a valuation v is the ring $\{f \in L \mid f = 0 \text{ or } v(f) \geq 0\}$. Conversely, for a maximal subring O with maximal ideal $m(O)$ generated by u, define the corresponding valuation $v(f) \in \mathbb{Z}$, $f \in L^*$, as the minimal $n \in \mathbb{Z}$ with $u^{-n} f \in O$.

In "*Sur les variétés algébriques et variétés qui s'en déduisent*" [123] André Weil took an "Absolute Riemann Surface" of a field extension L/K of transcendence degree one as the definition of a Riemann Surface, not explicitly referring to Ostrowski, and in particular, defined the genus, without using topological definitions. Moreover, he proved some of the main theorems such as the Riemann–Roch Theorem.

In the case of positive characteristic of the ground field K Weil entered new territory with the computation of the generating series of the number of points with coordinates in finite fields and proposed a series of conjectures, the so-called "Weil conjectures", that would extend his results to higher dimension. Alexandre

---

[7] Named after Alexander Markowich Ostrowski (1893–1986), a Ukrainian, educated in Germany, and later Professor at the University of Basel.

[8] József Kürschák (1864–1933), Hungarian mathematician.

Grothendieck outlined an approach to these conjectures, and Pierre Deligne, partially following this approach, gave a confirmative answer to the conjectures, see the article by Nicholas Katz in [59].

Chapter 13
Line Bundles and Cohomology

13.1 Divisors and Line Bundles

Let S be a connected Riemann surface. A divisor D on S is a function $n\colon S \to \mathbb{Z}$, $p \in S \mapsto n_p \in \mathbb{Z}$, with finite support. The support of a divisor D is the set of points at which n has value $n_p \neq 0$. A divisor can be written as a formal finite combination

$$D = \sum_{p \in S} n_p \cdot p$$

with integral coefficients of points in S.

Given two divisors D_1 and D_2 on S their sum $D_1 + D_2$ is again a divisor. The set of all divisors on S together with the sum operation is an abelian group attached to the surface S. This construction of an abelian group works for any set.

Considering divisors on Riemann surfaces is interesting mainly for the following reason: for each meromorphic function $f \neq 0$ on S is defined a divisor $(f)\colon S \to \mathbb{Z}$ by putting $(f)_p = \operatorname{order}_p(f)$.

How much information about a meromorphic function f is stored in its divisor (f)? We start with a non-promising example. The functions $f \equiv 1$ and $g = e^z$ on $\mathbb{C}$ have equal divisors $0 = (f) = (g)$. The functions f and g seem to have little in common! The divisor (f) informs us about the possible values of the function f: If $(f)_p > 0$ then $f(p) = 0$, if $(f)_p = 0$ then $f(p) \neq 0$, if $(f)_p < 0$ then $f(p) = \infty$.

The next example is more promising. Let S be a compact connected Riemann surface and f, g two meromorphic functions, both $\neq 0$, and such that $(f) = (g)$ holds. The ratio function f/g is meromorphic, $f/g \neq 0$, and from $(f/g) = (f) - (g) = 0$ it follows that f/g is holomorphic and nowhere 0 on S, hence constant $f/g \equiv \lambda \in \mathbb{C} \setminus \{0\}$, since S is compact. We can conclude from the equality of divisors $(f) = (g)$ that the functions f, g only differ by a multiplicative factor: $g = \lambda f, \lambda \in \mathbb{C}, \lambda \neq 0$.

A basic question about divisors is: which divisors appear as the divisor of a meromorphic function?

In the case $S = \mathbb{C}$ the answer is that every divisor appears as the divisor of a meromorphic function. Indeed, for the divisor D with support $\{p\}$, $p \in S$, and value $v \in \mathbb{Z}$ we have $D = ((z - z(p))^v)$. For a general divisor $D = \sum_{p \in S} n_p \cdot p$ we have

N. A'Campo, *Topological, Differential and Conformal Geometry of Surfaces*, Universitext,
https://doi.org/10.1007/978-3-030-89032-2_13

$D = (\prod_{p \in S}(z - z(p))^{n_p})$. Uniqueness up to a multiplicative factor does not hold: for instance, $\tilde{D} = (e^z \prod_{p \in S}(z - z(p))^{n_p})$ also holds. The function $\sin(z)$ has infinitely many zeros in $\mathbb{C}$, so $(\sin(z))$ is not a finite combination of points.

On a compact connected Riemann surface S not every divisor appears as the divisor of a meromorphic function. One main reason, but not the only one as we will see later, is given by the following theorem.

The degree of a divisor D on a compact surface is defined as $\deg(D) = \sum_{p \in S} D(p)$. The degree is additive: $\deg(D_1 + D_2) = \deg(D_1) + \deg(D_2)$. The degree is monotone: if D_1 is pointwise smaller than or equal to D_2, written $D_1 \leq D_2$, then $\deg(D_1) \leq \deg(D_2)$. The partial order relation $\leq$ and the addition of divisors are compatible in the sense that for three divisors A, B, C with $A \leq B$ it follows that $A + C \leq B + C$. A divisor D is called positive if $0 \leq D$ holds. These definitions and properties concerning degree, addition and order are set-theoretic. Not set-theoretic is the following important property of divisors of meromorphic functions on compact Riemann surfaces.

Theorem 13.1. *For a meromorphic function $f \neq 0$ on a compact surface S the equality $\deg((f)) = 0$ holds.*

Proof. Let $p_1, p_2, \ldots, p_n$ be the points of the support of the divisor (f) and let $z_1, z_2, \ldots, z_n$ be local coordinates in neighborhoods of the points p_l that are centered at p_l. Let $B_1, B_2, \ldots, B_n$ be a system of pairwise disjoint discs given by $|z_l| < r$ for $r > 0$ and small. Put $S_r = S \setminus \cup B_l$. The 1-differential form $\alpha = \frac{df}{f}$ is of type $(1, 0)$ and is holomorphic on S_r. Hence $d\alpha = 0$ on S_r. In each disk B_l at p_l the form α has a Laurent expansion with a pole of order 1

$$\alpha = \left(\frac{(f)_{p_l}}{z_l} + \text{higher-order terms} \cdots \right) dz_l.$$

By Stokes' Theorem and computation of residues we have

$$0 = \int_{S_r} d\alpha = \int_{\partial S_r} \alpha = -2\pi \sqrt{-1} \deg((f)).$$

This finishes the proof. $\square$

The basic question becomes: which divisors of degree 0 are divisors of a meromorphic function?

Two divisors D_1 and D_2 on a compact Riemann surface S are called linearly equivalent, and denoted by $D_1 \sim D_2$, if the difference $D_2 - D_1$ is the divisor of a meromorphic function. This definition of equivalence of divisors uses the structure of a Riemann surface on S. This equivalence relation is compatible with the basic equivalence: if $D_1 \sim D_2$, i.e. $D_2 - D_1 = (g)$, the divisor D_1 is equal to the divisor of a meromorphic function if and only if the divisor D_2 has this property. Indeed, let (f) be a meromorphic function with $D_1 = (f)$, then $(gf) = D_2 - D_1 + D_1 = D_2$. Similarly, if $D_2 = (f)$ one has $D_1 = (f/g)$.

The following definition associates to each divisor D a vector space $L(D)$ of meromorphic functions on S. The "L" comes from the German term *linearer Raum* for vector space. Let $L^*(D)$ be the set of meromorphic functions f on S with $f \neq 0$ and $-D \leq (f)$. The set $L(D)$ is obtained from $L^*(D)$ by adjoining the 0-function. One checks that $L(D)$ is a linear subspace of the space of meromorphic functions on S. The subspace $L(D)$ depends on the divisor, but if two divisors D_1, D_2 are equivalent, then the spaces $L(D_1)$ and $L(D_2)$ are isomorphic as vector spaces. Indeed, let the meromorphic function g provide this equivalence, i.e. $(g) = D_2 - D_1$, then for $f_2 \in L^*(D_2)$ we have $f_1 = g f_2 \in L^*(D_1)$ since

$$-D_1 = -D_2 + (g) \leq (f_2) + (g) = (g f_2) = (f_1).$$

Hence the multiplication m_g by g provides an isomorphism $m_g \colon L(D_2) \to L(D_1)$.

The quantity $l(D) = \dim(L(D))$, which is a natural number or eventually ∞, is constant on equivalence classes of divisors.

The second source for divisors on a Riemann surface S are the divisors (α) of meromorphic 1-differential forms α of type $(1, 0)$. Let $\alpha \neq 0$ be a meromorphic 1-form of type $(1, 0)$. In a local coordinate z at $p \in S$ the form α is written as $f dz$, where f is a local function with a Laurent expansion $f = \sum_{k=\nu}^{+\infty} a_k z^k$ with $\nu \in \mathbb{Z}$, $a_k \in \mathbb{C}$, $a_\nu \neq 0$. The value of the divisor (α) at the point p is defined as the integer ν of the above expansion.

Let α_1, α_2 be non-zero meromorphic 1-differential forms of type $(1, 0)$. Then the fraction $\frac{\alpha_2}{\alpha_1}$ is a meromorphic function f. One has the equality of divisors $(\alpha_2) = (\alpha_1) + (f)$, which shows that the divisors $(\alpha_1), (\alpha_2)$ are equivalent. In other terms, the divisors of non-zero meromorphic 1-differential forms of type $(1, 0)$ on S form an equivalence class of divisors. This class of divisors is called the *canonical class* of the Riemann surface S and is denoted by $K = K_S$.

Let D be a divisor on S. Let $I^*(D)$ be the space of meromorphic 1-differential forms $\omega \neq 0$ with $D \leq (\omega)$. Again as for $L^*(D)$, by adjoining the 0-form to $I^*(D)$ one gets a vector space $I(D)$ of meromorphic 1-differential forms. If the divisors D, D' are linearly equivalent, say $D' - D = (f)$, the spaces $I(D)$ and $I(D')$ are isomorphic vector spaces by the map $\omega \in I(D) \mapsto f \omega \in I(D')$. It follows that the quantity $i(D) = \dim(I(D))$ that is attached to a divisor D on a compact Riemann surface S is constant on the equivalence classes of the divisors D.

In the next section we will study the quantities $l(D), i(D)$ on compact Riemann surfaces S.

An important geometric object that is associated to a divisor D on a Riemann surface S is the corresponding holomorphic line bundle L_D on S. The line bundle L_D is in fact a holomorphic vector bundle on S with fiber dimension 1.

A holomorphic bundle E of dimension n on a complex manifold M of dimension m is a pair (E, π), where E is a complex manifold of dimension $m+n$ and $\pi \colon E \to M$ is a holomorphic surjective map without critical values. The manifold E is called the total space and the map π is called the projection of the fiber bundle. The fibers $E_p = \pi^{-1}(p)$, $p \in M$, are hence complex manifolds of dimension $n = \text{Dim}(E) - \text{Dim}(M)$. Such a pair (E, π) is a holomorphic vector bundle if moreover locally on M a condition is satisfied that compares the map π to a cartesian projection with fibers the space $\mathbb{C}^n$. More precisely, required is the existence of a covering $(U_\alpha)_{\alpha \in A}$ of M by open subsets U_α such that for each restriction π_α of π to $E_\alpha = \pi^{-1}(U_\alpha)$ there exists a holomorphic bijection $\phi_\alpha \colon E_\alpha \to U_\alpha \times \mathbb{C}^n$. The system $(U_\alpha, \phi_\alpha)_{\alpha \in A}$ is called an atlas of the fiber bundle . In order to be a holomorphic vector bundle the atlas is assumed to have the following properties.

(i) For each $p \in U_\alpha$, ϕ_α maps the fiber $\pi^{-1}(p) = \pi_\alpha^{-1}(p) \subset E_\alpha$ to the fiber $\{p\} \times \mathbb{C}^n \subset U_\alpha \times \mathbb{C}^n$.

(ii) For $\alpha, \beta \in A$ with $U_\alpha \cap U_\beta \neq \emptyset$ the composition of appropriate restrictions

$$\phi_\beta \circ \phi_\alpha^{-1} : (U_\alpha \cap U_\beta) \times \mathbb{C}^n \to (U_\beta \cap U_\alpha) \times \mathbb{C}^n$$

are holomorphic and linearly map fibers to fibers.

One may assume, possibly after refining the covering of the manifold M, that each U_α is the domain of a coordinate system $z_\alpha : U_\alpha \to \mathbb{C}^m$. The compositions

$$z_\alpha \times \mathrm{Id}_{\mathbb{C}^n} : U_\alpha \times \mathbb{C}^n \to \mathbb{C}^{m+n}$$

together with the gluing maps $\phi_{\beta\alpha} = \phi_\beta \circ \phi_\alpha^{-1}$ (re)define the complex manifold E.

A section of a fiber bundle (E, π) over M is a map $s : M \to E$ with $\pi \circ s = \mathrm{Id}_M$. One could say that sections are fields with values in fiber bundles. Vector fields on a manifold are sections of the tangent bundle. Most fields that we have encountered so far are sections of vector bundles.

After this intermezzo on holomorphic vector bundles we give the construction of the holomorphic line bundle L_D of a divisor D on a Riemann surface S.

Let D be a divisor on a Riemann surface S. Let $\{p_1, p_2, \ldots, p_k\}$ be the support of D. Let $\{z_1, z_2, \ldots, z_k\}$ be holomorphic local coordinates on S with $z_i(p_i) = 0$. Let $\{U_0, U_1, U_2, \ldots, U_k\}$ be the covering of S given by $U_0 = S \setminus \{p_1, p_2, \ldots, p_k\}$ and $U_i = \{p \in S \mid |z_i(p)| < r\}$, $1 \leq i \leq k$, $r > 0$, and moreover so small that $U_i^* = U_0 \cap U_i$, $1 \leq i \leq k$, are the only non-empty intersections. The total space of the holomorphic line bundle L_D is obtained by gluing the chart $U_0 \times \mathbb{C}$ to the union of the charts $U_i \times \mathbb{C}$ with the gluing maps

$$\phi_{i0} : (p, u) \in U_i^* \times \mathbb{C} \subset U_0 \times \mathbb{C} \mapsto (p, z_i^{D(p_i)} u) \in U_i^* \times \mathbb{C} \subset U_0 \times \mathbb{C}.$$

The Cartesian projections $U_i \times \mathbb{C} \to U_i$ glue together to a projection of a line bundle $L_D \to S$.

Note that the section with constant value 1 over U_0 extends to a meromorphic section s_D over S that has zeros and poles of order $D(p_i)$ at the points p_i. Hence the divisor (s_D) is precisely the divisor D. More generally, for every meromorphic function $f \in L(D)$, hence with $-D \leq (f)$, the section $f s_D$ on U_0 extends to a holomorphic section over S. Conversely, given a holomorphic section s of L_D, then the function $f = \frac{s}{s_D}$ is holomorphic on U_0 with meromorphic extension over S such that $-D \leq (f)$ holds, i.e. $f \in L(D)$. The map $s \mapsto \frac{s}{s_D}$ is an isomorphism of vector spaces from the space of holomorphic sections of L_D to the space $L(D)$.

The vector space of holomorphic sections over S of the holomorphic line bundle L is denoted by $\Gamma(S, L)$ or by $\Gamma(L)$. For a divisor D the vector spaces $L(D)$ and $\Gamma(L_D), \Gamma(S, L_D)$ are equal. In order to compute the dimension of the space of sections, it is of great help to provide the space of section with a cohomological interpretation. This is the concern of the next sections.

13.2 Čech and Dolbeault Cohomology

We start by changing the notation and denote by $\check{H}^0(S, L_D) = \Gamma(S, L_D)$ the space of sections. Also denote by $\check{H}^0(U, L_D)$ the space of holomorphic sections of L_D that are only defined in the open subset U of S and admit moreover an extension to S as a global smooth section of L_D.

The cohomology that we are going to explain is due to Eduard Čech.[1] Let $\mathcal{U} = (U_\alpha)_{\alpha \in A}$ be an open covering of S. A 0-chain c (with respect to the given covering) of holomorphic sections of L_D is defined as a system of holomorphic sections $c_\alpha \in \check{H}^0(U_\alpha, L_D)$ over U_α. Moreover, we assume for practical reasons that the holomorphic sections on an open subset have (many) smooth global extensions. Often and ambiguously we keep the same notation for such an extension. We write $c = (s_\alpha)$ for short. A very natural question concerning a 0-chain of sections is whether one can glue them together to a global section defined on S. The answer is that the gluing to a globally defined section is possible if and only the restrictions of c_α and of c_β to $U_\beta \cap U_\alpha$ are equal sections. A 1-chain b of holomorphic sections of L_D is an alternating system $b_{\beta\alpha} \in \check{H}^0(U_\beta \cap U_\alpha, L_D) = \Gamma(U_\beta \cap U_\alpha, L_D)$ of holomorphic sections over $U_\beta \cap U_\alpha$. Alternating means that $b_{\beta\alpha} = -b_{\alpha\beta}$ holds. The differential $\check{d} = \check{d}^0$ of a 0-chain $c = (s_\alpha)$ is the obstruction to gluing the locally given 0-chain of sections to a global section, hence and by definition, it is the 1-chain $\check{d}c$ given by

$$(\check{d}c)_{\alpha\beta} = -c_\alpha + c_\beta,$$

where more precisely the second member $-c_\alpha + c_\beta$ is the difference of the restrictions to $U_\beta \cap U_\alpha$ of the section c_α on U_α and c_β on U_β.

The spaces of 0-chains, 1-chains, ... have a natural vector space structure. The maps $\check{d}^0, \check{d}^1, \ldots$ are linear. The following observations are important:

$$c \text{ glues to a global section} \Leftrightarrow \check{d}^0 c = 0,$$

$$\check{H}^0(S, L_D) = \mathrm{Ker}(\check{d}^0).$$

A 2-chain a of holomorphic sections of L_D consists of an alternating system of holomorphic smoothly extendable sections $a = (a_{\gamma\beta\alpha})$ on the triple intersections $U_\gamma \cap U_\beta \cap U_\alpha$. Alternating means that a permutation of indices results in a sign change according to the signature of the permutation. The differential $\check{d} = \check{d}^1$ of a 1-chain b is the system $\check{d}b$ given by

$$(\check{d}b)_{\gamma\beta\alpha} = b_{\beta\alpha} - b_{\gamma\alpha} + b_{\gamma\beta},$$

where the right-hand side is an alternating sum of restrictions to $U_\gamma \cap U_\beta \cap U_\alpha$.

Important is the fact that $\check{d}^1 \circ \check{d}^0 = 0$, which allows to define the first Čech-cohomology group of S with coefficients in L_D as the quotient group

$$\check{H}^1(S, L_D) = \mathrm{Kernel}(\check{d}^1)/\mathrm{Image}(\check{d}^0).$$

---

[1] Eduard Čech (1893–1960), Czech mathematician.

The value of the group $\check{H}^1(S, L_D)$ may depend upon the chosen covering $\mathcal{U}$. We say that $\mathcal{U}$ is fine enough if for every finer covering $\mathcal{U}'$ the value of $\check{H}^1(S, L_D)$ and also of $\check{H}^k(S, L_D)$ is given by the covering $\mathcal{U}$. In the sequel we will work with coverings that are fine enough.

A k-chain a of holomorphic sections of L_D consists of an alternating system of holomorphic smoothly extendable sections $a = (a_{\alpha_0 \alpha_1 \cdots \alpha_k})$ on $U_{\alpha_0} \cap U_{\alpha_1} \cap \cdots \cap U_{\alpha_k}$. The co-boundary of a k-chain a is the $(k+1)$-chain $\check{d}^k a$ obtained as the restriction of a signed alternating sum of the sections given by a on the intersections of depth 1 less. The higher Čech cohomology groups are

$$\check{H}^k(S, L_D) = \text{Kernel}(\check{d}^k)/\text{Image}(\check{d}^{k-1}), k = 0, 1, 2, \ldots$$

We have already encountered the de Rham cohomology. In fact the de Rham cohomology is a tool that uses results of differential calculus for the computation of a special Čech cohomology on differentiable manifolds. Let X be a topological space. The singular Čech cohomology $\check{H}^*(X, \mathbb{R})$ with real coefficients of a topological space X is obtained by charging open sets U in X with locally constant real functions. In the case that the space X is a differentiable manifold, the locally constant real functions on an open set U in X are the functions $f: U \to \mathbb{R}$ with $df = 0$. So the space of locally constant real functions on U is the kernel of the map $d: \Omega^0(U, \mathbb{R}) \to \Omega^1(U, \mathbb{R})$. The following exercise states the de Rham theorem for the 0th- and 1th-cohomology groups.

Exercise 13.2. Prove that for a manifold X the Čech cohomology with values in the space of locally constant real functions $\check{H}^*(X, \mathbb{R})$ and the de Rham cohomology $H^*_{\text{dR}}(X)$ are naturally isomorphic, $* = 0, 1$. Hint: use differentiable partitions of unity and the fact that closed 1-differential forms are locally exact.

The Dolbeault cohomology, that we will introduce soon, is a tool that uses holomorphic calculus for the computation of the Čech cohomology $\check{H}^*(X, L)$, where X is a complex manifold and where L is a holomorphic vector bundle over X. Here we study only the case where X is a Riemann surface and L is a holomorphic line bundle.

We recall the following about functions and derivatives of functions in the case of real dimension 1. Let $f: \mathbb{R} \to \mathbb{R}$ be a (smooth) function. The derivative f' measures how far the function f is from being constant. It is important that for any given smooth function $g: \mathbb{R} \to \mathbb{R}$ one can find a function f with $f' = g$. Instead of working with the derivative f' please work with the differential $df = f'dx$. The differential df measures how far the function f is from being locally constant. Important is that given a real-valued 1-differential form $\alpha = g dx \in \Omega^1(U, \mathbb{R})$ on an open subset U of $\mathbb{R}$ one can find a function f on U with $df = \alpha$.

What about the case of complex dimension 1? Is there an operator $f \mapsto Pf$ such that for a function $f: U \to \mathbb{C}$ on U open in $\mathbb{C}$ the object Pf measures how far the function f is from being holomorphic? The answer is "yes", the operator is the Cauchy–Riemann operator $f \mapsto d''f$. The object $d''f \in \Omega^{0,1}(\mathbb{C}, \mathbb{C})$ is a 1-differential form of type $(0, 1)$ on $\mathbb{C}$ with values in $\mathbb{C}$. As in the real 1-dimensional case, the following important result holds.

Theorem 13.3 (Dolbeault Lemma). *Given a (smooth) 1-differential form $\alpha \in \Omega^{0,1}(U, \mathbb{C})$ or $\beta \in \Omega^{1,0}(U, \mathbb{C})$ on an open convex subset U of $\mathbb{C}$, there exist complex-valued functions f, g on U with $d''f = \alpha$ or $d'g = \beta$.*

Proof. For the first step, see the next theorem and exercise, and solve for a function k on U the equation $d'd''k = d'\alpha$. Observe that

$$d(d''k - \alpha) = d'(d''k - \alpha) = 0$$

since $d - d' = d''$ and d'' of the form $d''k - \alpha \in \Omega^{0,1}(U, \mathbb{C})$ vanishes. The closed 1-form $d''k - \alpha$ being exact on the convex open subset U, there exists a function h on U with $dh = d''k - \alpha$. Observe that $dh = d''h$, hence, putting $f = k - h$, we have

$$d''f = d''(k - h) = d''k - d''h = d''k - dh = \alpha.$$

The case for $\beta \in \Omega^{1,0}(U, \mathbb{C})$ is similar, first solve $d''d'k = d''\beta$. □

Theorem 13.4. *Given a smooth complex-valued function g on an open subset U of $\mathbb{C}$, there exists a smooth function f with $(\frac{\partial^2}{\partial x^2} + \frac{\partial^2}{\partial y^2})f = g$.*

Proof. For real $R > 0$ let U_R be the subset of U of points $p \in \mathbb{C}$ with $|p| < R$ such that the disk with center p and radius $\frac{1}{R}$ belongs to U. Let $q_c \in U$ be a system of points that equips each connected component c of U with a base point. Let g_R on $\mathbb{C}$ be a smooth doubly periodic function with periods $4R$ and $4Ri$, with $g_R(p) = g(p), p \in U_R$, and with $\int_{-2R}^{2R}\int_{-2R}^{2R} g_R dxdy = 0$. Using Fourier series, solve the equation $(\frac{\partial^2}{\partial x^2} + \frac{\partial^2}{\partial y^2})F_R = g_R$ for a function F_R on $\mathbb{C}$. Let $LF_R: U \to \mathbb{C}$ be the locally constant function on U with $LF_R(q_c) = F_R(q_c)$. Denote by f_R the restriction to U of $F_R - LF_R$ and put $f = \lim_{R\to\infty} f_R$, where the limit is in the sense of stationary convergence on compact subsets in U. □

Exercise 13.5. Adapt the above in order to prove: For every smooth function $g: U \to \mathbb{R}$ on an open subset U in $\mathbb{R}^n$ there exists a smooth function $f: U \to \mathbb{R}$ solving the Laplace operator equation $\Delta(f) = (\sum_{i=1}^n \frac{\partial^2}{\partial x_i^2})(f) = g$.

Exercise 13.6. Given a smooth complex-valued function g on an open subset U of $\mathbb{C}$, solve for a smooth function f on U the equation $d'd''f = gdz \wedge d\bar{z}$. Hint: solve $(\frac{\partial^2}{\partial x^2} + \frac{\partial^2}{\partial y^2})f = \frac{i}{2}g$.

We have seen that the Cauchy–Riemann operator $d'': \Omega^0(U, \mathbb{C}) \to \Omega^{0,1}(U, \mathbb{C})$, for U open in $\mathbb{C}$ or more generally for U open in a Riemann surface S, has as kernel the holomorphic functions on U. Now we ask for an operator $s \mapsto Ps$ that has as kernel the holomorphic sections of a holomorphic line bundle $\Gamma(U, L)$. Such an operator will be constructed now.

Let L be a holomorphic line bundle on a Riemann surface S. More precisely, suppose that the line bundle L is on S given by local charts $\phi_\alpha: i_\alpha^* L \to U_\alpha \times \mathbb{C}$ with holomorphic transition maps $\phi_{\beta\alpha}: U_\alpha \cap U_\beta \times \mathbb{C} \to U_\beta \cap U_\alpha \times \mathbb{C}$ of the form $(p, u) \mapsto (p, m_{\beta\alpha}(p)u)$ with $m_{\beta\alpha}: U_\alpha \cap U_\beta \to \mathbb{C}^*$ being holomorphic, and henceforth $d''m_{\beta\alpha} = 0$. A smooth section s on an open subset U of S is in the chart $\phi_\alpha\colon i_\alpha^* L \to U_\alpha \times \mathbb{C}$ given by a function $s_\alpha\colon U_\alpha \to \mathbb{C}$. On the overlap of charts $U \cap U_\alpha \cap U_\beta$ the expression transforms to $s_\beta = m_{\beta\alpha} s_\alpha$. It follows for the d''-differentials that $\mathrm{d}'' s_\beta = m_{\beta\alpha} \mathrm{d}'' s_\alpha$ holds, hence the expressions in the charts glue together as a section of $\Omega^{0,1}(U, L)$, which will be denoted by $\bar{\partial}_L s$. The resulting operator

$$\bar{\partial} = \bar{\partial}_L\colon \Omega^0(U, L) \to \Omega^{0,1}(U, L)$$

is called the Dolbeault operator. The kernel of the operator $\bar{\partial}_L\colon \Omega^0(U, L) \to \Omega^{0,1}(U, L)$ has as elements precisely the holomorphic sections $\Gamma(U, L)$ of the holomorphic line bundle over U.

The following definition and theorem is part of a more general theory due to Pierre Dolbeault[2] that holds for complex manifolds, holomorphic vector bundles and higher cohomology groups. For a Riemann surface S and holomorphic line bundle L the Dolbeault cohomology groups $H_{\mathrm{Dolb}}^*(S, L), * = 0, 1$, are defined as the kernel and cokernel of $\bar{\partial}_L\colon \Omega^0(U, L) \to \Omega^{0,1}(U, L)$.

Theorem 13.7. *For a compact Riemann surface S with holomorphic line bundle L the Čech cohomology groups and Dolbeault cohomology groups are naturally isomorphic: $H_{\mathrm{Dolb}}^*(S, L) \cong \check{H}^*(S, L), * = 0, 1$.*

Proof. From the definitions we have $H_{\mathrm{Dolb}}^0(S, L) \cong \check{H}^0(S, L)$. We propose first a definition of a linear map $w\colon H_{\mathrm{Dolb}}^1(S, L) \to \check{H}^1(S, L)$. Let $\omega \in \Omega^{0,1}(S, L)$ be a smooth form of type $(0, 1)$. The form ω is $\bar{\partial}$-closed since on a complex manifold of dimension 1 the group $\Omega^{0,2}(S, L)$ is trivial. The class $[\omega]$ of ω modulo $\bar{\partial}\Omega^{0,0}(S, L]$ is a Dolbeault cohomology class in $H_{\mathrm{Dolb}}^1(S, L)$. Let (U_α) be an open covering of S. We may assume that for each α we have a complex coordinate z_α on U_α such that $z_\alpha(U_\alpha)$ is open and convex in $\mathbb{C}$. The Dolbeault Lemma 13.3 applies, and provides a system of smooth local sections (f_α) of L with $\bar{\partial} f_\alpha = \omega$ on (U_α). In the intersections $U_\alpha \cap U_\beta$, the difference of restrictions of these sections $h_{\alpha\beta} = -f_\alpha + f_\beta$ are holomorphic and define a 1-Čech chain of holomorphic sections. This chain is a cycle and defines an element $w([\omega]) \in \check{H}^1(S, L)$.

In two steps of the proposed construction of the map w a choice was involved. First the class $[\omega]$ in $H_{\mathrm{Dolb}}^1(S, L)$ may have another representative $\omega' \in \Omega^{0,1}(S, L)$, second the Dolbeault Lemma might provide another system (f_α') of local solutions to the $\bar{\partial}$-equation.

Concerning the first choice, let $[\omega]$ and $[\omega']$ be equal in $H_{\mathrm{Dolb}}^1(S, L)$, which means $\omega' - \omega = \bar{\partial} k, k \in \Omega^{0,0}(S, L]$. Hence if (f_α) is a system as above for ω, the $(f_\alpha + k)$ will be a system for ω'. The differences in the pairwise intersections computed with (f_α) and with $(f_\alpha + k)$ agree, hence $w([\omega]) = w([\omega'])$.

Concerning the second choice, let (f_α') be a second system with $\bar{\partial} f_\alpha' = \omega$. The differences $k_\alpha = f_\alpha - f_\alpha'$ satisfy $\bar{\partial} k_\alpha = 0$ and define a 0-Čech chain $k = (k_\alpha)$ of holomorphic sections. The 1-chains of differences $h_{\alpha\beta} = f_\alpha - f_\beta$ and $h_{\alpha\beta}' = f_\alpha' - f_\beta'$ differ by the $\check{\mathrm{d}} k$, hence the classes of h and h' agree in $\check{H}^1(S, L)$.

We conclude that the linear map w is well defined and moreover canonical and also natural.

<hr>

[2] Pierre Dolbeault (1924–2015), French mathematician, was Professor in Poitiers and Paris.

The linear map w is injective. Indeed, let $\omega \in \Omega^{0,1}(S, L)$ be such that $w([\omega]) = 0$. Let $f_\alpha \in \Omega^{0,0}(S, L)$ be a smooth section of L such that on U_α the equation $\bar{\partial} f_\alpha = \omega$ holds. From $w([\omega]) = 0$ one deduces that the 1-Čech chain $(h_{\alpha\beta}) = (f_\alpha - f_\beta)$ of holomorphic sections is the $\bar{\partial}$ image of a 0-Čech chain (h_α) of holomorphic sections of L. The 0-Čech chain $(f_\alpha - h_\alpha)$ has vanishing Čech boundary, hence defines a global section $f - h$ of $\Omega^{0,0}(S, L)$ with $\bar{\partial}(f - h) = \omega$. It follows that $[\omega] = 0$ in $H^1_{\mathrm{Dolb}}(S, L)$, proving the injectivity of w.

The map w is surjective. Indeed, let $f \in \check{H}^1(S, L)$ be represented by the 1-Čech chain with vanishing Čech differential given by a system of holomorphic local sections $f_{\alpha\beta}$ on $U_\alpha \cap U_\beta$ of L. Choose a smooth partition of unity (χ_α) with Support$(\chi_\alpha) \subset U_\alpha$ and choose smooth extensions to S of the section $(f_{\alpha\beta})$. By the formula $g_\alpha = \sum_\gamma \chi_\gamma f_{\alpha\gamma}$ we define a 0-Čech chain by a system of smooth sections (g_α) on U_α with Čech differential $((\check{\mathrm{d}}g)_{\alpha\beta}) = (f_{\alpha\beta})$ since

$$(\check{\mathrm{d}}g)_{\alpha\beta} = -g_\alpha + g_\beta$$

$$= \sum_\gamma \chi_\gamma(-f_{\alpha\gamma} + f_{\beta\gamma})$$

$$= \sum_\gamma \chi_\gamma f_{\alpha\beta} = f_{\alpha\beta}.$$

The 0-Čech chain $\bar{\partial}g$ of smooth local sections of $\Omega^{0,1}(S, L)$ given by $(\bar{\partial}g_\alpha)$ has trivial Čech differential. Indeed

$$(\check{\mathrm{d}}(\bar{\partial}g))_{\alpha\beta} = \bar{\partial}(-g_\alpha + g_\beta) = \bar{\partial}f_{\alpha\beta} = 0.$$

Hence the 0-chain $(\bar{\partial}g_\alpha)$ defines a global smooth section of $\bar{\partial}g \in \Omega^{0,1}(S, L)$. One checks $w([g]) = f$. $\qquad\square$

13.3 Computations of Cohomology

As an important case and example, consider the case of the holomorphic cotangent bundle $K = T^*S$ to a Riemann surface S of genus g. The group $H^0_{\mathrm{Dolb}}(S, K) = \check{H}^0(S, K)$ consists of the space of holomorphic 1-differential forms of type $(1, 0)$, which is a complex vector space of dimension g by the previous Theorem 12.2 of Riemann that characterizes the genus analytically.

Theorem 13.8. *Let S be a Riemann surface of genus $g \geq 0$. The group $H^1_{\mathrm{Dolb}}(S, K) = \check{H}^1(S, K)$ is the complex vector space $\Omega^{1,1}(S, \mathbb{C})/\bar{\partial}\Omega^{1,0}(S, \mathbb{C})$, which is of dimension 1.*

Proof. Observe $\Omega^{0,0}(S, K) = \Omega^{1,0}(S, \mathbb{C})$ and $\Omega^{0,1}(S, K) = \Omega^{1,1}(S, \mathbb{C})$. From

$$d''d'\Omega^{0,0}(S, \mathbb{C}) \subset d''\Omega^{1,0}(S, \mathbb{C}) = d''\Omega^{0,0}(S, K) \subset \Omega^{1,1}$$

and

$$d''\Omega^{0,0}(S,K) = d\Omega^{1,0}(S,\mathbb{C}) \subset d\Omega^1(S,\mathbb{C}) \neq \Omega^2(S,\mathbb{C})$$

it follows that

$$d''d'\Omega^{0,0}(S,\mathbb{C}) = d''\Omega^{1,0}(S,\mathbb{C}) = d''\Omega^{0,0}(S,K)$$

since the sub-space $d''d'\Omega^0(S,\mathbb{C})$ in $\Omega^{1,1}(S,\mathbb{C}) = \Omega^2(S,\mathbb{C})$ has co-dimension 1 by the Laplace inversion property of the Laplacian Δ_J acting on complex-valued smooth functions. Also it follows that

$$H^1_{\mathrm{Dolb}}(S,K) = \check{H}^1(S,K) = \Omega^{1,1}(S,\mathbb{C})/d''d'\Omega^0(S,\mathbb{C})$$

is of dimension 1.   □

A further important example is the holomorphic trivial bundle $\theta = S \times \mathbb{C}$ on a Riemann surface of genus g. A section is just a holomorphic function on S, and it follows that $H^0_{\mathrm{Dolb}}(S,\theta) = \check{H}^0(S,\mathbb{C})$ is the 1-dimensional space of complex constant functions on S.

Theorem 13.9. *Let S be a Riemann surface of genus $g \geq 0$. The cohomology group $H^1_{\mathrm{Dolb}}(S,\theta) = \check{H}^1(S,\theta)$ is a complex vector space of dimension g.*

Proof. Observe that $\Omega^*(S,\theta) = \Omega^*(S,\mathbb{C})$ and the space of complex 1-differential forms decomposes as a direct sum $\Omega^1(S,\mathbb{C}) = \Omega^{1,0}(S,\mathbb{C}) \oplus \Omega^{0,1}(S,\mathbb{C})$.

The Dolbeault first cohomology group $H^1_{\mathrm{Dolb}}(S,\theta)$ is the quotient of $\Omega^{0,1}(S,\mathbb{C})$ by the image of the map $d'' = \bar\partial^{0,0}\colon \Omega^{0,0}(S,\mathbb{C}) \to \Omega^{0,1}(S,\mathbb{C})$. For $\alpha \in \Omega^{0,1}(S,\mathbb{C})$ we denote by $[\alpha]_{\mathrm{Dolb}}$ its class in $H^1_{\mathrm{Dolb}}(S,\theta)$. The integral $\int_S d\alpha$ vanishes by Stokes' Theorem, hence $d\alpha$ is in the image of the Laplace operator $\Delta = d \circ d''\colon \Omega^{0,0}(S,\mathbb{C}) \to \Omega^{1,1}(S,\mathbb{C})$. We modify the form α to $\beta = \alpha - d''f$ with $\Delta f = d\alpha$. Since the function f with $\Delta f = d\alpha$ is well defined up to a constant function, the map $\pi_c\colon \alpha \mapsto \beta$ is a projection of the space $\Omega^{0,1}(S,\mathbb{C})$ onto its subspace $\Omega^{0,1}_c(S,\mathbb{C})$ of closed 1-forms of type $(0,1)$. Note $[\alpha]_{\mathrm{Dolb}} = [\pi_c(\alpha)]_{\mathrm{Dolb}}$. Moreover, the map $\beta \in \Omega^{0,1}_c(S,\mathbb{C}) \mapsto [\beta]_{\mathrm{Dolb}} \in H^1_{\mathrm{Dolb}}(S,\theta) = \check{H}^1(S,\theta)$ is an isomorphism.

For the closed form $\beta \in \Omega^{0,1}_c(S,\mathbb{C})$ we denote by $[\beta]_{\mathrm{dR}} \in H^1_{\mathrm{dR}}(S,\mathbb{C})$ its class in the first de Rham cohomology group with complex coefficients. The map $\beta \in \Omega^{0,1}_c(S,\mathbb{C}) \to [\beta]_{\mathrm{dR}} \in H^1_{\mathrm{dR}}(S,\mathbb{C})$ is injective.

We have already studied the space $\Omega^{1,0}_c(S,\mathbb{C})$ of complex closed 1-differential forms of type $(1,0)$ in the proof of Riemann's Theorem 12.2 on holomorphic 1-forms. Remember $\mathrm{Hol}(S) = \Omega^{1,0}_c(S,\mathbb{C})$ is a complex vector space of dimension g. The complex conjugation map $\beta \in \Omega^{0,1}_c(S,\mathbb{C}) \mapsto \bar\beta \in \Omega^{1,0}_c(S,\mathbb{C})$ is an isomorphism of vector spaces. Again, the map $\bar\beta \in \Omega^{1,0}_c(S,\mathbb{C}) \to [\bar\beta]_{\mathrm{dR}} \in H^1_{\mathrm{dR}}(S,\mathbb{C})$ is injective.

It follows that the spaces $H^1_{\mathrm{Dolb}}(S,\theta) = \check{H}^1(S,\theta)$, $\Omega^{0,1}_c(S,\mathbb{C})$, $\Omega^{1,0}_c(S,\mathbb{C})$ are isomorphic complex vector spaces of dimension g.   □

With the notation of the preceding proof, denote by $H^{1,0}_{\mathrm{dR}}(S,\mathbb{C})$ and $H^{0,1}_{\mathrm{dR}}(S,\mathbb{C})$ the images of $\Omega^{1,0}_c(S,\mathbb{C})$ and $\Omega^{0,1}_c(S,\mathbb{C})$ in $H^1_{\mathrm{dR}}(S,\mathbb{C})$. The following is a first glimpse of Hodge theory.[3] For more, see Claire Voisin's book [122].

---

[3] Initiated by Scottish mathematician William Vallance Douglas Hodge (1903–1975).

Theorem 13.10 (Hodge decomposition). *The group $H_{\mathrm{dR}}^1(S, \mathbb{C})$ is the direct sum of $H_{\mathrm{dR}}^{1,0}(S, \mathbb{C})$ and $H_{\mathrm{dR}}^{0,1}(S, \mathbb{C})$.*

Proof. The group $H_{\mathrm{dR}}^1(S, \mathbb{R})$ is a real vector space of dimension $2g$, so $H_{\mathrm{dR}}^1(S, \mathbb{C})$ is a complex vector space of dimension $2g$. The subspaces $H_{\mathrm{dR}}^{1,0}(S, \mathbb{C})$ and $H_{\mathrm{dR}}^{0,1}(S, \mathbb{C})$, both of complex dimension g, have only the origin in common. The direct sum decomposition

$$H_{\mathrm{dR}}^1(S, \mathbb{C}) = H_{\mathrm{dR}}^{1,0}(S, \mathbb{C}) \oplus H_{\mathrm{dR}}^{0,1}(S, \mathbb{C})$$

follows. $\qquad\square$

13.4 More General Computation of Cohomology

Up to now we have computed the Čech and Dolbeault cohomology for the trivial line bundle θ and the canonical line bundle K on a compact connected Riemann surface. Both bundles are very special and we have used special properties for the computation of their cohomologies.

The cohomologies $H_{\mathrm{Dolb}}^*(S, L) = \check{H}^*(S, L)$ for a general holomorphic line bundle can be brought under control, mainly due to the following theorem. A Fredholm map[4] is a map $A \colon E \to F$ between Fréchet spaces, which is linear, continuous, with closed image and with finite-dimensional kernel and co-kernel.

Theorem 13.11. *Let S be a compact Riemann surface and let L be a holomorphic line bundle. The Dolbeault operator $\bar{\partial}_L \colon \Omega^{0,0}(S, L) \to \Omega^{0,1}(S, L)$ is a Fredholm map.*

Let S be a compact connected Riemann surface and let L be a holomorphic line bundle on S. The spaces of smooth sections $\Omega^{a,b}(S, L)$, $0 \le a, b \le 1$, are Fréchet spaces in a natural way given by a system of norms $\|s\|_{S,L,a,b,C^k}$, $k \in \mathbb{N}$. These norms are unique up to norm equivalence. The actual construction of the norms depends on the choice of a bundle atlas (U_α, ϕ_α), $\alpha \in A$, with finite many charts and the choice of a system (U_α') of open subsets $U_\alpha' \subset U_\alpha$ such that the open sets U_α' still cover the surface S, but in U_α have a compact closure $\bar{U}_\alpha'$. A section s of $\Omega^{a,b}(S, L)$ corresponds to a system of smooth functions or differential forms $s_\alpha \in \Omega^{a,b}(\phi_\alpha(U_\alpha), \mathbb{C})$. On the image $\phi_\alpha(U_\alpha) \subset \mathbb{C}$ complex and real coordinate functions $z_\alpha = x_\alpha + iy_\alpha$ allow us to express s_α as $s_\alpha = f_\alpha(x_\alpha + iy_\alpha)\mathrm{d}z_\alpha^a \wedge \mathrm{d}\bar{z}_\alpha^b$, where f_α is a complex function. The norm $\|s\|_{C^k} = \|s\|_{S,L,a,b,C^k}$ on $\Omega^{a,b}(S, L)$ is defined by taking the maximum over $\alpha \in A$ and $i, j \in \mathbb{N}$ with $i + j \le k$ of the absolute value of the partial derivative

$$\left| \frac{\partial^{i+j} f_\alpha}{\partial^i x_\alpha \partial^j y_\alpha}(p) \right|, \quad p \in \bar{U}_\alpha'.$$

A subset P of $\Omega^{a,b}(S, L)$ is C^k-bounded if for $k \in \mathbb{N}$ and some constant M_k the estimate

---

[4] Named after Swedish mathematician Erik Ivar Fredholm (1866–1927).

$$\|s\|_{C^k} \leq M_k, \quad s \in P,$$

holds. A subset P of $\Omega^{a,b}(S, L)$ is bounded if P is C^k-bounded for every $k \in \mathbb{N}$. The system of constants M_k may grow arbitrarily.

The norms $\|s\|_{C^k}$, $k \in \mathbb{N}$, are not equivalent but satisfy the inequalities

$$\|s\|_{C^k} \leq \|s\|_{C^{k+1}}.$$

The completion $\Omega^{a,b}_{C^k}(S, L)$ of $\Omega^{a,b}(S, L)$ with respect to the norm $\|s\|_{C^k}$ is a Banach space. The above norm inequalities induce continuous injective linear maps

$$N_{k+1} \colon \Omega^{a,b}_{C^{k+1}}(S, L) \to \Omega^{a,b}_{C^k}(S, L).$$

The maps N_{k+1} are compact, i.e. the image of a bounded set has compact closure, by the Arzelà–Ascoli theorem.[5]

The Fréchet topology on $\Omega^{a,b}(S, L)$ is the coarsest topology such that the inclusions

$$N_k \colon \Omega^{a,b}(S, L) \to \Omega^{a,b}_{C^k}(S, L)$$

are continuous. The Fréchet space $\Omega^{a,b}(S, L)$ together with the system of norms $\|s\|_{C^k}$ and the continuous linear inclusion N_k is a nuclear space, as introduced by Alexander Grothendieck, see [43].

In the space $\mathbb{R}^n$ with some norm $\| \cdot \|_{\mathbb{R}^n}$, by the Heine–Borel theorem[6] a closed subset C is compact if and only if the norm $\| \cdot \|_{\mathbb{R}^n}$ is bounded on C. In the nuclear Fréchet space $\Omega^{a,b}(S, L)$ a similar criterion holds. A closed subset C in $\Omega^{a,b}(S, L)$ is compact if and only if the subset C is bounded, i.e. is $\| \cdot \|_{C^k}$-bounded for all $k \in \mathbb{N}$. More generally, nuclear spaces among Fréchet spaces are designed by Grothendieck as spaces that share many topological properties with finite-dimensional real or complex vector spaces.

Proof. The first part shows that the dimension of the kernel of $\bar{\partial}_L$ has finite dimension. This part only uses $\| \cdot \|_{C^0}$- and $\| \cdot \|_{C^1}$-estimates. The kernel of $\bar{\partial}_L$ is precisely the space $\Gamma(S, L) \subset \Omega^{0,0}(S, L)$ of holomorphic sections of L.

The space $\Gamma(S, L)$ is complete for the C^0-norm, since the limit of a uniformly convergent sequence of holomorphic sections has as limit a holomorphic section. Hence the normed space $(\Gamma(S, L), \| \cdot \|_{C^0})$ is a Banach space.

From the Cauchy integral formula for the derivative of a holomorphic function, it follows that a C^0-bounded set in $\Omega^{0,0}(S, L)$ is also C^1-bounded and together with the Arzelà–Ascoli Theorem one concludes that the closed C^0-norm unit ball in $(\Gamma(S, L)$ is compact.

The closed unit ball in $(\Gamma(S, L), \| \cot \|_{C^0})$ being compact, one concludes by the Riesz theorem that the vector space $\Gamma(S, L)$ is of finite dimension.

The second part shows that the co-kernel of $\bar{\partial}_L$ has finite dimension and uses all $\| \cdot \|_{C^k}$-estimates.

---

[5] Named after Italian mathematicians Cesare Arzelà (1847–1912) and Giulio Ascoli (1843–1896).
[6] Named after German mathematician Eduard Heine (1821–1881) and French mathematician Félix Edouard Justin Émile Borel (1871–1956).

Let $(U_\alpha, z_\alpha)_{\alpha \in A}$ be a finite atlas for the Riemann surface S, such that the image of each coordinate function $z_\alpha(U_\alpha)$ covers in $\mathbb{C}$ the disk D_3 of radius 3. Moreover, we assume that a trivialization of the line bundle over U_α is given. Further we assume that the open sets $U'_\alpha = \{p \in U_\alpha \mid |z_\alpha(p)| < 1\}$ cover S. Let $(\phi_\alpha)_{\alpha \in A}$ be a smooth partition of unity with $\mathrm{Supp}(\phi_\alpha) \subset U'_\alpha$. Let $(\psi_\alpha)_{\alpha \in A}$ be a system of smooth functions with $\mathrm{Supp}(\psi_\alpha) \subset U''_\alpha = \{p \in U_\alpha \mid |z_\alpha(p)| < 2\}$ and with $\psi_\alpha \equiv 1$ on U'_α.

The above preparations allow the construction of an operator

$$k \colon \Omega^{(0,1)}(S, L) \to \Omega^{(0,0)}(S, L)$$

such that composition

$$\bar{\partial}_L \circ k \colon \Omega^{(0,1)}(S, L) \to \Omega^{(0,1)}(S, L)$$

is of the form $\mathrm{Id} + K$, where

$$K \colon \Omega^{(0,1)}(S, L) \to \Omega^{(0,1)}(S, L)$$

is a compact operator. The construction of k is as follows. For $\omega \in \Omega^{(0,1)}(S, L)$, using the trivialization of L above U_α, put

$$\omega_\alpha = z_{\alpha *} \phi_\alpha \omega \in \Omega^{(0,1)}(\mathbb{C}, \mathbb{C}).$$

Let $f_\alpha \in \Omega^{(0,0)}(\mathbb{C}, \mathbb{C})$ solve the local Poisson equation

$$\mathrm{d}'\mathrm{d}''(f_\alpha) = \mathrm{d}'\omega_\alpha = \mathrm{d}'(z_{\alpha *} \phi_\alpha \omega), \quad f_\alpha = G * (*\mathrm{d}'(z_{\alpha *} \phi_\alpha \omega)).$$

In the last formula for f_α there are three stars, the first means convolution of functions on $\mathbb{C}$ with the elementary solution G of the Poisson equation, the second is the Hodge star map $* \colon \Omega^{(1,1)}(\mathbb{C}, \mathbb{C}) \to \Omega^{(0,0)}(\mathbb{C}, \mathbb{C})$, the third is the push forward $z_{\alpha *} \colon \Omega^{(0,1)}(U_\alpha, \mathbb{C}) \to \Omega^{(0,1)}(\mathbb{C}, \mathbb{C})$ of forms via the coordinate function z_α. Finally, put

$$k(\omega) = \sum_\alpha \psi_\alpha z_\alpha^* f_\alpha = \sum_\alpha \psi_\alpha z_\alpha^*(G * (*\mathrm{d}'(z_{\alpha *} \phi_\alpha \omega))) \in \Omega^{(0,0)}(S, L).$$

One has

$$\bar{\partial}_L \circ k(\omega) = \omega + \sum_\alpha \bar{\partial}_L(\psi_\alpha) \wedge z_\alpha^*(G * (*\mathrm{d}'(z_{\alpha *} \phi_\alpha \omega)))$$

with

$$K(\omega) = \sum_\alpha \bar{\partial}_L(\psi_\alpha) \wedge z_\alpha^*(G * (*\mathrm{d}'(z_{\alpha *} \phi_\alpha \omega))).$$

The Fréchet norms $\| \cdot \|_{k+1}$ of the solutions f_α and also of the right-hand sides $*\mathrm{d}'(z_{\alpha *} \phi_\alpha \omega)$ of the local Poisson equations are controlled up to multiplicative constants by the norms $\| \cdot \|_k$ of ω. Since the operator K consists of wedging with $\bar{\partial}_L(\psi_\alpha)$, with similar control of the Fréchet norms, it follows that K is compact by the nuclearity of the system of C^k-norms on $\Omega^{(0,1)}(S, L)$.

Clearly, the image of $\bar{\partial}_L$ contains the image of Id + K. Hence the image of $\bar{\partial}_L \colon \Omega^{(0,0)}(S, L) \to \Omega^{(0,1)}(S, L)$ is of finite codimension and closed since the image of Id + K is of finite codimension and closed.   □

The Dolbeault differential $\bar{\partial}_L$ being a Fredholm operator one concludes:

Theorem 13.12. *Let S be a compact Riemann surface and let L be a holomorphic line bundle on S. The Dolbeault cohomology groups $H^*_{\mathrm{Dolb}}(S, L)$ are of finite dimension.*   □

The Dolbeault differential $\bar{\partial}_L = \bar{\partial}_{J,L} \colon \Omega^{(0,0)}(S, L) \to \Omega^{(0,1)}(S, L)$ depends upon the J-structure on the surface S and upon the holomorphic line bundle L. The space of J-structures on a surface of genus g is path connected. The space of holomorphic line bundles of degree d is also path connected. The index of a connected continuous family of Fredholm operators is constant. It follows that the index of $\bar{\partial}_{J,L}$ depends only on the topology of the data, hence only on the genus g of S and the degree d of L. The precise formula is given by the Riemann–Roch Theorem, that we will study later (see Section 15.1).

13.5 Roch's Inequality

Let L be a holomorphic line bundle on a Riemann surface S. There is a natural bilinear map

$$B \colon H^1_{\mathrm{Dolb}}(S, L) \times H^0_{\mathrm{Dolb}}(S, K_S \otimes L^{-1}) \to \mathbb{C}$$

induced by the integration map

$$\int_S \colon \Omega^{1,0}(S, L^{-1}) \times \Omega^{0,1}(S, L) \to \mathbf{C}$$

$$(\alpha, \omega) \mapsto \int_S \alpha \wedge \omega \in \mathbb{C}.$$

Indeed, for $\omega = \bar{\partial}_L(\sigma)$, $\beta \in \Omega^{0,0}(S, L^{-1})$ and $\alpha \in \Omega^{1,0}(S, L^{-1})$ it follows from Stokes' Theorem that

$$\int_S \alpha \wedge \omega = \int_S \alpha \wedge \bar{\partial}_L(\sigma) = \int_S \bar{\partial}_{K_S \otimes L^{-1}}(\alpha) \wedge \sigma = 0,$$

showing that the map B is well defined. The map B provides a duality. This duality is in fact non-degenerate as we will see later in Section 15.3 on Serre duality. The proof of the following theorem shows only that $H^1_{\mathrm{Dolb}}(S, L)$ separates $H^0_{\mathrm{Dolb}}(S, K_S \otimes L^{-1})$.

Theorem 13.13. *Let L be a holomorphic line bundle on a Riemann surface S. The following inequality holds:*

$$\dim(H^1_{\mathrm{Dolb}}(S, L)) \geq \dim(H^0_{\mathrm{Dolb}}(S, K_S \otimes L^{-1})).$$

Proof. For

$$\alpha \in H^0_{\text{Dolb}}(S, K_S \otimes L^{-1}) \subset \Omega^{0,0}(S, K_S \otimes L^{-1}) = \Omega^{1,0}(S, L^{-1}), \; \alpha \neq 0,$$

let $p \in S$ be such that $\alpha_p \neq 0$. Let $U \subset S$ be an open neighborhood of p in S, together with a local coordinate $z \colon U \to \mathbb{C}$ and a trivialization ϕ of the restriction of the line bundles L^{-1} and L. In this trivialization ϕ of L^{-1}, the form α has on U the expression $a dz$, where $a \colon U \to \mathbb{C}$ is a holomorphic function with $a(p) \neq 0$. Let $f \colon U \to \mathbb{C}$ be a smooth function with compact support such that $\int_U f a dz \wedge d\bar{z} = 1$. Using the trivialization ϕ for L the form $f d\bar{z}$ on U extends by 0 to $\omega \in \Omega^{0,1}(S, L)$ such that $\int_S \alpha \wedge \omega = 1$. It follows that the bilinear duality map B induces a linear injective correlation map

$$C \colon H^0_{\text{Dolb}}(S, K_S \otimes L^{-1}) \to (H^1_{\text{Dolb}}(S, L))^*.$$

The claimed inequality follows. $\qquad\qquad\qquad\qquad\qquad\qquad\qquad\qquad\square$

13.6 Line Bundles, Degree and Exact Čech Cohomology Sequences

Let S be a Riemann surface. For an open subset U of S, denote by $O(U)$ the group of holomorphic functions on U and by $O^*(U)$ the group of nowhere vanishing holomorphic functions on U. The group law on $O(U)$ is by addition of values, as on $O^*(U)$ the law is by multiplication of values. Both groups are abelian. We denote by $H^*(S, O)$ and $H^*(S, O^*)$ the corresponding Čech cohomologies of those sheaves. Finally, let $\mathbb{Z}$ be the sheaf of locally constant functions with values in $\mathbb{Z}$. So, $\mathbb{Z}(U)$ is the additive abelian group of $\mathbb{Z}$-valued continuous functions on U. The map $E \colon O(U) \to O^*(U)$, $f \mapsto e^{2\pi i f}$, is part of a short exact sequence of sheaves on S

$$0 \to \underline{\mathbb{Z}} \to O \to O^* \to 1.$$

The corresponding long exact sequence of Čech cohomology groups is

$$0 \to \check{H}^0(S, \underline{\mathbb{Z}}) \to \check{H}^0(S, O) \to \check{H}^0(S, O^*)$$
$$\to \check{H}^1(S, \underline{\mathbb{Z}}) \to \check{H}^1(S, O) \to \check{H}^1(S, O^*) \to \check{H}^2(S, \underline{\mathbb{Z}}) \to 0.$$

For a Riemann surface S of genus g all groups are known or have an interesting interpretation. Most interesting is that the group $\check{H}^1(S, O^*)$ parametrizes the space of all holomorphic line bundles on S. Indeed it follows directly from the definition of holomorphic line bundle and Čech cohomology that the group $\check{H}^1(S, O^*)$ parametrizes the space of isomorphy classes of holomorphic line bundles on S. Moreover, the map

$$\check{H}^1(S, O^*) \to \check{H}^2(S, \underline{\mathbb{Z}}) = \mathbb{Z}$$

assigns to a line bundle its degree, provided the identification $\check{H}^2(S, \mathbb{Z}) = \mathbb{Z}$ is chosen such that the oriented fundamental cycle $[S] \in \check{H}^2(S, \mathbb{Z})$ of the oriented surface S maps to $1 \in \mathbb{Z}$. The group $\check{H}^1(S, O)$ is isomorphic to the group of the first Dolbeault cohomology $H^1_{\mathrm{Dolb}}(S, \theta)$ with values in the trivial line bundle. Hence, $\check{H}^1(S, O) = \mathbb{C}^g$ holds. After replacing terms by known values in the long exact cohomology sequence, the sequence becomes

$$0 \to \mathbb{Z} \to \mathbb{C} \to \mathbb{C}^* \to \mathbb{Z}^{2g} \to \mathbb{C}^g \to \check{H}^1(S, O^*) \to \mathbb{Z} \to 0.$$

The image Γ_S of $\mathbb{Z}^{2g} \to \mathbb{C}^g$ is a lattice of maximal real rank $2g$. Hence $\mathbb{C}^g / \Gamma_S$ is compact and homeomorphic to the torus $(S^1)^{2g}$ of real dimension $2g$. From this we obtain

Theorem 13.14. *The topological group $\check{H}^1(S, O^*)$ and hence the space of isomorphy classes of line bundles on S are isomorphic to $(S^1)^{2g} \times \mathbb{Z}$.*   □

The group $\check{H}^1(S, O^*)$ is called the Picard group[7] $\mathrm{Pic}(S)$ of the Riemann surface S.

Exercise 13.15. Show that the group law on $\check{H}^1(S, O^*)$ corresponds to the tensor product of holomorphic line bundles. Hint: use $(fg) = (f) + (g)$.

Denote by $\mathrm{Pic}_d(S)$ the space of holomorphic line bundles of degree $d \in \mathbb{Z}$. The space $\mathrm{Pic}_0(S)$ is a subgroup of $\check{H}^1(S, O^*)$. Topologically it is a real torus $(S^1)^{2g}$ of dimension $2g$. It has a natural complex structure since $\mathrm{Pic}_0(S) = \mathbb{C}^g / \Gamma_S$. This complex structure is special, as we will see later. It is a very special *abelian variety*.

Let S be a Riemann surface and let $p \in S$ be a point. We denote by θ_p the line bundle to the divisor $D = p$ and by 1_p a non-zero holomorphic section of θ_p that vanishes at p.

Let L be a holomorphic line bundle on S and define L_p to be the line bundle $L_p = L \otimes \theta_p$. For each open subset $U \subset S$ define the linear map $m_p(U) \colon \Gamma(L, U) \to \Gamma(L_p, U)$ by $s \in \Gamma(L, U) \mapsto s1_p \in \Gamma(L_p, U)$. The maps $m_p(U)$ are injective. A map $m_p(U)$ is surjective if and only if $p \notin U$. If $p \in U$, the cokernel $S_p(U)$ of $m_p(U)$ is isomorphic as a vector space to the 1-dimensional fiber $F_p(L)$ of the line bundle L over p and if $p \notin U$ the cokernel $S_p(U)$ is the 0-dimensional vector space. The assignment $U \mapsto S_p(U)$ defines a sheaf S_p on S. Sheaves of this type are called *sky scraper* sheaves.

The maps $m_p(U)$ define in fact a morphism m_p of sheaves between the sheaves $\Gamma(L)$ and $\Gamma(L_p)$ of holomorphic sections of the line bundles L and L_p. The kernel of m_p is the sheaf assignment $U \mapsto \{0\}$, the cokernel of m_p is the sheaf S_p.

The short exact sequence of sheaves

$$0 \to \Gamma(L) \to \Gamma(L_p) \to S_p \to 0$$

leads to a long exact sequence in Čech cohomology:

$$0 \to \check{H}^0(S, L) \to \check{H}^0(S, L_p) \to \check{H}^0(S, S_p)$$
$$\to \check{H}^1(S, L) \to \check{H}^1(S, L_p) \to \check{H}^1(S, S_p) \to 0$$

---

[7] Named after French mathematician Charles Émile Picard (1856–1941).

and also to the additivity of Euler characteristics of the three sheaves. In particular:

Theorem 13.16 (Additivity of Euler Characteristics).

$$\dim(\check{H}^0(S, L_p)) - \dim(\check{H}^1(S, L_p))$$
$$= (\dim(\check{H}^0(S, L)) - \dim(\check{H}^0(S, L))) + 1.$$

Proof. One computes:

$$\dim(\check{H}^0(S, S_p)) - \dim(\check{H}^1(S, S_p)) = 1 - 0 = 0. \qquad \square$$

13.7 Intermezzo: Global Infinitesimal Deformations of Locally Rigid Structures

We have seen several examples of locally rigid structures σ on smooth manifolds M: J-fields on surfaces, surfaces with Riemannian metrics of fixed Gaussian curvature, integrable J-fields on manifolds, symplectic manifolds and manifolds equipped with a volume form. We should not forget smooth manifolds which have no such extra locally rigid structure. In this section we follow the seminal work of Kodaira and Spencer on infinitesimal deformations of a structure on a compact manifold.

A deformation of the structure σ on M is a family σ_t, $t \in [0, \epsilon]$, $\epsilon > 0$, $\sigma = \sigma_0$, of structures of the same type. For instance, a smooth family J_t of J-fields on a surface. One way of thinking of a deformation of (M, σ) on a compact manifold M is as follows. First, choose a smooth finite atlas $(U_\alpha, \phi_\alpha)_{\alpha \in A}$ with charts $\phi_\alpha : U_\alpha \to (E, \sigma)$ where (E, σ) is the model space for the given type of structure and with transition maps $\phi_{\beta\alpha} : \phi_\alpha(U_\alpha \cap U_\beta) \to \phi_\beta(U_\beta \cap U_\alpha)$ that preserve the structure σ. The deformation will not consist in deforming the structure inside each chart, but will keep the structure inside each chart fixed and deform instead the gluing maps $\phi_{\beta\alpha}$. Second, choose for each $\alpha \in A$ an open subset $U'_\alpha \subset U_\alpha$ such that $\bar{U}'_\alpha \subset U_\alpha$ and $\bigcup_{\alpha \in A} U'_\alpha = M$ hold. Third, choose on $U_\alpha \cap U_\beta$ a time-dependent vector field $X_{\alpha\beta} = X_{\alpha\beta;t}$ such that the corresponding flow $\Phi_t^{X_{\alpha\beta}}$ preserves the structure σ. Choose $\epsilon > 0$ such that for all $t \in [0, \epsilon]$, all $\alpha, \beta \in A$, the inclusions $\Phi_t^{X_{\alpha\beta}}(U'_\alpha \cap U'_\beta) \subset U_\alpha \cap U_\beta$ hold. Finally, for $t \in [0, \epsilon]$ glue $p \in U'_\alpha$ to $q = \Phi_t^{X_{\alpha\beta}}(p)$ if $q \in U'_\beta$.

To ensure that this gluing defines a symmetric relation we need the property $X_{\alpha\beta} = -X_{\beta\alpha}$. Let $\phi_{\beta\alpha,t}$ be the corresponding gluing maps. In order to get a manifold we also need the transitivity property $\phi_{\gamma\alpha,t} = \phi_{\gamma\beta,t} \circ \phi_{\beta\alpha,t}$, which is more difficult to achieve. However, at first order at $t = 0$ this will be achieved if the above choice of vector fields satisfies $X_{\alpha\beta;0} - X_{\alpha\gamma;0} + X_{\beta\gamma;0} = 0$. The given initial data $X_{\alpha\beta;0}$ that extend to a family $X_{\alpha\beta;t}$ with both properties are called integrable.

Sometimes, the deformation σ_t induced by the system $(X_{\alpha\beta;t})$ of vector fields will in fact be induced by a flow Ψ_t on M. So $\sigma_t = \Psi_t.\sigma$. This happens if there exist σ-preserving vector fields $X_\alpha = X_{\alpha;t}$ on U_α such that $X_{\alpha\beta} = X_\beta - X_\alpha$ holds on $U_\alpha \cap U_\beta$.

Let $(TM)_\sigma$ be the sheaf of locally defined vector fields that generate σ-preserving flows.

The datum $Z = (Z_{\alpha\beta}) = (X_{\alpha\beta;0})$ defines a Čech cocycle with values in $(TM)_\sigma$. The corresponding cohomology class $KS(\sigma_t) = [Z] \in \check{H}^1(M, (TM)_\sigma)$ is called the Kodaira[8]–Spencer class of the deformation [61].

We will not study the Kodaira–Spencer class, but show by examples that this theory completes the work of Darboux, Ehresmann and Moser.

The case $\sigma = C^\infty$: By the Theorem of Ehresmann the structure of being a smooth compact manifold M has no infinitesimal deformation, clearly in accordance with $\check{H}^1(M, (TM)_\sigma) = 0$. The sheaf $(TM)_\sigma$ admits smooth partitions of unity, so all the Čech cohomology groups $\check{H}^*(M, (TM)_\sigma)$, $* > 0$, vanish, see [122].

The case when σ is given by a J-field on a surface: The sheaf $(TS)_\sigma$ is the sheaf of holomorphic vector fields on (S, J). On a compact surface S of genus g, the group $\check{H}^1(S, (TS)_\sigma)$ is a complex vector space of dimension 0 if $g = 0$, of dimension 1 if $g = 1$, of dimension $3g - 3$ if $g \geq 2$ by the Riemann–Roch theorem, confirming that those are the dimensions of the corresponding Teichmüller spaces. The group $\check{H}^1(S, (TS)_\sigma)$ is dual to the group $\check{H}^0(S, K_S^2)$ of holomorphic quadratic differentials on (S, J).

The case when σ is given by a symplectic form ω on a compact smooth manifold M: In this case the sheaf $(TM)_\sigma$ is the sheaf of smooth vector fields X such that locally the contraction $i_X\omega$ is the differential of a smooth function. In other words, the vector field X is locally the symplectic gradient $\mathrm{grad}_\omega(f)$ of a function f, i.e. locally $i_Y df = \omega(X, Y)$ holds. So we have a short exact sequence $0 \to A \to B \to (TM)_\sigma \to 0$ of sheaves, A being the sheaf of locally constant real functions, B the sheaf of smooth functions. The long exact sequence in cohomology shows the isomorphism $\check{H}^1(M, (TM)_\sigma) \sim H_{dR}^2(M)$ confirming with Darboux and Moser that the only way of deforming the symplectic structure, up to symplectomorphism, is by deforming the periods. This means by families of symplectic forms $\omega_t = \omega + t\eta$, where η is a closed form, t small, and $0 \neq [\eta]$ in $H_{dR}^2(M)$.

13.8 Hyperelliptic Curves

We have seen that compact Riemann surfaces of gonality 1 are isomorphic to the complex projective line $\mathbb{P}^1(\mathbb{C})$. Compact connected Riemann surfaces S of gonality 2 are called *hyperelliptic surfaces* if the genus exceeds 1. In the case of genus 1 the surface is called an *elliptic surface*. In the proof of the Riemann Existence Theorem 12.4 we used the surface $S_g \subset \mathbb{P}^2(\mathbb{C})$, $g > 0$, given by the inhomogeneous equation $y^2 = x^{2g+1} - 1$. The surface S_g is of genus g and the coordinate function x induces a holomorphic map $x \colon S_g \to \mathbb{P}^1(\mathbb{C})$ of degree 2. More generally, surfaces $S_P \subset \mathbb{P}^2(\mathbb{C})$ given by the inhomogeneous equation $y^2 = P(x)$, where $P(X)$ is a polynomial in X of degree $2g + 1$, $g > 0$, with $2g + 1$ distinct roots, is an elliptic or hyperelliptic

---

[8] Named after 小平邦彦 Kunihiko Kodaira (1915–1997), Japanese mathematician, and Donald Clayton Spencer (1912–2001), American mathematician.

surface of genus g. We may normalize the polynomial, choosing it as being monic and having $0, 1$ as roots. Such normalized polynomials depend upon $2g - 1$ constants.

Theorem 13.17. *Every hyperelliptic or elliptic surface S of genus g is isomorphic to a curve $S_P \subset \mathbb{P}^2(\mathbb{C})$ with inhomogeneous equation $y^2 = P(x)$, where $P(X)$ is a normalized polynomial of odd degree $2g + 1$.*

Proof. Let $f: S \to \mathbb{P}^1(\mathbb{C})$ be holomorphic of degree 2. At each critical point $p \in S$ of f the local expression of f is $f(q) = z_p(q)^2$, where z_p is a local coordinate on S at p.

The number k of critical values equals the number of critical points. Indeed, it follows that the Euler characteristic $\chi(S) = 2 - 2g$ equals $k + 2(2 - k)$, hence $k = 2g + 2$.

Post composing f with an automorphism of $\mathbb{P}^1(\mathbb{C}) = \mathbb{C} \cup \{\infty\}$ allows us to choose f such that $0, 1, \infty$ are among the critical values. Let $P(X)$ be the monic polynomial having as roots the $k - 1 = 2g + 1$ critical values in $\mathbb{C}$.

The Riemann surfaces S and S_P are isomorphic. The following construction of an isomorphism $\phi: S \to S_P$ uses as input the degree two mappings $f: S \to \mathbb{P}^1(\mathbb{C})$ and $y: S_P \to \mathbb{P}^1(\mathbb{C})$, which have the same set C of critical values. Remember that C has an even number $2g + 2$ of elements. Think of $\mathbb{P}^1(\mathbb{C})$ as the round two-sphere S^2 with great circles as geodesics. Choose $b \in \mathbb{P}^1(\mathbb{C})$ such that no great circle through b intersects C in two or more points. In particular $b \notin C$. Choose $p \in S$ with $f(p) = b$ and choose $q \in S_P$ with $y(q) = b$. Let $\bar{p} \in S$ and $\bar{q} \in S_P$ be such that $p \neq \bar{p}$, $f(\bar{p}) = b$ and $q \neq \bar{q}$, $y(\bar{q}) = b$.

Define $\phi(p) = q$ and $\phi(\bar{p}) = \bar{q}$.

Every great circle E through b with $E \cap C = \emptyset$ divides the set C into two sets C'_E and C''_E of equal parity. The circle E is called even/odd if the number of elements of C'_E is even/odd. The preimages by f or y in S or S_P of E are the disjoint union of two/one simply closed curves on S or S_P according to E being even/odd.

If a great circle E through b intersects C, then the preimages by f or y are the disjoint union of two intervals, both copies of $E \setminus C$.

Extend ϕ on $S \setminus f^{-1}(C)$ by extending ϕ for each great circle E through b to $\phi_E: f^{-1}(E) \to y^{-1}(E)$ such that $y \circ \phi_E = f$ holds. The extensions ϕ_E of ϕ over E exist uniquely.

The obtained extension is a holomorphic bijective map

$$\phi: S \setminus f^{-1}(C) \to S \setminus y^{-1}(C)$$

and finally, extends to a holomorphic bijection $\phi: S \to S_P$ with $y \circ \phi = f$ by Riemann's Removable Singularities Theorem. □

Exercise 13.18. Let S be the Riemann surface of genus $g \geq 1$ defined by the hyperelliptic equation $y^2 = P(x)$, $\deg(P) = 2g + 1$. Points in the support of the divisor (P) correspond to a set W of $2g + 2$ points on S. Show that compositions $h_{x(p),x(q)} \circ x$ of the Green function $h_{x(p),x(q)}$, $p \neq q \in W$, on $\mathbb{P}^1(\mathbb{C})$ and the coordinate x are Green functions on S. Construct special Green functions on the surface defined by $y^2 = x(x^{2g} - 1)$. Construct many special Green functions on the surfaces defined by $y^2 = x^3 - 1$ and $y^2 = x(x^4 - 1)$.

Chapter 14
Moduli Spaces and Teichmüller Spaces

14.1 Teichmüller Spaces as Smooth Manifolds

We have already encountered a moduli space: the hyperbolic plane $\mathbb{H}_I$ is the space of all surjective ring homomorphism $\mathbb{R}[X] \rightarrow F$ up to an equivalence relation, where F is a field isomorphic to the field of complex numbers. From the first instance of its birth $\mathbb{H}_I$ was just a set, but surprisingly this set carries in a natural way interesting structures such as a Riemannian metric, a J-field, a geometry that satisfies the first 4 Euclidean axioms and a new version of the fifth axiom.

The hyperbolic three space $\mathbb{H}^3$ is the moduli space for all Riemannian metrics of curvature $+1$ in the conformal class of a fixed conformal 2-sphere.

Let S be a compact connected oriented surface of genus g. The space $\mathbb{J}(TS)$ of J-fields on S is a topological space, see Section 5.6. The topology is given by a metric. On $\mathbb{J}(TS)$ acts the group $\mathrm{Diff}^+(S)$ of orientation-preserving diffeomorphisms of S. The quotient space $\mathbb{M}_g = \mathbb{J}(TS)/\mathrm{Diff}^+(S)$ is called the moduli space in genus g. The quotient space $\mathbb{T}_g = \mathbb{J}(TS)/\mathrm{Diff}_0^+(S)$ of $\mathbb{J}(TS)$ by the group $\mathrm{Diff}_0^+(S)$ of diffeomorphisms that are homotopic to the identity is called the Teichmüller space in genus g. These are quick definitions that do not yet take into account the rich amount of structure that is carried by these spaces.

The Uniformization Theorem gives an alternative definition of the moduli and Teichmüller spaces $\mathbb{M}_g$, $\mathbb{T}_g$, $g \geq 2$, namely let $\mathrm{Hyp}(S)$ be the space of all fields of scalar products $p \in S \mapsto g_p \in \mathrm{Sym}^2_{>0}(T_p(S))$ which have constant Gaussian curvature equal to -1. The groups $\mathrm{Diff}^+(S)$ and $\mathrm{Diff}_0^+(S)$ act on $\mathrm{Hyp}(S)$ and the moduli and Teichmüller space are the quotient spaces $\mathbb{M}_g = \mathrm{Hyp}(S)/\mathrm{Diff}^+(S)$ and $\mathbb{T}_g = \mathrm{Hyp}(S)/\mathrm{Diff}_0^+(S)$, respectively. In short, for the case $g \geq 2$ one takes profit from the fact that the spaces $\mathbb{J}(TS)$, $Z(S)$ and $\mathrm{Hyp}(S)$ are naturally identified.

The Uniformization Theorem defines an identification between $\mathbb{J}(TS)$ and $\mathrm{Hyp}(S)$, which we use in order transport the topology of $\mathbb{J}(TS)$ to $\mathrm{Hyp}(S)$. The moduli space and Teichmüller space are equipped with the quotient topology.

An orientation-preserving diffeomorphism ϕ of S that maps each curve c of a marking such that $\phi(c)$ is isotopic to c, is isotopic to the identity of S.

The next theorem defines a canonical structure of a smooth manifold on the Teichmüller spaces $\mathbb{T}_g$, $g \geq 2$.

N. A'Campo, *Topological, Differential and Conformal Geometry of Surfaces*, Universitext, https://doi.org/10.1007/978-3-030-89032-2_14

Theorem 14.1. *Let S be an oriented compact smooth surface of genus $g \geq 2$. A marking $m = (P, P')$ on S defines a homeomorphism*

$$\pi_m \colon \mathbb{T}_g \to\,]-\infty, +\infty[^P \times\,]-\infty, +\infty[^{P'} = \mathbb{R}^{6g-6}.$$

Given two markings m, n, the composition $\pi_n \circ \pi_m^{-1}$ is smooth. The family of maps $\pi_m \colon \mathbb{T}_g \to\,]-\infty, +\infty[^P \times\,]-\infty, +\infty[^{P'} = \mathbb{R}^{6g-6}$, considered as a smooth atlas, define a natural smooth structure on the Teichmüller space $\mathbb{T}_g$.

Proof. Let $(S, (P, P'))$ be a pair consisting of a smooth surface S of genus $g \geq 2$ and a marking $m = (P, P')$ on it. A Riemannian metric $\mu \in \mathrm{Hyp}(S)$ defines a new marking $m_\mu = (\bar{P}, \bar{P}')$ by replacing each curve c in m by the closed geodesic $\bar{c}$ for the metric μ that is isotopic to c.

Denote by $l_P(\mu)$ the string of numbers

$$l_P(\mu) = (\log(\mathrm{Length}_\mu(\bar{c}_i))_{1 \leq i \leq 3g-3} \in \mathbb{R}^{3g-3}.$$

A pair of curves $\bar{c}_i, \bar{c}'_i$ intersect in one or two points. Let C_i be a connected component of the complement of $\cup_j \bar{c}_j$, i.e. a pair of pants, having $\bar{c}_i$ twice or once in its closure. Let $s_{1,i}, s_{2,i}, s_{3,i}, s_{4,i}$ be four sectors that appear near those intersection points and that belong to C_i. We order these sectors such that the angles of sector $s_{1,i}$ and $s_{3,i}$ increase (strictly) from 0 to π by positive Dehn twisting along the curve $\bar{c}_i$. We denote by $\theta_\mu(s_{1,i}), \theta_\mu(s_{3,i}) \in\,]0, \pi[$ the angles measured with respect to the metric μ and by $\theta_{P,P'}(\mu)$ the string of real numbers

$$\theta_{P,P'}(\mu) = \left(\arctan \left(\frac{1}{2} \left(\theta_\mu\left(s_{1,i}\right) + \theta_\mu\left(s_{3,i}\right) \right) - \frac{\pi}{2} \right) \right)_{1 \leq i \leq 3g-3} \in \mathbb{R}^{3g-3}.$$

The maps l_P and $\theta_{P,P'}$ are constant along orbits of the group $\mathrm{Diff}_0(S)$, hence define a map $\pi_m = l_P \times \theta_{P,P'} \colon \mathbb{T}_g \to \mathbb{R}^{6g-6}$.

Using hyperbolic trigonometry one proves that π_m is a homeomorphism. Again by hyperbolic trigonometry one proves that the composition $\pi_n \circ \pi_m^{-1}$ is smooth, see [37], hence the family (π_m) is a natural smooth atlas on the Teichmüller space $\mathbb{T}_g$.

$\square$

14.2 The Space $\mathbb{J}_\mu(TS_g)$ as a Symplectic Product

For $g \geq 2$ and S_g a differentiable surface of genus g, from the identification of $\mathbb{J}(TS_g)$ and $\mathrm{Hyp}(S_g)$ by the Uniformization Theorem we deduce a smooth identification of the quotient spaces $\mathbb{J}(TS_g)/\mathrm{Diff}_0(S_g) = \mathrm{Hyp}(S_g)/\mathrm{Diff}_0(S_g)$. So, conformal methods and hyperbolic methods provide a combined method of study of the Teichmüller space $\mathbb{T}_g$.

The quotient map $\pi_{P,P'} \colon \mathrm{Hyp}(S_g) \to \mathbb{R}^{6g-6}$ previously constructed from a marking (P, P') of S_g is a submersion. It follows that the quotient map $\pi \colon \mathbb{J}(TS_g) \to \mathbb{T}_g$ is a submersion too.

The Uniformization Theorem defines a map $v\colon \mathbb{J}(TS_g) \to \mathrm{Vol}_\chi(S_g)$. Indeed for $J \in \mathbb{J}(TS_g)$ define $v(J) \in \mathrm{Vol}_\chi(S_g)$ to be the unique volume form v such that $g(v, J)$ is the hyperbolic metric in the conformal class of J. By unicity, it follows that the map v is $\mathrm{Diff}_0(S_g)$-equivariant. The map v is surjective, since the action of the group $\mathrm{Diff}_0(S_g)$ on $\mathrm{Vol}_\chi(S_g)$ is transitive by Moser's Global Rigidity Theorem 6.18. For each $J \in \mathbb{J}(TS_g)$ there exists a $\phi \in \mathrm{Diff}_0(S_g)$ with $\phi_* J \in \mathbb{J}_\mu(TS_g)$. Indeed, choose ϕ such that $\phi^* v(J) = \mu$ holds.

For $\mu \in \mathrm{Vol}_\chi(S_g)$ define $\mathbb{J}_\mu(TS_g) = v^{-1}(\mu) \subset \mathbb{J}(TS_g)$. Define $\pi_\mu\colon \mathbb{J}_\mu(TS_g) \to \mathbb{T}_g$ to be the restriction of π, which is surjective by the previous remark. The map π_μ is the quotient map for the action of $\mathrm{Diff}_{\mu,0}(S_g)$ on $\mathbb{J}_\mu(TS_g)$. Here $\mathrm{Diff}_{\mu,0}(S_g)$ is the group of isotopies that preserve the volume form μ.

The 2-form

$$\omega^{\circ,\mu}(H, K) = \int_{[S_g]} \omega^\circ_J(H, K)\mu$$

restricts on $\mathbb{J}_\mu(S_g)$ to a symplectic form. It is clearly non-degenerate since the function $\omega^\circ_J(H, J \circ H)$ is positive on an open non-empty subset in S_g if $0 \neq H \in T_J \mathbb{J}_\mu(S_g)$. The form is closed since the integrand is a closed form with values in the function space $C^\infty(S, \mathbb{R})$ and since the volume form μ satisfies $L_F \mu = 0$ for all vector fields F on $\mathbb{J}_\mu(S_g)$.

Theorem 14.2. *The fibers of π_μ are $\mathrm{Diff}_{\mu,0}(S_g)$-orbits, which are $\omega^{\circ,\mu}$-symplectic submanifolds in $(\mathbb{J}_\mu(TS_g), \omega^{\circ,\mu})$. The J-field J° on $\mathbb{J}_\mu(TS_g)$ leaves invariant the tangent spaces of the $\mathrm{Diff}_{\mu,0}(S_g)$-orbits.*

Proof. Let H be a tangent vector at $J \in \mathbb{J}_\mu(TS_g)$ to the $\mathrm{Diff}_{\mu,0}(S_g)$-orbit through J. The tangent vector H can be realized as the speed vector $\dot{\Phi}^Z(0)(J)$ of the flow Φ^Z on $\mathbb{J}(TS_g)$ that is induced from a flow ϕ^Z of a vector field Z on the surface S_g. It follows that

$$H = H^Z = \frac{\mathrm{d}}{\mathrm{d}t}\phi^Z(-t) \circ J_{\phi^Z(t)} \circ \phi^Z(t)_{|t=0}.$$

Remember, H is a field of endomorphism of TS_g that anti-commutes with J. We compute the field H locally on S_g. The computation simplifies on a local coordinate chart $U \subset \mathbb{R}^2$ on which J is constantly given by the matrix $\left(\begin{smallmatrix} 0 & -1 \\ 1 & 0 \end{smallmatrix}\right)$. Indeed, in the chart U, the field H is given by

$$H = -\mathrm{D}Z \circ J + J \circ \mathrm{D}Z,$$

where $\mathrm{D}Z$ is the differential of $Z\colon U \to \mathbb{R}^2$. It follows that the speed vector $\dot{\Phi}^{J(Z)}(0)(J)$ of the flow $\Phi^{J(Z)}$ is the tangent vector

$$J \circ H = -\mathrm{D}(J(Z)) \circ J + J \circ \mathrm{D}(J(Z))$$

and hence, that the J-field J° leaves invariant the tangent spaces to the $\mathrm{Diff}_0(S_g)$-orbits. It proves in fact that those orbits are J°-holomorphic submanifolds in $(\mathbb{J}_\mu(TS_g), J^\circ)$.

The restriction of $\omega^{\circ,\mu}$ to a $\mathrm{Diff}_{\mu,0}(S_g)$-orbit is a closed two form. This form is non-degenerate since for tangent vectors $H = H^Z \neq 0$, $H^{J(Z)} = J^\circ(H)$ to such an orbit we have that the function

$$p \in S_g \mapsto (\omega^\circ(H, J^\circ(H)))_p = \text{Trace } H_p \circ H_p$$

is non-negative and positive at $p \in S_g$ if $H_p \neq 0$, so $\omega^{\circ,\mu}(H, J^\circ(H)) > 0$.   □

Let τ_μ be the distribution given by

$$J \in \mathbb{J}_\mu(TS_g) \mapsto \tau_{\mu,J} = (T_J \mathbb{J}_\mu(TS_g))^{\omega^{\circ,\mu,\perp}}.$$

So, $\tau_{\mu,J} = \{H \in T_J \mathbb{J}_\mu(TS_g) \mid \omega_J^{\circ,\mu}(H, Z) = 0, \ Z \in T_J \text{Orb}_\mu(J)\}$ is a linear subspace of $T_J \mathbb{J}_\mu(TS_g)$ on which the restriction R of the differential $(D\pi_\mu)_J$ is an isomorphism $R \colon \tau_J \to T_{\pi(J)} \mathbb{T}_g$. The distribution τ_μ is $\text{Diff}_\mu(S_g)$-invariant and J°-invariant.

Theorem 14.3. *The distribution τ_μ satisfies the Frobenius integrability condition. The restriction of π_μ to each leaf $\Lambda \subset \mathbb{J}_\mu(TS_g)$ of the corresponding foliation Φ is a diffeomorphism $\pi_\Lambda \colon \Lambda \to \mathbb{T}_g$. The foliation Φ is $\text{Diff}_\mu(S_g)$-invariant. The tangent space of each leaf is J°-invariant. The restriction of $\omega^{\circ,\mu}$ to a leaf is a symplectic form.*

Proof. For vector fields X, Y on $\mathbb{T}_g$ let $X^{\tau_\mu}, Y^{\tau_\mu}$ be vector fields on $\mathbb{J}_\mu(TS_g)$ in the distribution τ_μ that map by $D\pi_\mu$ to the fields X, Y. The Frobenius integrability condition (Theorem 2.23) is satisfied if for all fields X, Y as above the vector field $[X^{\tau_\mu}, Y^{\tau_\mu}]_{\text{Lie}}$ is in the distribution τ_μ. The condition will be satisfied if for all vector fields Z on S_g the equality $\omega^{\circ,\mu}(\dot{\Phi}^Z, [X^{\tau_\mu}, Y^{\tau_\mu}]_{\text{Lie}}) = 0$ holds. Here Φ^Z is, as above, the flow

$$(t, J) \in \mathbb{R} \times \mathbb{J}_\mu(TS_g)) \mapsto \phi^Z(t)_* J \in \mathbb{J}_\mu(TS_g)$$

on $\mathbb{J}_\mu(TS_g)$ generated by the vector field Z and its flow ϕ^Z on S_g. The needed equality follows from

$$
\begin{aligned}
0 = {}& d\omega^{\circ,\mu}(\dot{\Phi}^Z, X^{\tau_\mu}, Y^{\tau_\mu}) \\
= {}& \dot{\Phi}^Z \omega^{\circ,\mu}(X^{\tau_\mu}, Y^{\tau_\mu}) - X^{\tau_\mu} \omega^{\circ,\mu}(\dot{\Phi}^Z, Y^{\tau_\mu}) + Y^{\tau_\mu} \omega^{\circ,\mu}(\dot{\Phi}^Z, X^{\tau_\mu}) \\
& - \omega^{\circ,\mu}([\dot{\Phi}^Z, X^{\tau_\mu}]_{\text{Lie}}, Y^{\tau_\mu}) + \omega^{\circ,\mu}([\dot{\Phi}^Z, Y^{\tau_\mu}]_{\text{Lie}}, X^{\tau_\mu}) \\
& - \omega^{\circ,\mu}([X^{\tau_\mu}, Y^{\tau_\mu}]_{\text{Lie}}, \dot{\Phi}^Z)
\end{aligned}
$$

since in last sum the first five terms vanish. In fact, the form $\omega^{\circ,\mu}$ and the vector fields X^{τ_μ}, Y^{τ_μ} are $\text{Diff}_\mu(S_g)$ invariant, so the first term $\dot{\Phi}^Z \omega^{\circ,\mu}(X^{\tau_\mu}, Y^{\tau_\mu})$ and the Lie brackets in terms 4 and 5 vanish. The other two vanish by the definition of the distribution τ_μ.

The foliation by the free $\text{Diff}_\mu(S_g)$ orbits and the foliation Φ generated by the distribution τ_μ are supplementary. Each orbit and each leaf intersect in one point.

The restrictions of $\omega^{\circ,\mu}$ and of J° induce a Kähler structure on each leaf. The foliation, including the leaf-wise Kähler structure, is $\text{Diff}_\mu(S_g)$-invariant. The leaf-wise restrictions of π transport the Kähler structure to a natural Kähler structure on the Teichmüller space $\mathbb{T}_g$ that is a mapping class group invariant.   □

As a main corollary:

Theorem 14.4. *The pair $(J^\circ, \omega^{\circ,\mu})$ defines a natural Kähler structure on $\mathbb{J}_\mu(TS)$. If S is compact of genus $g \geq 2$ the structure $(J^\circ, \omega^{\circ,\mu})$ induces a natural Kähler structure on the Teichmüller space $\mathbb{T}_g$ and on the moduli space M_g.*   □

In the above proof it is shown that the space $\mathbb{J}_\mu(TS_g)$, $g \geq 2$, is homeomorphic to the product of the group $\mathrm{Diff}_{\mu,0}(S_g)$ with its smooth Fréchet topology and the Teichmüller space $\mathbb{T}_g$. More precisely:

Theorem 14.5. *Assume* $g \geq 2$. *The* $\mathrm{Diff}_{\mu,0}(S_g)$-*orbit decomposition of the free action on* $\mathbb{J}_\mu(TS_g)$ *is a foliation by* $\omega^{\circ,\mu}$-*symplectic leafs. The* $\omega^{\circ,\mu}$-*orthogonal distribution* τ_μ *is integrable and defines a foliation by* $\omega^{\circ,\mu}$-*symplectic leafs. Both foliations define on* $(\mathbb{J}_\mu(TS_g), \omega^{\circ,\mu})$ *a symplectic product decomposition that is symplectomorphic to* $\mathrm{Diff}_{\mu,0}(S_g) \times \mathbb{T}_{S_g}$.

14.3 The Space $\mathbb{J}(TS)$ as a Product With Three Factors

The space $\mathrm{Vol}_\chi(S)$ is a convex set, in particular the points ω_0, ω_1 are connected by the segment $(1-t)\omega_0 + t\omega_1$. The map $v : \mathbb{J}(TS) \to \mathrm{Vol}_\chi(S)$ is a trivial fibration. The method of Moser, at the price of the choice of a base point $v \in \mathrm{Vol}_\chi(S)$ and $J_0 \in \mathbb{J}_v(TS)$, yields a trivialization $\mathbb{J}(TS) \overset{v}{=} \mathbb{J}_v(TS) \times \mathrm{Vol}_\chi(S)$. The field J_0 allows us to make a uniform choice of the 1-form α, see Moser's proof, by putting $\alpha_\mu = d(\Delta_{J_0}^{-1}(\mu - v))$ for $\mu \in \mathrm{Vol}_\chi(S)$.

The product decomposition of Fréchet manifolds follows

$$\mathbb{J}(TS_g) \overset{v}{=} \mathbb{J}_v(TS_g) \times \mathrm{Vol}_\chi(S_g) = \mathrm{Diff}_{v,0}(S_g) \times \mathbb{T}_g \times \mathrm{Vol}_\chi(S_g).$$

Define on $\mathbb{J}(TS)$ the 2-differential form $\omega^{\circ,v}$ by

$$\omega_J^{\circ,v(J)}(H,K) := \int_{[S]} \omega_J^\circ(H,K)v(J), H, K \in T_J\mathbb{J}(TS).$$

The restriction of the non-degenerate 2-differential form $\omega^{\circ,v}$ on $\mathbb{J}(TS_g)$ to the free $\mathrm{Diff}_0(S_g)$ orbits is non-degenerate too. It follows that the $\omega^{\circ,v}$-perpendicular distribution τ to these orbits is contained in the tangent spaces to the fibers of the map v. Moreover, at each point $J \in v^{-1}(\mu)$ the equality $\tau_J = \tau_{\mu,J} \subset T_J\mathbb{J}_\mu(TS_g)$ holds. It follows that the distribution τ is integrable with leafs that are copies of $\mathbb{T}_g$.

The two transversal foliations by $\mathrm{Diff}_0(S_g)$ orbits and τ-leafs give a Cartesian product decomposition on $\mathbb{J}(TS_g)$.

The spaces $\mathbb{J}(TS_g)$, $\mathbb{T}_g$ and $\mathrm{Vol}_\chi(S_g)$ are contractible. Hence, the spaces $\mathbb{J}_\mu(TS_g), \mu \in \mathrm{Vol}_\chi(S)$, are contractible. So, in particular, see [9, 24, 35, 66], we have the following theorem.

Theorem 14.6 (Rheinhold Baer, Jean Cerf, J.C. Earle and J. Eells, Yves Ladegaillerie). *The groups* $\mathrm{Diff}_0(S_g)$, $\mathrm{Diff}_{v,0}(S_g)$, $g \geq 2$, *with their smooth Fréchet topologies are contractible.* □

Exercise 14.7. Let S be a connected surface. Let δ be an oriented tangent direction at a point $p \in S$. Show that the group $\mathrm{Stab}_{\mathrm{Diff}_0(S)}(\delta)$ is contractible.

14.4 The Geometry of Tangent Vectors to a Teichmüller Space

Fields H of endomorphisms of the tangent bundle TM of a manifold M can be interpreted as 1-differential forms on M with values in TM. So, $H \in \Omega^1(M, TM)$. Consequently, a tangent vector $H \in T_J(\mathbb{J}(TS))$ to the space $\mathbb{J}(TS)$ of J-fields on the surface S can be interpreted as $H \in \Omega^1(S, TS)$. Such a field H is special since pointwise $J \circ H + H \circ J = 0$ holds. So, more precisely $H \in \Omega_J^{0,1}(S, TS)$ holds.

Let S be a smooth surface of genus $g \geq 2$. Tangent vectors h at $J \in \mathbb{T}_g$ to the Teichmüller space $\mathbb{T}_g$ of S can be lifted along the $\mathrm{Diff}_0(S)$-orbit of J as a vector field H^τ with values in the distribution $\tau = \omega^{\circ, \mu, \perp}$ on $T(\mathbb{J}(TS))$ of the previous section. The distribution τ is $\mathrm{Diff}_0(S)$-invariant, hence h can be represented uniquely by a closed 1-differential form $H^\tau \in \Omega_J^{0,1}(S, TS)$.

Locally, after having chosen a complex coordinate z for (S, J), the field H^τ can be written as $\theta \frac{\partial}{\partial z} d\bar{z}$, where θ is a complex anti-holomorphic function, since from $d(\theta \frac{\partial}{\partial z} d\bar{z}) = (\frac{\partial}{\partial z} \theta) \frac{\partial}{\partial z} dz \wedge d\bar{z} = 0$ follows $d_J' \theta = 0$.

In local isothermal coordinates $z = x + yi$ the real 2×2 matrix of H^τ on the basis $(\frac{\partial}{\partial x}, \frac{\partial}{\partial y})$ of TS is symmetric and traceless $\left(\begin{smallmatrix} a & b \\ b & -a \end{smallmatrix} \right)$, where a, b are real functions on S.

The two representations of H^τ are related by $a = \mathrm{Re}(\theta)$, $b = \mathrm{Im}(\theta)$. From $dH^\tau = 0$ follows that the functions a, b are J-harmonic, and that the anti-Cauchy–Riemann equations $\frac{\partial}{\partial x} a = \frac{\partial}{\partial y} b$, $\frac{\partial}{\partial y} a = -\frac{\partial}{\partial x} b$ hold.

The 2×2 matrix $\left(\begin{smallmatrix} a & b \\ b & -a \end{smallmatrix} \right)$ of H^τ has two real J-orthogonal eigendirections with eigenvalues $\pm \sqrt{a^2 + b^2}$.

The eigendirections integrate on S to two J-perpendicular 1-foliations with singularities. From $dH^\tau = 0$ follows that the foliations are measured in the sense of William Thurston,[1] see [37].

Exercise 14.8. The above measured foliations $F_\pm$, as discovered by Thurston, have isolated singularities. At a singular point p of F there exists a local holomorphic coordinate z, centered at p, such that the tangent space $T_q F_\pm$, $q \neq p$, is the real span of the vectors $u \in T_q S$ with $\pm z(q)^k (dz(u))^2 \in \mathbb{R}_{>0}$, $k = 1, 2, \cdots$.

The deformation corresponding to tH^τ of J_p at p in S stretches the the infinitesimal J_p circle in $T_p S$ to an ellipse by factors $e^{+t\sqrt{a(p)^2 + b(p)^2}}$ and $e^{-t\sqrt{a(p)^2 + b(p)^2}}$ in the directions of the corresponding eigendirections.

Riemann's Theorem 12.2 about the equality of genus and the dimension of the space $\mathrm{Hol}(S)$ of holomorphic differentials, i.e. the subspace of closed forms in $\Omega_J^{1,0}(S, \mathbb{C})$, has a companion:

Theorem 14.9. *The space $K(S, J)$ of closed forms in $\Omega_J^{0,1}(S, TS)$ on a Riemann surface (S, J) of genus $g \geq 2$ has complex dimension $3g - 3 = \mathrm{Dim}(\mathbb{T}_g)$.* $\square$

As explained above, the space $K(S, J)$ can be identified with the tangent space $T_J \mathbb{T}_g$ at J to the Teichmüller space. The space of quadratic differentials $L(K_{(S,J)}^{\otimes 2})$ is identified with the dual space $(T_J \mathbb{T}_g)^*$. The contraction map C,

---

[1] William Paul Thurston (1946–2012), American mathematician.

$$(f\mathrm{d}z^2, \theta\tfrac{\partial}{\partial z}\mathrm{d}\bar{z}) \in L(K^{\otimes 2}_{(S,J)}) \times K(S,J) \xmapsto{\ C\ } f\theta\mathrm{d}z \wedge \mathrm{d}\bar{z} \in \Omega_J^{1,1}(S,\mathbb{C})$$

locally obtained by the contraction $C\colon \mathrm{d}z(\tfrac{\partial}{\partial z}) \mapsto 1$ and composed with the integration $\int_S$,

$$f\theta\mathrm{d}z \wedge \mathrm{d}\bar{z} \in \Omega_J^{1,1}(S,\mathbb{C}) \xmapsto{\ \int_S\ } \int_S f\theta\mathrm{d}z \wedge \mathrm{d}\bar{z} \in \mathbb{C}$$

induces a non-degenerate duality

$$B = \int_S \circ\, C\colon L(K^{\otimes 2}_{(S,J)}) \times K(S,J) \to \mathbb{C}$$

between $\mathbb{C}$-vector spaces of equal finite dimension $3g-3$. The duality B is compatible with the hermitian Petersson scalar product[2] [93] on $L(K^{\otimes 2}_{(S,J)})$ and the Hermitian structure underlying $(\omega^{\circ,\perp}, J^{\circ})$. In conclusion

Theorem 14.10. *The Kähler structure[3] $(\omega^{\circ,\perp}, J^{\circ})$ on $\mathbb{T}_g$, $g \geq 2$, is equal to the Weil–Petersson Kähler structure.*

See the seminal papers [127, 128] by Scott Wolpert on the metric study of Teichmüller and moduli spaces.

---

[2] Named after German mathematician Hans Petersson (1902–1984).
[3] Named after German mathematician Erich Kähler (1906–2000).

Chapter 15
Dimensions of Spaces of Holomorphic Sections

15.1 The Riemann–Roch Theorem

The goal of this section is the Riemann–Roch Theorem on compact connected Riemann surfaces. We first formulate the classical Riemann–Roch Theorem:

Theorem 15.1 (Classical Riemann–Roch). *Let S be a connected compact Riemann surfaces of genus g. Let D be a divisor on S. The quantities $l(D), i(D)$ are natural numbers and the following holds*

$$l(D) - i(D) = \deg(D) - g + 1.$$

Moreover, $i(D) = l(-D + K_S)$ holds.

Here is the statement of the cohomological Riemann–Roch Theorem:

Theorem 15.2 (Cohomological Riemann–Roch). *Let S be a connected compact Riemann surfaces of genus g. Let L be holomorphic line bundle on S with Dolbeault operator $\bar{\partial}_L$. Then*

$$\mathrm{index}(\bar{\partial}_L) = \dim(H^0_{\mathrm{Dolb}}(S, L)) - \dim(H^1_{\mathrm{Dolb}}(S, L)) = \deg(L) - g + 1.$$

The first version concerns divisors D, the second version concerns holomorphic line bundles L on Riemann surfaces S. To a divisor D on S we have associated a holomorphic line bundle L_D. So far we have not yet shown the converse: every holomorphic line bundle L on S is obtained from a divisor D on S. In order to do so we need to prove that every holomorphic line bundle L on S has a meromorphic section. This fact will follow from the cohomological Riemann–Roch Theorem.

Since for a divisor D on S we have $\deg(D) = \deg(L_D)$, we observe that the right-hand sides in both Riemann–Roch Theorems coincide. Further, remember that by definition $l(D) = \dim(H^0_{\mathrm{Dolb}}(S, L_D))$ holds.

The main differences are manifested by the terms $i(D) = l(-D + K_S)$ and $\dim(H^1_{\mathrm{Dolb}}(S, L(D)))$ and also that in the cohomological Riemann–Roch Theorem the left-hand side has the interpretation of an Euler characteristic in the Dolbeault cohomology and also as the index of a Fredholm operator. Both interpretations will

© The Author(s), under exclusive license to Springer Nature Switzerland AG 2021
N. A'Campo, *Topological, Differential and Conformal Geometry of Surfaces*, Universitext,
https://doi.org/10.1007/978-3-030-89032-2_15

be important for the proofs and for later generalizations to holomorphic manifolds and holomorphic bundles of higher dimensions.

Proof (Cohomological Riemann–Roch). As explained above, it suffices to prove for each pair (g, d) the cohomological Riemann–Roch Theorem for only one Riemann surface of genus $g \in \mathbb{N}$ and only one holomorphic line bundle L of degree $d \in \mathbb{Z}$ on it. For the cases $g \geq 0$, $d = 0$, one can take for L the trivial bundle θ on any Riemann surface S of genus g. Observe,

$$\dim(H^0_{\text{Dolb}}(S, L)) = 1, \ \dim(H^1_{\text{Dolb}}(S, L)) = g, \ \deg(L) = 0$$

so the cohomological Riemann–Roch Theorem

$$\dim(H^0_{\text{Dolb}}(S, \theta)) - \dim(H^1_{\text{Dolb}}(S, \theta)) = 0 - g + 1$$

holds in this case.

Let $p \in S$ be a point and let θ_{dp} be the line bundle to the divisor $D = dp$, $d \in \mathbb{Z}$. So $\theta_0 = \theta$. From the additivity of Euler characteristics in short exact sequences one deduces that the cohomological Riemann–Roch holds for (S, θ_d) if and only if it does for $(S, d + 1)$. It follows by up-down induction on d with anchor at $d = 0$ that the cohomological Riemann–Roch holds for all $d \in \mathbb{Z}$.   □

Proof (Classical Riemann–Roch). Let L_D be a holomorphic line bundle to the divisor D on a Riemann surface S of genus g. Recall the definition $l(D) = \dim(H^0_{\text{Dolb}}(S, L_D))$. From the Roch inequality

$$\dim(H^1_{\text{Dolb}}(S, L_D)) \geq \dim(H^0_{\text{Dolb}}(S, K_S - L_D)) = l(K_S - D)$$

and the cohomological Riemann–Roch theorem for L_D follows the inequality

$$l(D) - l(K_S - D) \geq l(D) - \dim(H^1_{\text{Dolb}}(S, L_D)) = \deg(D) - g + 1.$$

Applying this inequality to the line bundle $L_{K_S - D}$ yields

$$l(K_S - D) - l(D) \geq \deg(K_S - D) - g + 1$$

hence, remembering $\deg(K_S) = 2g - 2$, yields by "magic" the opposite inequality

$$l(D) - l(K_S - D) \leq -\deg(K_S - D) + g - 1 = \deg(D) - g + 1.$$

Both inequalities imply the equality of the classical Riemann–Roch theorem for the divisor D.   □

Weaker statements of both Riemann–Roch Theorems give the inequalities first proved by Riemann:

$$\dim(H^0_{\text{Dolb}}(S, L)) \geq \deg(L) - g + 1,$$

$$l(D) \geq \deg(D) - g + 1.$$

The excess term $i(D) = l(-D + K_S)$, called the irregularity, was discovered by Roch. The cohomological interpretation of the excess term as the equality

$$i(D) = l(-D + K_S) = \dim(H^1_{\text{Dolb}}(S, L_D))$$

was discovered by Serre. Putting all of this together, both Riemann–Roch Theorems become equivalent statements.

The classical Riemann–Roch Theorem dates back to the work of Riemann (1857) and Roch[1] (1865). The cohomological Riemann–Roch Theorem dates back to the work of Dolbeault and Serre (1953).

The classical proof is very interesting. We sketch some of its ingredients, which are clearly the forerunners of later concepts.

The first ingredient is the definition of a vector space of dimension $\deg(D)$ for a positive divisor D on the surface S. This classical definition is a forerunner of what nowadays would be called a *categorification of the degree*. For the case of positive divisors D the vector space in question is the space $PP(D)$ of principal parts.

The second ingredient is a linear map

$$R \colon L(D) \to PP(D)$$

with 1-dimensional kernel. This already implies the estimate

$$l(D) \le \deg(D) + 1$$

showing that the vector space $L(D)$ has finite dimension.

The third ingredient is a linear map

$$\text{Res}_D \colon \text{Hol}(S) \to PP(D)^*$$

that gives at most

$$i(D) = \dim(\text{Res}_D(\text{Hol}(S))) \le \dim(\text{Hol}(S)) = g$$

sufficient conditions for a principal part $\pi \in PP(D)$ to be realized as a principal part of a meromorphic function $f \in L(D)$.

From the inequality $i(D) \le \dim(\text{Hol}(S)) = g$ one obtains for positive divisors D the Riemann inequality:

$$l(D) \ge \deg(D) - g + 1.$$

The fourth ingredient uses more properties of the map Res_D and sharpens the Riemann inequality to the Riemann–Roch inequality

$$l(D) - i(D) \ge \deg(D) - g + 1.$$

The fifth ingredient extends the Riemann–Roch inequality with an inductive argument from positive divisors to all divisors.

---

[1] Gustav Adolph Roch (1839–1866), German mathematician.

The sixth ingredient uses the involution $D \mapsto -D + K_S$ on divisors and classes of divisors. This involution changes the quantity $l(D) - i(D)$ to its opposite $l(-D + K_S) - i(-D + K_S) = -l(D) + i(D)$. The involution also changes $\deg(D) - g + 1$ to its opposite $\deg(-D + K_S) - g + 1 = -\deg(D) + g - 1$ but does not flip the orientation of the inequality. It follows magically that the Riemann–Roch inequality is an equality.

A meromorphic function f on a Riemann surface S has locally at each of its poles $p \in S$ a Laurent expansion

$$f = \sum_{k=\mathrm{ord}_p(f)}^{+\infty} a_k z_p^k,$$

where z_p is a holomorphic local coordinate at p. The part of this expansion with negative exponents

$$PP_p(f) = \sum_{k=\mathrm{ord}_p(f)}^{-1} a_k z_p^k$$

is the principle part $PP_p(f)$ of f at p.

For a positive divisor D let $PP(D)$ be the vector space of systems of Laurent polynomials with negative exponents at each point $p \in \mathrm{Supp}(D)$ and orders at p not lower than $-D(p)$. Clearly, the dimension of $PP(D)$ equals $\deg(D)$. So, the construction of $PP(D)$ is a categorification of $\deg(D)$, in the case when the divisor D is positive.

Assume that D is a positive divisor. Let

$$R_D : L(D) \to PP(D)$$

be the map

$$f \in L(D) \mapsto (PP_p(f))_{p \in \mathrm{Supp}(D)} \in PP(D).$$

The kernel of R_D consists of holomorphic functions on S, hence its dimension is 1. Thus we have the inequality

$$l(D) \le \deg(D) + 1$$

This inequality will be an equality for every positive divisor D if and only if every $\pi \in PP(D)$ is in the image of the map $R : L(D) \to PP(D)$, which only occurs if $g = 0$.

Let

$$\mathrm{Res}_D : \mathrm{Hol}(S) \to (PP(D))^*$$

be the map

$$\omega \in \mathrm{Hol}(S) \mapsto \mathrm{Res}_D(\omega) \in (PP(D))^*$$

where for $\pi \in PP(D)$ one puts

$$\mathrm{Res}_D(\omega)(\pi) = \sum_{p \in \mathrm{Supp}(D)} \mathrm{Res}_p(\pi_p \omega).$$

The images of $R_D : L(D) \to PP(D)$ and $\mathrm{Res}_D : \mathrm{Hol}(S) \to (PP(D))^*$ are orthogonal to each other. Indeed, the natural pairing $\langle R_D(f), \mathrm{Res}_D(\omega) \rangle$ vanishes since

$$\langle R_D(f), \mathrm{Res}_D(\omega)\rangle = \sum_{p \in \mathrm{Supp}(D)} \mathrm{Res}_p(f\omega) = 0$$

by the Theorem of Residues.

By definition, one has the equality $\mathrm{Ker}(\mathrm{Res}_D) = I(D)$. It follows that the image of Res_D has dimension $g - i(D)$. From the orthogonality it follows that the dimension of the image of R_D is less than $\deg(D) - (g - i(D))$. In conclusion

$$l(D) \geq \deg(D) - (g - i(D)) + 1,$$

since the kernel of R_D has dimension 1. Reordering gives

$$l(D) - i(D) \geq \deg(D) - g + 1.$$

Exercise 15.3. For a Riemann surface S compute $\mathrm{Min}_{0 \leq D}\ \deg(D) - l(D) + 1$.

15.2 Consequences of the Riemann–Roch Theorem

Let L be a holomorphic line bundle of degree d on a Riemann surface S of genus g. Let D be a divisor of degree $g - d$. The bundle $L \otimes L_D$ is of degree g, hence by the Riemann–Roch Theorem $\dim(H^0_{\mathrm{Dolb}}(S, L \otimes L_D)) \geq 1$. It follows that $L \otimes L_D$ admits a holomorphic section $s \neq 0$. The holomorphic section s corresponds to a meromorphic section λ of L. Let $\Lambda = (\lambda)$ be the divisor of λ and L_Λ the corresponding line bundle. The following theorem states that every holomorphic line bundle comes from a divisor.

Theorem 15.4. *Let L be a holomorphic line bundle on a Riemann surface S. The bundle L admits a non-zero meromorphic section. Each such section λ defines an isomorphism of holomorphic line bundles $\lambda: L \to L_\Lambda$, $\Lambda = (\lambda)$.* $\qquad\square$

Let D be a divisor of degree $d \geq (2g - 1)$ on a Riemann surface of genus $g > 0$. One has $l(K_S - D) = 0$ since $\deg(K_S - D) \leq -1 < 0$. The equality $l(D) = d - g + 1$ follows. For high enough degree d the Riemann inequality becomes an equality.

As an important example, consider the kth tensor power $K_S^{\otimes k} = K_S \otimes \cdots \otimes K_S$, k factors, of the canonical bundle K_S. For $k > 0$, $g > 0$, one has $l(K_S^{\otimes k}) = k(2g - 2) - g + 1$. In particular, for $g \geq 2$ the dimension $l(K_S^{\otimes k})$ of the space of holomorphic sections of $K_S^{\otimes k}$ grows linearly with k and with g. Locally, a holomorphic section of $K_S^{\otimes k}$ with respect to a local coordinate $z: U \to \mathbb{C}$ has the form $f(\mathrm{d}z)^k$, where $f: U \to \mathbb{C}$ is a holomorphic function. Sections of $K_S^{\otimes k}$ are called k-differentials. One calls 1-differentials simply (holomorphic) differentials, and 2- and 3-differentials respectively quadratic and cubical differentials.

For $k = 3$, $g \geq 2$, one has $l(K_S^{\otimes 3}) = 5g - 5$. For $p \in S$ it follows from $l(K_S^{\otimes 3} \otimes \theta_{-p}) = 5g - 6$ that the subspace $Z_p = \{s \in L(K_S^{\otimes 3}) \mid s(p) = 0\}$ in $L(K_S^{\otimes 3})$ is of codimension 1. The so-called *tri-canonical* map

$$E: p \in S \mapsto Z_p \in \mathbb{P}((K_S^{\otimes 3})^*)$$

is holomorphic.

Theorem 15.5. *Let S be a Riemann surface of genus $g \geq 2$. The tri-canonical map $E: S \to \mathbb{P}((K_S^{\otimes 3})^*)$, $p \to Z_p$, is a holomorphic embedding of the Riemann surface S into the projective space of hyperplanes $\mathbb{P}((K_S^{\otimes 3})^*)$ in $L(K_S^{\otimes 3})$.*

Proof. The map E is injective. Indeed, assume $p, q \in S$, $p \neq q$. The intersection $Z_p \cap Z_q$ is the space of holomorphic section of the $K_S^{\otimes 3} \otimes \theta_{-p} \otimes \theta_{-q}$. Hence, $\dim(Z_p \cap Z_q) = 5g - 7 < \dim(Z_p)$ and $Z_p \neq Z_q$ follows.

The map E has a $\mathbb{C}$-linear differential $(DE)_p : T_p S \to T_{E(p)} \mathbb{P}((K_S^{\otimes 3})^*)$ of complex rank 1. For $p \in S$, let $z: U \to \mathbb{C}$ be a local coordinate at p. Let t be a holomorphic section of $K_S^{\otimes 3}$ with $t(p) \neq 0$ and let $T \subset L(K_S^{\otimes 3})$ be the span of t. Let $s_0, s_1, \ldots, s_{5g-7}$ be a basis of Z_p such that the first $5g - 8$ elements are sections of $K_S^{\otimes 3} \otimes \theta_{-2p}$. This is possible since by the Riemann–Roch Theorem $L(K_S^{\otimes 3} \otimes \theta_{-2p})$ is of codimension 1 in $Z_p = L(K_S^{\otimes 3} \otimes \theta_{-p})$. Let V be the open neighborhood of Z_p in $\mathbb{P}((K_S^{\otimes 3})^*)$ given by $V = \{Z \in \mathbb{P}((K_S^{\otimes 3})^*) \mid t \notin Z\}$. Think of $Z \in V$ as the graph of a linear map $m(Z): Z_p \to T$. The matrix coefficients $x(Z)$ of $m(Z)$ define a map $x: V \to \mathbb{C}^{5g-7}$ which is a coordinate chart on $\mathbb{P}((K_S^{\otimes 3})^*)$. For $Z \in V$ the number $x_i(Z)$ is such that $s_i + x_i(Z)t$ belongs to Z. Hence $x_i(Z) = \frac{-s_i(p)}{t(p)}$. It follows that the partial derivatives $\frac{\partial x_i(Z_p)}{\partial \bar{z}}$, $0 \leq i \leq 5g - 7$, vanish since the sections s_i and t are holomorphic. It follows that the map E is holomorphic. By the choice of the basis s_i of Z_p one has $\frac{\partial x_i(Z_p)}{\partial z} = 0$, $0 \leq i \leq 5g - 8$, and $\frac{\partial x_{5g-7}(Z_p)}{\partial z} \neq 0$. It follows that the $\mathbb{C}$-rank 1 of the differential of E is maximal.

One concludes that the tri-canonical map E is a holomorphic embedding of the Riemann surface S of genus g in a projective space of dimension $5g - 6$. □

The above proof shows that if a point q on S travels near and once around $p \in S$, then the image Z_q turns in $L(K_S^{\otimes 3})$ near and once around Z_p such that the instantaneous velocity center, *axe instantané de rotation*, $A_q = Z_p \cap Z_q$ in Z_p stays near $L(K_S^{\otimes 3} \otimes \theta_{-2p})$.

What remains from the above, if we work with holomorphic differentials, i.e. with the complex vector space $L(K_S)$ of dimension g?

Let S be a Riemann surface of genus $g \geq 1$. Again, for $p \in S$ the space Z_p is of codimension 1 in $L(K_S)$ by Riemann–Roch. So, the canonical map $E_S: S \to \mathbb{P}((K_S)^*) = \mathbb{P}^{g-1}(\mathbb{C})$ is well defined and holomorphic. If $g = 1$ the canonical map is constant since the space $\mathbb{P}^0(\mathbb{C})$ consists of one point only. The canonical map is not injective in general.

Theorem 15.6. *Let S be a Riemann surface of genus $g \geq 2$. The map $E_S: S \to \mathbb{P}((K_S)^*) = \mathbb{P}^{g-1}(\mathbb{C})$ is injective if and only if the surface S is not hyperelliptic. Moreover, if the surface S is not hyperelliptic, then the canonical map E_S is an embedding.*

Proof. Assume that the map E_S is not injective. Let $p, q \in S$, $p \neq q$, be such that $E_S(p) = E_S(q)$. Hence by classical Riemann–Roch $\dim L(\theta_p \otimes \theta_q) = 2$. Let $\alpha, \beta \in L(\theta_p \otimes \theta_q)$ be linear independent. The function

$$f := \frac{\alpha}{\beta} : S \to \mathbb{P}^1(\mathbb{C})$$

is holomorphic and of degree 2, so S is hyperelliptic.

Assume that the surface S is hyperelliptic. Let $f : S \to \mathbb{P}^1(\mathbb{C})$ be holomorphic of degree 2. By post composing with an automorphism of $\mathbb{P}^1(\mathbb{C})$ we may assume that ∞ is a regular value of f. Let p, q be the preimages of ∞. From 1, $f \in L(\theta_p \otimes \theta_q)$ follows $Z_p = Z_p \cap Z_q = Z_q$. So the map E_S is not injective.

Assume E_S injective. If for $p \in S$ we had dim $L(\theta_p^{\otimes 2}) = 2$, then the surface S would be hyperelliptic and hence E_S not injective, which is a contradiction. So for all $p \in S$ we have dim $L(\theta_p^{\otimes 2}) = 1$, which implies that the rank of the differential at p is maximal and that the injective map E_S is indeed an embedding. $\qquad\square$

15.3 The Birth of Serre Duality

Let L be a holomorphic line bundle on a Riemann surface S. Let f be a meromorphic section of L with divisor $D = (f)$. From the classical and cohomological Riemann–Roch Theorems follows that the dimension of the space $H^1(S, L)$ equals $l(K_S - D)$. Also, see Section 13.5 on Roch's inequality, it follows that the linear correlation map

$$C : H^0_{\text{Dolb}}(S, K_S \otimes L^{-1}) \to (H^1_{\text{Dolb}}(S, L))^*$$

is an isomorphism and that the bilinear pairing

$$B : H^1_{\text{Dolb}}(S, L) \times H^0_{\text{Dolb}}(S, K_S \otimes L^{-1}) \to \mathbb{C}$$

is non-degenerate. These facts hold much more generally, as the following (weak version) of the Serre Duality Theorem [112] shows. We state it without proof.

Theorem 15.7. Serre Duality. *Let $X \subset \mathbb{P}^N(\mathbb{C})$ be a smooth holomorphic sub-manifold of dimension $n > 0$ in projective space $\mathbb{P}^N(\mathbb{C})$. Let K_X be the sheaf of complex volume forms of type $(0, n)$ on X. Let L be a holomorphic line bundle on X. For all $a = 0, \ldots, n$, the integration map*

$$\int_X : \Omega^{0,a}(X, L) \times \Omega^{0,n-a}(X, K_X \otimes L^{-1}) \to \mathbb{C}$$

induces a non-degenerate pairing

$$B : H^a_{\text{Dolb}}(X, L) \times H^{n-a}_{\text{Dolb}}(X, K_X \otimes L^{-1}) \to \mathbb{C}. \qquad\square$$

If we have the Serre Duality Theorem at our disposal, the proofs of the Riemann–Roch Theorems simplify. In particular the magic reversal of inequalities disappears and the involution on divisors $D \to K - D$ induces a duality in the Dolbeault cohomology of line bundles on Riemann surfaces.

Chapter 16
The Teichmüller Curve and its Universal Property

Let S_g be a smooth compact surface of genus $g \geq 2$. The Teichmüller space $\mathbb{T}_g = \mathbb{J}(TS_g)/\mathrm{Diff}_0(S_g)$ for genus $g \geq 2$ is the target space of a proper holomorphic submersion $\pi_g \colon C_g \to \mathbb{T}_g$ such that for any point $t \in \mathbb{T}_g$ the fiber $C_{g,t}$ is a marked Riemann surface that represents the point t. The total space C_g is called the universal Teichmüller curve for genus $g \geq 2$.

The construction of the holomorphic manifold C_g is as follows. Let S be a compact smooth oriented surface of genus $g \geq 2$. The space $\mathbb{J}(TS)$ of all smooth J-fields on S is a complex Fréchet manifold. Put on $\mathbb{J}(TS) \times S$ the J-field that corresponds to the identity map $\mathbb{J}(TS) \to \mathbb{J}(TS)$. This J-field on $\mathbb{J}(TS) \times S$ is integrable by the Fibered Nijenhuis Criterion (Theorem 6.11).

The group $\mathrm{Diff}_0(S)$ of diffeomorphisms isotopic to the identity acts on $\mathbb{J}(TS)$ and by the diagonal action on $\mathbb{J}(TS) \times S$. Both actions are by holomorphic motions. The action on $\mathbb{J}(TS)$ is moreover free and hence a smooth quotient manifold $\mathbb{J}(TS)/\mathrm{Diff}_0(S)$ exists as a consequence of the Uniformization Theorem and by the hyperbolic theory of surfaces. It follows that the quotient $\mathbb{J}(TS) \times S/\mathrm{Diff}_0(S)$ also exists first as a smooth manifold, and second carries a holomorphic structure. Put $C_g = C(S) = \mathbb{J}(TS) \times S/\mathrm{Diff}_0(S)$.

From $C_g = C(S_g)$ a natural holomorphic map $\pi_g \colon C_g \to \mathbb{T}_g$ is induced by the projection map from $\mathbb{J}(TS) \times S$ to $\mathbb{J}(TS)$. As a smooth map it is a smooth trivial fibration. The map π_g can be considered as a family of marked Riemann surfaces of genus $g \geq 2$, namely as the family $((\pi^{-1})(t))_{t \in \mathbb{T}_g}$. This family is universal for holomorphic families of Riemann surfaces of genus $g \geq 2$, which as a family of smooth surfaces is trivial. More precisely, the following theorem due to Teichmüller [117] (see also [4]) holds:

Theorem 16.1 (Universal Property of the Teichmüller Curve). *Let $\pi_X \colon X \to B$ be a holomorphic map from a holomorphic manifold X to a connected holomorphic manifold B pointed by $b_0 \in B$. Assume that π_X is a smooth trivial fibration with fibers diffeomorphic to a surface of genus $g \geq 2$.*

Let $t_0 \in \mathbb{T}(S_g)$ be chosen together with a bi-holomorphic map $U_0 \colon \pi_X^{-1}(b_0) \to \pi_g^{-1}(t_0)$. Then U_0 extends uniquely to a holomorphic map $U \colon X \to \mathbb{C}(S_g)$ and there exists a unique holomorphic map $u \colon B \to \mathbb{T}(S_g)$ with $u(b_0) = t_0$ and $u \circ \pi_X = \pi_g \circ U$.

N. A'Campo, *Topological, Differential and Conformal Geometry of Surfaces*, Universitext, https://doi.org/10.1007/978-3-030-89032-2_16

Proof. The holomorphic structure on X is given by an integrable J-field J_X. Let $\phi: X \to B \times S$ be a smooth trivialization of the fiber bundle π_X. The push-forward J-field $J_{B \times S} = \phi_* J_X$ on $B \times S$ is integrable too. By the Fibred Nijenhuis Criterion 6.11 the field $J_{B \times S}$ is given by a holomorphic map $\sigma: B \to \mathbb{J}(TS)$.

$$
\begin{array}{ccccc}
X & \xrightarrow{\phi} & B \times S & \longrightarrow & \mathbb{J}(TS) \times S \\
\downarrow{\scriptstyle \phi_X} & & \downarrow{\scriptstyle \pi_{B \times S}} & & \downarrow{\scriptstyle \pi_{\mathbb{J}(TS) \times S}} \\
B & \xrightarrow{\mathrm{Id}_B} & B & \longrightarrow & \mathbb{J}(TS)
\end{array}
$$

Define the map $u: B \to \mathbb{T}(S_g)$ as the composition of σ followed by the quotient map $\mathbb{J}(TS) \to \mathbb{T}(S_g)$. The map u is holomorphic. In the sequel, we always denote by J' the restriction of a J-field J to a holomorphic sub-manifold arising in the context.

From the construction of the map u it follows that for all $b \in B$ the fibers $(S = \pi_{B \times S}^{-1}(b), J'_{B \times S})$ and $(S = \pi_g^{-1}(u(b)), \sigma'_g(u(b)))$ are isomorphic Riemann surfaces by a diffeomorphism that is isotopic to the identity of S. Moreover, the action of $\mathrm{Diff}_0(S)$ on $\mathbb{J}(TS)$ being free, this diffeomorphism is a unique holomorphic map

$$
V_b: (S = \pi_{B \times S}^{-1}(b), J'_{B \times S}) \to (S = \pi_g^{-1}(u(b)), \sigma'_g(u(b))).
$$

Define the $V: B \times S \to C_g$ by $V(b, s) = V_b(s)$. The map V is partially holomorphic from $(B \times S, J_{B \times S})$ to C_g. So V is holomorphic. Define

$$
U = V \circ \phi: X \to C_g,
$$

which is holomorphic too, since it is a composition of holomorphic maps. This proves the existence of a pair of maps (U, u), as claimed.

The uniqueness of u follows from the definition of the Teichmüller space $\mathbb{T}(S_g)$ as pairs (Σ, m) of a Riemann Surface Σ together with a marking $m: \Sigma \to S$. The uniqueness of U follows from the freeness of the action of $\mathrm{Diff}_0(S)$ on $\mathbb{J}(TS)$. $\quad\square$

The above theorem of Teichmüller was one of the two main motivations for Grothendieck: in his 20 lectures in the Henri Cartan seminar 1960–1961, [44], he uses Teichmüller spaces and the level 3 moduli spaces, genus > 1, as founding examples for his new theory of schemes. See Chap. 3 in [1].

Chapter 17
Riemann Surfaces and Algebraic Curves

17.1 Chow's Theorem

Let $\mathbb{P}^n(\mathbb{C})$ be the complex projective space of dimension n. A point L in $\mathbb{P}^n(\mathbb{C})$ represents a sub-vector space $L \subset \mathbb{C}^{n+1}$ of dimension one. For short we call $L \in \mathbb{P}^n(\mathbb{C})$ a line L through the origin in $\mathbb{C}^{n+1}$. The space $\mathbb{P}^n(\mathbb{C})$ can be covered by $n + 1$ affine charts $U_0, U_1, \ldots, U_n$. The chart U_i is the subset of lines L such that the restriction of the coordinate function z_i to L is not constant. On each line $L \in U_i$ exists a unique point p_L with $z_i(p_L) = 1$. The map $L \in U_i \mapsto p_L \in \mathbb{C}^{n+1}$ identifies U_i with the hyperplane $H_i = \{z_i \equiv 1\}$ in $\mathbb{C}^{n+1}$. Let $\pi_i \colon \mathbb{C}^{n+1} \to \mathbb{C}^n$ be the projection to the coordinate hyperplane $\{z_i \equiv 0\}$. The restriction to $U_i = H_i$ of π_i is a coordinate map $\phi_i \colon U_i \to \mathbb{C}^n$. The coordinate changes $\phi_{ij} \colon \phi_j(U_j) \to \phi_i(U_i)$ are holomorphic. The space $\mathbb{P}^n(\mathbb{C})$ with the charts (U_i, ϕ_i) is a complex compact connected manifold. Let $J = J_{\mathbb{P}^n(\mathbb{C})}$ be its J-field. An ideal I in the ring $\mathbb{C}[z_0, z_1, \ldots, z_n]$ that is generated by homogeneous polynomials is called homogeneous and its variety $Z(I)$ in $\mathbb{C}^{n+1}$ is a union of lines and defines an algebraic variety $V(I)$ in $\mathbb{P}^n(\mathbb{C})$.

The following theorem of Chow[1] gives a description of all smooth compact differential sub-manifolds $M \subset \mathbb{P}^n(\mathbb{C})$ having the property $J_{\mathbb{P}^n(\mathbb{C})}(TM) = TM$.

Theorem 17.1. *Let $M \subset \mathbb{P}^n(\mathbb{C})$ be a compact differential sub-manifold with $J_{\mathbb{P}^n(\mathbb{C})}(TM) = TM$. Then M is an algebraic variety in $\mathbb{P}^n(\mathbb{C})$ defined by a homogeneous ideal in $\mathbb{C}[z_0, z_1, \ldots, z_n]$.*

The following proof uses as the main tools the classical theorem of Liouville[2] that characterizes polynomials among entire holomorphic functions by a growth property at infinity, and the theorem of removable singularities of Riemann (more precisely, a locally bounded function $f \colon \mathbb{C}^n \to \mathbb{C}$ that is holomorphic in the complement of a sub-manifold of real dimension $\leq (2n - 2)$ is holomorphic).

The standard hermitian scalar product $(u, v) \in \mathbb{C}^n \mapsto u_1 \bar{v}_1 + u_2 \bar{v}_2 + \cdots + u_n \bar{v}_n \in \mathbb{C}$ on $\mathbb{C}^n$ induces a metric on $\mathbb{P}^{n-1}(\mathbb{C})$ such that the diameter is $\pi/2$. The distance between two points L, L' in $\mathbb{P}^{n-1}(\mathbb{C})$ is the angular distance between the corresponding complex lines in $\mathbb{C}^n$. The real part of the hermitian form is a scalar product

---

[1] Wei-Liang Chow, 周炜良 (1911–1995), Chinese mathematician.

[2] Joseph Liouville (1809–1882), French mathematician.

© The Author(s), under exclusive license to Springer Nature Switzerland AG 2021
N. A'Campo, *Topological, Differential and Conformal Geometry of Surfaces*, Universitext,
https://doi.org/10.1007/978-3-030-89032-2_17

on $\mathbf{C}^n = \mathbf{R}^{2n}$ with the following property: the real orthogonal subspace $H^\perp$ of a complex linear subspace H in $\mathbf{C}^n$ is a complex linear subspace in $\mathbf{C}^n$.

As a consequence one obtains an involution $L \in \mathbb{P}^1(\mathbf{C}) \mapsto L^\perp \in \mathbb{P}^1(\mathbf{C})$. For every $L \in \mathbb{P}^1(\mathbf{C})$ the points $L, L^\perp$ realize the diameter distance. We call $L^\perp$ the anti-podal of L.

Proof. The proof is in three steps. Step one concerns the case of real surfaces in $\mathbb{P}^2(\mathbb{C})$. Step two is the case of real codimension 2 sub-manifolds of $\mathbb{P}^n(\mathbb{C})$. Step three is the case of higher real codimensions in $\mathbb{P}^n(\mathbb{C})$.

Step 1. Let $S \subset \mathbb{P}^2(\mathbb{C})$ be a compact real sub-manifold of dimension 2 with $J_{\mathbb{P}^2(\mathbb{C})}(TS) = TS$. At every point $p \in S$ the tangent space T_pS is complex, hence there exists a unique projective line $L_p(S)$ in $\mathbb{P}^2(\mathbb{C})$ with $p \in L_p(S)$ and $T_p(L_p(S)) = T_pS$. For $p \in S$ let $p_S^\perp$ be the anti-podal of p in $L_p(S)$. The point $p_S^\perp$ is the intersection of the hyperplane $p^\perp$ with $L_p(S)$. Put $S^\perp = \{p_S^\perp \mid p \in S\}$. Let $PL(S) \subset \mathbb{P}^2(\mathbb{C}) \times \mathbb{P}^2(\mathbb{C})$ be the subspace of pairs (p, q) with $p \in S$ and $q \in L_p(S)$. Let $PL^*(S)$ be set of pairs $(p, q) \in PL(S)$ with $q \neq p$ and $q \neq p_S^\perp$. The map $(p, q) \in PL^*(S) \mapsto p \in S$ is a bundle map having fibers diffeomorphic to the punctured complex line $\mathbb{C}^*$.

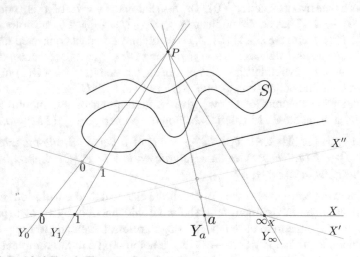

Fig. 17.1 Proof of Chow's Theorem, Step 1.

By Sard's Theorem 1.61 the set C of critical values of the map $E \colon (p, q) \in PL^*(S) \mapsto q \in \mathbb{P}^2(\mathbb{C})$ has Lebesgue measure 0. Hence, there exists a regular value $P \in \mathbb{P}^2(\mathbb{C})$ with $P \notin S \cup S^\perp$. For $p \in S$ let $L_{p,P} \subset \mathbb{P}^2(\mathbb{C})$ be the line passing through p, P. If $L_{p,P} = L_p(S)$ the rank of the differential $DE_{(p,P)}$ is 4. Since the rank of the restriction of $DE_{(p,P)}$ to $T_PL_{p,P}$ is 2, the rank of the map $p \in S \mapsto L_p(S) \in \mathbb{P}^{2*}(\mathbb{C})$ is maximal, i.e. 2. It follows that lines through P and tangent to S at some point in S

are isolated, hence form a finite set. So, all projective lines through P are transversal to S up to finitely many exceptions.

Let X be a projective line with $P \notin X$ and $X \not\subseteq S$. Choose three points $0_X, 1_X, \infty_X$ on X with $\infty_X \notin S$. Think of X as $\mathbb{C} \cup \{\infty\}$ in such a way that $0_X, 1_X, \infty_X$ correspond to $0, 1, \infty$. Let $X' \neq X$ be a line with $\infty_X \in X', X' \not\subseteq S$ and $P \notin X'$.

For $a \in X$ denote by Y_a the projective line through P and a. By the choice of the point P the lines Y_a at each point $q \in S \cap Y_a$ intersect the surface S transversally, or the surface S and the line Y_a have a contact of order 1 at q. It follows that the number d_a of intersection points (if counted twice at tangencies) of S and Y_a is independent of $a \in X$.

A line L through $s \in S$ and P intersects the line X in a point a_s, hence $d_{a_s} \geq 1$ and therefore $d_a \geq 1$, $a \in X$. On the line Y_a, $a \neq \infty_X$, lie three points a, a', P, where a' is the intersection point of X' with Y_a. Moreover there are $d = d_a$ intersection points $s_{1,a}, s_{2,a}, \ldots, s_{d,a}$. The numbering is freely chosen, and tangency points are repeated twice. Let $z_{1,a}, z_{2,a}, \ldots, z_{d,a}$, $a \neq \infty_X$, be the cross-ratios of $a, a', s_{j,a}, P$. Each of the d symmetric functions $\sigma_j(a)$ in the numbers $z_{1,a}, z_{2,a}, \ldots, z_{d,a}$, $a \neq \infty_X$, defines a function $\sigma_j : X \backslash \{\infty_X\} \to \mathbb{C}$. The functions σ_j are indeed holomorphic. The map $\pi : s \in S \mapsto a_s \in X = \mathbb{C} \cup \{\infty\}$ is a function that, near points $a \in X$ with $Y_a = L_{a,P}$, is transversal to S and that is locally bounded near the remaining finitely many points. If Y_a intersects S transversally, the intersections points $s_{1,a}, s_{2,x}, \ldots, s_{d,x}$ of S with Y_a are transversal and we can compute the differential $d\sigma_j$ at a by replacing S by the union of the complex tangent projective lines $L_{s_{j,a}}$ to S. It follows that $(d\sigma_j)_a$ is $\mathbb{C}$-linear. So σ_j is holomorphic at a. Finally, the function σ_j is holomorphic on $X \backslash \{\infty\}$ by the Riemann removable singularities theorem at the finitely many remaining points.

Since the lines X and X' intersect transversally at ∞_X the functions σ_j grow at ∞_X with bounds $|\sigma_j(a)| \leq A|x(a)|^j$. Here $x : X \backslash \{\infty_X\} \to \mathbb{C}$, $a \mapsto x(a)$, is the coordinate function given by the cross-ratio of $0_X, a, 1_X, \infty_X$. Hence the functions σ_j are polynomials of degree $\leq d$ in x by Liouville's theorem. Let $f(x, y)$ be the polynomial of degree d in x, y given by

$$y^d - \sigma_1(x)y^{d-1} + \sigma_2(x)y^{d-2} - \cdots + (-1)^d \sigma_d(x).$$

On the affine part $\mathbb{P}^2(\mathbb{C}) \backslash Y_\infty$ the manifold S is the zero set of f. It follows that S is the zero set in $\mathbb{P}^2(\mathbb{C})$ of the corresponding homogeneous polynomial $F(x, y, z)$.

Step 2. Let M be a compact manifold of real dimension $2n - 2$ in $\mathbb{P}^n(\mathbb{C})$, $n > 2$, with $J(TM) = TM$. Choose projective hyperplanes X, X' in $\mathbb{P}^n(\mathbb{C})$ such that the intersection of $X \cap X'$ with M is transversal. As in Step 1, choose a point $P \notin M$, $P \notin M^\perp$, $P \notin X \cup X'$ and not a critical value of the following map $E : PH^*(M) \to \mathbb{P}^n(\mathbb{C})$. Here $PH^*(M) \subset \mathbb{P}^n(\mathbb{C}) \times \mathbb{P}^n(\mathbb{C})$ is the set of pairs of points (p, q) with $p \in M$, $q \neq p$, $L_{p,q}$ at p tangent to M and q not anti-podal to p on $L_{p,q}$. The set $M^\perp$ comprises the points q for which a point $p \in M$ with $L_{p,q}$ tangent at p to M and p, q anti-podal exists. The map E is induced by $(p, q) \mapsto q$. The version of Sard's theorem needed here is beyond the case of equal dimensions, see Milnor's book [75]. Now as in Step 1, on $X \backslash X \cap X'$ construct functions σ_j. Observe, again by the reasoning as in Step 1, that the restrictions of σ_j to projective lines L in X are holomorphic with poles of order $\leq j$ at $X \cap X' \cap L$. It follows that the restriction of σ_j to $X \setminus X \cap X'$ is a multivariate polynomial of degree $\leq j$. The polynomial $f(x, y)$, $x = (x_1, x_2, \ldots, x_{n-1})$, given by

$$y^d - \sigma_1(x)y^{d-1} + \sigma_2(x)y^{d-2} - \cdots + (-1)^d \sigma_d(x)$$

on the affine part, that is spanned by P and $X \setminus X \cap X'$, is an equation for M. The corresponding homogeneous polynomial $F(x, y, z)$ in $(n+1)$ variables defines M as an algebraic projective variety.

Step 3. Let M be a compact manifold of real dimension $2m - 2$, $m < n$, in $\mathbb{P}^n(\mathbb{C})$ with $J(TM) = TM$. For every linear projective subspace Z of complex dimension $n - m$ and such that the real dimension of the projective join $Z * M$ is $2n - 2$, apply Step 2 to the projective join $Z * M$ and get an equation $F_{Z,M}(x, y, z)$ for it. The projective join $Z * M$ is the closure in $\mathbb{P}^n(\mathbb{C})$ of the union of the projective lines $L_{p,m} \subset \mathbb{P}^n(\mathbb{C})$ that pass through points $p \in Z$, $m \in M$, $p \neq m$. The projective join $Z * M$ is not a submanifold, but Step 2 still works. The homogeneous ideal I spanned by all (in fact by finitely many) such polynomials $F_{Z,M}(x, y, z)$ defines M as an algebraic projective variety. $\square$

The original proof of Chow of 1949 is rather complicated. His proof uses diophantine analysis in the ring of convergent power series and the subring of power series that are algebraic over the ring of polynomials. The proof is by contradiction: assume that the holomorphic closed subvariety X in projective space $\mathbb{P}^N$ is not algebraic. Then for large d there exists an algebraic subvariety Y of complementary dimension of degree d, that intersects X in an isolated point p and the intersection multiplicity $m_p(X, Y)$ exceeds the topological bound $[X] \cdot [Y]$. The proof is local and also works for analytically closed subvarieties X [25].

In 1956 a new proof using cohomological methods, commonly known under the abbreviation GAGA, was given by Serre [113].

The proof given above only uses elementary theorems from the theory of holomorphic functions in one variable. Here the proof is written for closed holomorphic submanifolds, but it can also be adapted in order to cover the case of a reduced holomorphic cycle.

17.2 Riemann Surfaces as Projective Curves

Chow's Theorem applies to the image $E(S)$ of the tri-canonical map $E: S \to \mathbb{P}((L(K_S^{\otimes 3}))^*)$. The tri-canonical map is canonical, hence the group of holomorphic symmetries of S is realized by a group of algebraic symmetries of $E(S)$ that extends canonically to a group of projective motions.

Theorem 17.2. Let S be a Riemann surface of genus $g \geq 2$. The image $E(S)$ of the tri-canonical map E is a smooth projective algebraic curve in a projective space $\mathbb{P}^{5g-6}(\mathbb{C})$ of dimension $5g - 6$. $\square$

The tri-canonical map is canonical, hence commutes with symmetries of the Riemann surface S. More precisely:

Theorem 17.3. *Let S be a Riemann surface of genus g $\geq$ 2. The group of holomorphic symmetries of S is realized by the group of algebraic symmetries of E(S) that extends canonically to the group of projective motions of* $\mathbb{P}((L(K_S^{\otimes 3}))^*)$. □

For non-hyperelliptic surfaces S of genus $g \geq 2$ the canonical map $E\colon S \rightarrow \mathbb{P}^{g-1}(\mathbb{C})$ is an embedding that commutes with the symmetries of the Riemann surface S.

In contrast, a holomorphic embedding $f \times g \times h\colon S \rightarrow \mathbb{P}^3(\mathbb{C})$ given by three meromorphic functions that separate points and tangent vectors is very rarely canonical. Indeed, the construction of three such functions depends on choices. Nevertheless, this embedding shows in a rather elementary way that each compact connected Riemann surfaces is isomorphic to an algebraic curve in projective 3-space.

Chapter 18
The Jacobian of a Riemann Surface

18.1 Vector Spaces Attached to a Riemann Surface

Let $S = (S, J)$ be a Riemann surface of genus $g \geq 1$. Three vector spaces of equal real dimension are attached to the surface S, namely the $\mathbb{C}$-vector spaces of complex dimension g of J-holomorphic, resp. $(-J)$-holomorphic, differentials $\mathrm{Hol}(S, J)$, $\mathrm{Hol}(S, -J)$, and the space of J- or $(-J)$-harmonic real-valued 1-differential forms $\mathrm{Harm}(S, J)$, which are real vector spaces of real dimension $2g$.

We recall the definitions of the preceding spaces. J-holomorphic differentials $\omega \in \mathrm{Hol}(S, J)$ are in local z-charts expressed as $f\mathrm{d}z$, where f is a J-holomorphic function. Elements $\omega \in \mathrm{Hol}(S, -J)$ are in local z-charts expressed as $f\mathrm{d}\bar{z}$, where f is a $(-J)$-holomorphic function. Elements $\alpha \in \mathrm{Harm}(S, J)$ are real differential 1-forms α, which in a local J-holomorphic coordinate $z = x + iy$ are represented as $\alpha = a\mathrm{d}x + b\mathrm{d}y$, where a, b are real functions. The forms α and $\alpha \circ J$ are closed, i.e. $-a_y + b_x = 0$ and $-b_y - a_x = 0$ hold. Hence, by differentiating once more it follows that the functions a, b satisfy $a_{xx} + a_{yy} = b_{xx} + b_{yy} = 0$.

The operation J°, $\alpha \mapsto \alpha \circ J$, is a linear complex structure on the real vector space $\mathrm{Harm}(S, J)$. This structure extends to a complex structure $J^\circ_{\mathbb{C}} = J^\circ \otimes_{\mathbb{R}} \mathrm{Id}_{\mathbb{C}}$ on $\mathrm{Harm}(S, J) \otimes_{\mathbb{R}} \mathbb{C}$. The multiplication m_i on $\mathbb{C}$ induces on the space $\mathrm{Harm}(S, J) \otimes_{\mathbb{R}} \mathbb{C}$ a second complex structure again denoted by m_i. The structures $J^\circ_{\mathbb{C}}$ and m_i coincide on the $+i$ eigenspace E_{+i} of $J^\circ_{\mathbb{C}}$ and are opposite on the $-i$ eigenspace E_{-i}. The complex conjugation c on $\mathbb{C}$ induces an involution $I_c = \mathrm{Id}_{\mathrm{Harm}(S,J)} \otimes_{\mathbb{R}} c$ on $\mathrm{Harm}(S, J) \otimes_{\mathbb{R}} \mathbb{C}$. The involution I_c exchanges the summands in the decomposition $\mathrm{Harm}(S, J) \otimes_{\mathbb{R}} \mathbb{C} = E_{+i} \oplus E_{-i}$. It follows that both summands are of complex dimension g. The eigenspaces $E_{+1} = \mathrm{Harm}(S, J) \otimes_{\mathbb{R}} \mathbb{R}$ and $E_{-1} = \mathrm{Harm}(S, J) \otimes_{\mathbb{R}} i\mathbb{R}$ of the involution I_c are real subspaces of real dimension $2g$. The four subspaces $E_{+1}, E_{+i}, E_{-1}, E_{-i}$ of half real dimension $2g$ are pairwise transversal in the ambient space of real dimension $4g$. The cross-ratio is special and evaluates in the above order to $-\mathrm{Id}_{E_{+1}}$. Indeed, $\alpha \in E_{+1} \mapsto \alpha - i\alpha \circ J \in E_{+i} \mapsto -i\alpha \circ J \in E_{-1} \mapsto -\alpha - i\alpha \circ J \in E_{-i} \mapsto -\alpha \in E_{+1}$.

After the identification $\mathrm{Harm}(S, J) = \mathrm{Harm}(S, J) \otimes_{\mathbb{R}} \mathbb{R} = E_{+1}$, the bijective mappings

$$\kappa^\pm : \alpha \in \mathrm{Harm}(S, J) = E_{+1} \mapsto \frac{1}{2}(\alpha \mp i\alpha \circ J) \in E_{\pm i}$$

© The Author(s), under exclusive license to Springer Nature Switzerland AG 2021
N. A'Campo, *Topological, Differential and Conformal Geometry of Surfaces*, Universitext,
https://doi.org/10.1007/978-3-030-89032-2_18

are of special interest.

The subspaces E_{+i} and E_{-i} are clearly the spaces $\mathrm{Hol}(S, J)$ and $\mathrm{Hol}(S, -J)$ in $\mathrm{Harm}(S, J) \otimes_{\mathbb{R}} \mathbb{C}$. This yields the decomposition as a $\mathbb{C}$-vector space

$$\mathrm{Harm}(S, J) \otimes_{\mathbb{R}} \mathbb{C} = \mathrm{Hol}(S, J) \oplus \mathrm{Hol}(S, -J).$$

The map $[\cdot] : \mathrm{Harm}(S, J) \to H^1_{\mathrm{dR}}(S, \mathbb{R})$, $\alpha \mapsto [\alpha]$, is an isomorphism. This map $[\cdot]$ allows us to identify $\alpha \in \mathrm{Harm}(S, J)$ and $[\alpha] \in H^1_{\mathrm{dR}}(S, \mathbb{R})$ and moreover extends to an isomorphism $[\cdot] : \mathrm{Harm}(S, J) \otimes_{\mathbb{R}} \mathbb{C} \to H^1_{\mathrm{dR}}(S, \mathbb{C})$.

The complex structures J° and $J^\circ_{\mathbb{C}}$ are transported by the map $[\cdot]$ to complex structures, again denoted by the same symbols, on $H^1_{\mathrm{dR}}(S, \mathbb{R})$ and $H^1_{\mathrm{dR}}(S, \mathbb{C})$. Moreover the involution I_c is transported to an involution on $H^1_{\mathrm{dR}}(S, \mathbb{C})$, and on the direct sum decomposition, where it exchanges the summands.

Finally, one obtains the decomposition of the first cohomology group with complex coefficients as a direct sum of complex sub-vector spaces

$$H^1_{\mathrm{dR}}(S, \mathbb{C}) = \mathrm{Hol}(S, J) \oplus \mathrm{Hol}(S, -J)$$

and as a direct sum of real sub-vector spaces

$$H^1_{\mathrm{dR}}(S, \mathbb{C}) = E_{+1} \oplus E_{-1}$$

together with operations $J^\circ_{\mathbb{C}}$ and I_c. Also recall the above special identification $\kappa^- : H^1_{\mathrm{dR}}(S, \mathbb{R}) \to \mathrm{Hol}(S, -J)$ given by $[\alpha] \mapsto \kappa^-([\alpha]) = \frac{1}{2}(\alpha + i\alpha \circ J) \in \mathrm{Hol}(S, -J)$ for $\alpha \in \mathrm{Harm}(S, J) = H^1_{\mathrm{dR}}(S, \mathbb{R})$. The linear identification κ^- is (J°, m_i)-holomorphic. Indeed, $\kappa^-([\alpha \circ J]) = i\kappa^-([\alpha])$. The identification $\kappa^+ : H^1_{\mathrm{dR}}(S, \mathbb{R}) \to \mathrm{Hol}(S, J)$ is (J°, m_i) anti-holomorphic.

Be aware that the formula $\kappa^-([\alpha]) = \frac{1}{2}(\alpha + i\alpha \circ J)$ does not hold for a non-harmonic representative α of its cohomology class $[\alpha]$. Indeed, $\frac{1}{2}(\alpha + i\alpha \circ J)$ is a holomorphic 1-form only if the closed form α is harmonic. For a non-harmonic closed form α we proceed in two steps. First make α harmonic $\alpha_{\mathrm{Harm}} = \alpha - df$, where f is the function $f = \Delta_J^{-1}(d(\alpha \circ J))$. Second put $\kappa^\pm([\alpha]) = \frac{1}{2}(\alpha_{\mathrm{Harm}} \mp i\alpha_{\mathrm{Harm}} \circ J)$.

18.2 The Period Matrix and Riemann's Bilinear Relations

Of special interest are the spaces $\mathrm{Hol}(S, J)$ and $\mathrm{Hol}(S, -J)$ of holomorphic and anti-holomorphic differentials. The $\mathbb{C}$-bilinear map w on

$$H^1_{\mathrm{dR}}(S, \mathbb{C}) = \mathrm{Holo}(S, J) \oplus \mathrm{Holo}(S, -J)$$

given by

$$w : (\omega, \eta) \mapsto w(\omega, \eta) = \int_S \omega \wedge \eta$$

is a non-degenerate pairing. This pairing allows the interpretation of elements η in $\mathrm{Hol}(S, -J)$ as linear forms w_η on $\mathrm{Hol}(S, J)$ by putting $w_\eta(\omega) = \int_S \eta \wedge \omega$, $\omega \in$

$\mathrm{Hol}(S, J)$. In fact the map $\eta \mapsto w_\eta$ identify the spaces $\mathrm{Hol}(S, -J)$ and the dual space $\mathrm{Hol}(S, J)^*$.

Riemann's first bilinear relation claims that both spaces of holomorphic or anti-holomorphic differentials are maximal w-isotropic subspaces. Clearly, these spaces are isotropic, since in a local z-coordinate these differentials are $f\,dz$ or $f\,d\bar{z}$. Moreover they are maximal isotropic, since an isotropic subspace E strictly containing $\mathrm{Hol}(S, J)$ would intersect $\mathrm{Hol}(S, -J)$ non-trivially. For $0 \neq \eta \in E \cap \mathrm{Hol}(S, -J)$ the form $\bar{\eta}$ will belong to $\mathrm{Hol}(S, J) \subset E$, so from $\int_S \eta \wedge \bar{\eta} = 0$ follows the contradiction $\eta = 0$.

Riemann's second bilinear relation claims that $\frac{i}{2} w(\eta, \bar{\eta}) = \frac{i}{2} \int_S \eta \wedge \bar{\eta}$, $\eta \in \mathrm{Hol}(S, J)$, is real and positive for $\eta \neq 0$. Clearly, locally $\eta = f\,dz$, $z = x + iy$. So locally, the integrand $\frac{i}{2} \eta \wedge \bar{\eta}$ is equal to $\frac{i}{2} f \bar{f}\,dz \wedge d\bar{z} = f\bar{f}\,dx \wedge dy$.

The first cohomology $H^1(S, \mathbb{Z})$ with integral coefficients is a lattice Γ in $H^1_{\mathrm{dR}}(S, \mathbb{R}) = \mathbb{R}^{2g}$. It follows that the quotient is diffeomorphic to the torus $(S^1)^{2g}$. Via the identification $\mathrm{Harm}(S, J) = H^1_{\mathrm{dR}}(S, \mathbb{R})$ the cohomology group $H^1_{\mathrm{dR}}(S, \mathbb{R})$ is equipped with the complex structure J°. Via the maps $\kappa^\pm$ the $\mathbb{C}$-bilinear map w defines a Hermitian structure h_{H^1} on $H^1_{\mathrm{dR}}(S, \mathbb{R})$ by $h_{H^1}([\alpha], [\beta]) = w(\kappa^+([\alpha]), \kappa^+([\beta])) = \int_S \kappa^+([\alpha]), \kappa^-([\beta]))$. The structures J° and h_{H^1} are invariant under Γ translations, hence descend to the torus $H^1_{\mathrm{dR}}(S, \mathbb{R})/H^1(S, \mathbb{Z})$.

Our next concern is how to store the structures hidden in the triple

$$(H^1(S, \mathbb{Z}), H^1_{\mathrm{dR}}(S, \mathbf{R}), J^\circ)$$

of a lattice $H^1(S, \mathbb{Z})$ in the real vector space $H^1_{\mathrm{dR}}(S, \mathbf{R})$ that is equipped with a linear complex structure J°.

As before, in the construction of the Teichmüller space, we need again a marking on the surface S. This time we need a *symplectic marking SM* consisting of a system of $2g$ simply closed curves on S. More precisely, it is a system $SM = (a_1, \ldots, a_g, b_1, \ldots, b_g)$ of simply closed oriented curves on the oriented surface S, where we think of the curves as regularly parametrized by the oriented circle S^1, with the following property:

The curves are pairwise disjoint, except the curves a_k, b_k, $k = 1, 2, \ldots, g$, that intersect transversally in one point such that the oriented tangent vectors at the intersection point of a_k and b_k are an oriented frame.

Now let α_k, β_k be a system of closed real 1-differential forms with support near the curves in SM and with, as the only non-zero periods, the periods $\int_{a_k} \beta_k = -\int_{b_k} \alpha_k = 1$, $k = 1, \ldots, g$. Recall the expansion

$$[\omega] = \sum_{k=1}^{g} \int_{a_k} \omega\,[\beta_k] - \int_{b_k} \omega\,[\alpha_k]$$

for real or complex closed 1-forms ω.

Observe for $\omega \in \mathrm{Hol}(S, J)$ the equalities

$$\int_{a_k} \omega = \int_S \alpha_k \wedge \omega = \int_S \kappa^+(\alpha_k) \wedge \omega,$$

$$\int_{b_k} \omega = \int_S \beta_k \wedge \omega = \int_S \kappa^+(\beta_k) \wedge \omega.$$

It follows that the (J°, m_i) linear map κ^+ sends the above triple to

$$(\kappa^+(H^1(S, \mathbb{Z}), \mathrm{Hol}(S, J)^*, m_i).$$

Let $\omega_1, \omega_2, \ldots, \omega_g \in \mathrm{Hol}(S, J)$ be the $\mathbb{C}$-basis with $\int_{a_k} \omega_h = \delta_h^k$.
Now the information in the triple

$$(H^1(S, \mathbb{Z}), H^1_{\mathrm{dR}}(S, \mathbf{R}), J^\circ)$$

is stored by $\Omega = \Omega(S, J)$, the complex $(g \times g)$-matrix $\Omega_{h,k} = \int_{b_h} \omega_k$, which is called Riemann's Period Matrix.

The chosen forms ω_k form a basis in the complex vector space $\mathrm{Hol}(S, J)$ and the linear forms $u_k \colon \omega \mapsto u_k(\omega) = \int_{a_k} \omega$ are the corresponding linear coordinates. The matrix Ω relates the period-vector $u(\omega) = (u_1(\omega), \ldots, u_g(\omega))$ to the period-vector $v(\omega) = (\int_{b_1} \omega, \ldots, \int_{b_g} \omega)$ by $v(\omega) = \Omega u(\omega)$.

The Period Matrix Ω has two properties that follow from the bilinear relations.
First property: Ω is symmetric, i.e. $\Omega_{h,k} = \Omega_{k,h}$.
Indeed, by the above, we expand

$$[\omega_h] = [\beta_h] - \sum_i \int_{b_i} \omega_h, [\alpha_i]$$

$$[\omega_k] = [\beta_k] - \sum_i \int_{b_i} \omega_k [\alpha_i]$$

and by the first bilinear relation we conclude that

$$0 = \int_S \omega_h \wedge \omega_k = -\Omega_{h,k} + \Omega_{k,h}.$$

Hence $\Omega = X + iY$, where X, Y are real symmetric $(g \times g)$-matrices.
Second property: the imaginary part of Ω, i.e. the real symmetric matrix Y, is positive definite. Indeed, with above notation of u, v, note for $\omega \in \mathrm{Hol}(S, J)$ the expansions

$$[\omega] = \sum_k u_k(\omega)[\beta_k] - v_k(\omega)[\alpha_k],$$

$$[\bar{\omega}] = \sum_k \bar{u}_k(\omega)[\beta_k] - \bar{v}_k(\omega)[\alpha_k].$$

Hence, for all $0 \neq \omega \in \mathrm{Hol}(S, J)$, the integral $\frac{i}{2} \int_S \omega \wedge \bar{\omega}$ is real and

$$0 < \frac{i}{2} \int_S \omega \wedge \bar{\omega} = \frac{i}{2} \sum_k (u_k(\omega)\bar{v}_k(\omega) - \bar{u}_k(\omega)v_k(\omega))$$

$$= \frac{i}{2}(\langle u(\omega), \bar{\Omega}\bar{u}(\omega)\rangle - \langle \bar{u}(\omega), \Omega u(\omega)\rangle)$$

$$= \langle \mathrm{Re}(u(\omega)), Y\mathrm{Re}(u(\omega))\langle,$$

showing the claim.

The above triple can be reconstructed from a matrix $\Omega \in \mathrm{Mat}(g, \mathbb{C})$, $g \geq 1$, that satisfies the Riemann bilinear relations. It is the triple $(L_\Omega, \mathbb{C}^g, m_i)$, where L_Ω is the lattice $\mathbb{Z}^g + \Omega(\mathbb{Z}^g)$.

For example in the case $\Omega = \Omega(S, J)$, the $\mathbb{C}$-linear map $E : \mathbb{C}^g \to \mathrm{Hol}(S, -J)$ that sends the ith standard basis vector e_i to the linear form $\omega \in \mathrm{Hol}(S, J) \mapsto \int_{a_i} \omega \in \mathbb{C}$ will send the lattice L_Ω to the above lattice $\kappa^-(H^1(S, \mathbb{Z}))$ in $\mathrm{Hol}(S, -J)$.

18.3 The Jacobian Jac(S)

Let $S = (S, J)$ be a Riemann surface of genus $g \geq 1$. The *Jacobian* of S is the torus $\mathrm{Jac}(S) = H^1_{\mathrm{dR}}(S, \mathbb{R})/H^1(S, \mathbb{Z})$ equipped with the complex structure induced by J° on $H^1_{\mathrm{dR}}(S, \mathbb{R}) = \mathrm{Harm}(S, J)$. Using the above notations κ^- and Γ one has the equivalent definition $\mathrm{Jac}(S) = \mathrm{Hol}(S, -J)/\kappa^-(\Gamma)$ with the complex structure induced from m_i on the $\mathbb{C}$-vector space $\mathrm{Hol}(S, -J)$.

Let Γ be any lattice in $\mathbb{C}^g$. The complex torus $T = \mathbb{C}^g/\Gamma$ is a complex manifold. Its function theory is very poor in general, for instance non-constant meromorphic functions on T in general do not exist. As a consequence, T cannot be embedded in a projective space $\mathbb{P}^N(\mathbb{C})$. Entire functions $f(z)$ with two real independent periods do not exist. In contrast, the meromorphic Theta function introduced by Jacobi

$$z \mapsto \vartheta_\tau(z) = \sum_{n \in \mathbb{Z}} e^{\pi i \tau n^2 + 2\pi i n z}, \; z \in \mathbb{C}, \; \tau \in \mathbb{C}_+,$$

is 1-periodic and weakly τ-periodic. The Jacobian varieties of Riemann surfaces are very special:

Theorem 18.1 (Riemann). *The Jacobian $Jac(S)$ of a Riemann surface $S = (S, J)$ admits a holomorphic embedding into a projective space.*

This theorem follows from the more general statement:

Theorem 18.2 (Riemann). *Let $\Omega \in \mathrm{Mat}(g, \mathbb{C})$, $g \geq 1$, satisfy Riemann's bilinear relations. Then there exists a holomorphic embedding of $T_\Omega = \mathbb{C}^g/L_\Omega$ into a projective space.*

Proof. For the proof of this theorem L_Ω-periodic meromorphic functions $f : \mathbb{C}^g \to \mathbb{C}$ have to be constructed.

A first Ansatz (with $\langle n, z \rangle = n_1 z_1 + \cdots + n_g z_g$) is

$$\vartheta(z, \Omega) = \sum_{n \in \mathbb{Z}^g} e^{2\pi i \langle n, z \rangle}.$$

Convergence is problematic. The following weighting, as used by Jacobi for the case of one complex variable, helps out

$$\vartheta(z, \Omega) = \sum_{n \in \mathbb{Z}^g} e^{\pi i \langle n, \Omega(n) \rangle} e^{2\pi i \langle n, z \rangle}$$

and defines the basic Riemann Theta function. The first factor decays exponentially very fast as $\langle n, n \rangle$ grows by Riemann's second bilinear relation, i.e. from the estimates

$$i \langle n, \Omega(n) \rangle \le -a \langle n, n \rangle, \ e^{\pi i \langle n, \Omega(n) \rangle} \le e^{-a\pi \langle n, n \rangle}, \ a > 0$$

and the second factor grows less fast, for all $z \in K \subset \mathbb{C}^g$, K compact, there exists a real number b_K with

$$|e^{2\pi i \langle n, z \rangle}| \le e^{b_K \sqrt{\langle n, n \rangle}}.$$

So this sum defines a holomorphic function on $\mathbb{C}^g$. This function is clearly $\mathbb{Z}^g$-periodic:

$$\vartheta(z + n, \Omega) = \vartheta(z, \Omega)$$

holds for $n \in \mathbb{Z}^g$. A weak form of $\Omega(\mathbb{Z}^g)$-periodicity holds:

$$\vartheta(z + \Omega(m), \Omega) = e^{-\pi i \langle m, \Omega(m) \rangle - 2\pi i \langle m, z \rangle} \vartheta(z, \Omega).$$

The correcting factor does not vanish, so the divisor $\Theta = (\vartheta(z, \Omega))$ is invariant under translations of elements in L_Ω. So this construction does not yet yield a meromorphic function on the torus $\mathbb{C}^g / L_\Omega$ but a divisor, the so-called Theta divisor, on this torus.

The weak periodicity follows from Riemann's first bilinear relation, i.e. from the symmetry ${}^t\Omega = \Omega$,

$$\vartheta(z + \Omega(m), \Omega)$$

$$= \sum_{n \in \mathbb{Z}^g} e^{\pi i \langle n, \Omega(n) \rangle} e^{2\pi i \langle n, z + \Omega(m) \rangle}$$

$$= \sum_{n \in \mathbb{Z}^g} e^{\pi i \langle n, \Omega(n) \rangle} e^{2\pi i \langle n, z \rangle}$$

$$= \sum_{n \in \mathbb{Z}^g} e^{\pi i (\langle n, \Omega(n) \rangle + 2\langle n, \Omega(m) \rangle + \langle m, \Omega(m) \rangle - \langle m, \Omega(m) \rangle - 2i \langle m, z \rangle)} e^{2\pi i \langle n+m, z \rangle}$$

$$= e^{-\pi i \langle m, \Omega(m) \rangle - 2\pi i \langle m, z \rangle} \sum_{n \in \mathbb{Z}^g} e^{\pi i \langle n+m, \Omega(n+m) \rangle} e^{2\pi i \langle n+m, z \rangle}$$

$$= e^{-\pi i \langle m, \Omega(m) \rangle - 2\pi i \langle m, z \rangle} \vartheta(z, \Omega).$$

Let $a \in \mathbb{Q}^g$ be a division vector with $la \in \mathbb{Z}^g$. The meromorphic function on $\mathbb{C}^g$

$$\vartheta_a(z) = \frac{\vartheta(z + a, \Omega)^l}{\vartheta(z, \Omega)^l}$$

is L_Ω-periodic. Indeed, it is clearly $\mathbb{Z}^g$-periodic. For $\Omega(m)$ the two correcting factors cancel:

$$\vartheta(z + a + \Omega(m), \Omega)^l = e^{-l\pi i\langle m,\Omega(m)\rangle - 2l\pi i\langle m,z+a\rangle}\vartheta(z + a, \Omega)^l$$
$$= e^{-l\pi i\langle m,\Omega(m)\rangle - 2l\pi i\langle m,z\rangle}\vartheta(z + a, \Omega)^l,$$

since $e^{2l\pi i\langle m,a\rangle} = 1$. As before

$$\vartheta(z + \Omega(m), \Omega)^l = e^{-l\pi i\langle m,\Omega(m)\rangle - 2l\pi i\langle m,z\rangle}\vartheta(z, \Omega)^l.$$

The restriction of ϑ to the real line $(t, 0, \ldots, 0)$, $t \in \mathbb{R}$, is not a constant function since it is 1-periodic and its Fourier expansion has many non-zero terms. It follows that for $l >> 0$ and many l-division vectors a the meromorphic function $\vartheta_a(z, \Omega)$ is not constant, so must have zeros and poles since the torus is compact. It follows also that the divisor $\Theta = (\vartheta(z, \Omega))$ has non-empty support.

We repeat the construction with the modified symplectic marking $SM = b_1, b_2, \ldots, b_g, -a_1, -a_2, \ldots, -a_g$. Then the Period Matrix changes to $-\Omega^{-1}$. We get a new basic function $\vartheta(\Omega(z), -\Omega^{-1})$. Also for l-division vectors we obtain the functions $\vartheta_b(\Omega(z), -\Omega^{-1})$.

For given $l >> 0$ we have l^g l-division vectors $a \in \mathbb{Q}^q$ with $0 \le a_k < 0$ and also l^g l-division vectors $b = \Omega(a)$. Those division vectors $a+b$ define an ϵ_l, $\lim_{l\to\infty} \epsilon_l = 0$, dense set in the torus. The corresponding l^{2g} meromorphic functions $\vartheta_{a,b}(z, \Omega) = \vartheta_a(z, \Omega)\vartheta_b(\Omega(z), -\Omega^{-1})$ separate the points of $\mathbb{C}^g/L_\Omega$ and define a holomorphic map $E_l : \mathbb{C}^g/L_\Omega \to \mathbb{P}^{l^g-1}$. This map is for $l >> 0$ an embedding. Hence, by Chow's Theorem 17.1, the torus $T_\Omega = \mathbb{C}^g/L_\Omega$ is a projective algebraic variety. $\square$

Theorem 18.3. *The group laws on T_Ω and on $\mathrm{Jac}(S, J)$ are algebraic maps. The laws of the inverse are algebraic maps.*

Proof. Choose an embedding $E_l : T_\Omega \to \mathbb{P}^N$. The graph of the addition is a holomorphic sub-variety in $\mathbb{P}^N \times \mathbb{P}^N \times \mathbb{P}^N$ and hence by Chow's Theorem 17.1 an algebraic sub-variety. The same argument works for the inverse map. $\square$

In conclusion, the complex tori T_Ω and Jacobians of a Riemann surface are complex *abelian varieties*. A complex abelian variety is a complex connected algebraic compact group. All complex abelian varieties are isomorphic to a complex torus $\mathbb{C}^g/L_\Omega$, where $\Omega \in \mathrm{Mat}(g, \mathbb{C})$ satisfies the Riemann bilinear relations.

Exercise 18.4. Compute the period matrix Ω and lattice L_Ω for the elliptic curve defined by $y^2 = -x^3 + 1$.

Exercise 18.5. Compute the period matrix Ω and lattice L_Ω for the Bolza curve defined by $y^2 = -x^5 + x$.

18.4 The Abel–Jacobi Map

One important link between a Riemann surface (S, J) of genus $g \ge 1$ and $\mathrm{Jac}(S)$ is the Abel–Jacobi map. For a path $\gamma : [0, 1] \to S$ the path integral

$$\omega \in \mathrm{Hol}(S, J) \mapsto \int_\gamma \omega \in \mathbb{C}$$

defines a linear form $\int_\gamma \in \mathrm{Hol}(S, J)^*$. For two paths γ_1, γ_2 with common endpoints from p to q the difference $\int_{\gamma_1} - \int_{\gamma_2}$ belongs to the lattice $\kappa^-(H^1(S, \mathbb{Z}))$. Hence, the class $[\int_\gamma] \in \mathrm{Hol}(S, J)^* / \kappa^-(H^1(S, \mathbb{Z})) = \mathrm{Jac}(S, J)$ depends only upon the end points $p = \gamma(0), q = \gamma(1)$. Denote this class by $\int_p^q \in \mathrm{Jac}(S, J)$.

The *Abel–Jacobi map* based at $b \in S$ is

$$u_b : S \to \mathrm{Jac}(S, J), \ q \in S \mapsto u_b(q) = \int_b^q \in \mathrm{Jac}(S, J).$$

Associated, and independent of a base point, is the group homomorphism

$$u : \mathrm{Div}^0(S) \to \mathrm{Jac}(S, J).$$

For a degree 0 divisor $D = c_1 p_1 + \cdots + c_N p_N$ with $\sum_{n=1}^N c_n = 0$ define

$$u(D) = \sum_{n=1}^N c_n u_b(p_n).$$

This expression in fact does not depend on the base point b.

The Abel–Jacobi maps u_b, u play an important role. Observe that these maps are holomorphic. Indeed, let $\omega_1, \dots, \omega_g$ be a basis of $\mathrm{Hol}(S, J)$ and let $z_1, \dots, z_g \in \mathrm{Hol}(S, J)^*$ be the dual coordinates. The map is holomorphic since its differential $h \in T_p S \mapsto (Du_b)_p(h) = \omega_{1,p}(h) z_1, \dots, \omega_{g,p} z_g$ is clearly (J_p, m_i)-linear. Moreover, $(Du_b)_p \neq 0$, $p \in S$, since at every point p one of the forms $\omega_{j,p}$ does not vanish.

Theorem 18.6 (Abel). *A divisor $D \in \mathrm{Div}^0(S)$ is principal if and only if its image $u(D) \in \mathrm{Jac}(S, J)$ vanishes. Equivalently, two effective divisors $D_1, D_2 \in \mathrm{Div}_{\mathrm{eff}}^E$ of equal degree $E > 0$ are linearly equivalent if and only if $u(D_1) = u(D_2)$ holds.*

Proof. Assume $D_1, D_2 \in \mathrm{Div}_{\mathrm{eff}}^E(S)$ are linearly equivalent, i.e. $D_1 - D_2 = (f)$ for a non-constant holomorphic map $f : S \to \mathbb{P}^1(\mathbb{C})$. Define $u^f : \mathbb{P}^1(\mathbb{C}) \to \mathrm{Jac}(S, J)$ by $u^f(t) = u(D_t)$, where D_t is the divisor weighted sum of points in $f^{-1}(t)$, $t \in \mathbb{P}^1(\mathbb{C})$, the weighting being by multiplicity. Observe $D_0 = D_1$, $D_\infty = D_2$. The pull back by u^f of the forms $\mathrm{d}z_k$ on $\mathrm{Jac}(S, J)$ are holomorphic differential forms $u^{f*}\mathrm{d}z_k$, $k = 1, \dots, g$, on $\mathbb{P}^1(\mathbb{C})$ which vanish by the analytic characterization of the genus. It follows that the map u^f is constant, hence $u(D_1) = u(D_2)$.

Assume $u(D_1) = u(D_2)$. We write $D_1 = p_1 + p_2 + \cdots + p_E$, $D_2 = q_1 + q_2 + \cdots + q_E$. We have already seen for $p, q \in S$ the real harmonic 1-form $\omega_{p,q} = \mathrm{d}f_{p,q}$, the differential of the Green function $f_{p,q}$, such that its meromorphic completion $\mu_{p,q} = \omega_{p,q} - i\omega_{p,q} \circ J$ has only simple poles at p, q with opposite residues $+1, -1$, and local periods $\pm 2\pi i$. So, all periods $\int_\gamma \mu_{p,q}$ are an integral multiple of $2\pi i$.

Let $\mu_{p,q;a}$ be

$$\mu_{p,q;a} = \mu_{p,q} - \sum_{k=1}^g \left(\int_{a_k} \mu_{p,q} \right) \omega_k.$$

The a-periods of $\mu_{p,q;a}$ vanish. We will compute the b-periods of $\mu_{p,q;a}$ for the case when the points p, q do not lie on the curves a_k and b_k of the symplectic marking. Using a suitable isotopy we may assume that the marking is disjoint from the support of the divisors.

Cutting the surface S along the curves of the marking produces a surface $S_{a,b}$ with boundary. The surface $S_{a,b}$ is diffeomorphic to the 2-sphere minus g squares. Each square V_k is oriented and its boundary ∂V_k consists of 4 pieces that correspond to the oriented curves $a_k, b_k, -a_k, -b_k$. We connect the squares by an embedded system of paths $c_1, c_2, \ldots, c_{g-1}$, cut also along them and get the surface $S_{a,b,c}$, which is homeomorphic to a polygon with $6g - 2$ sides.

Let B_p, B_q be small balls with centers p, q. The restrictions of the forms $\mu_{p,q}, \omega_k$ to $S_{a,b,c} \setminus B_p \cup B_q$ are of type $(1, 0)$, hence $\mu_{p,q} \wedge \omega_k = 0$ on $S' = S_{a,b,c} \setminus B_p \cup B_q$. The closed form ω_k has on $S_{a,b,c}$ a primitive $g \colon s \in S_{a,b,c} \mapsto \int_{s_0}^{s} \in \mathbb{C}$. By the Green–Stokes Theorem we get

$$0 = \int_{S'} \mu_{p,q} \wedge \omega_k = \int_{\partial S'} g\mu_{p,q}.$$

Observe, $\int_{\partial V_l} g\mu_{p,q} = 0$, $l \neq k$, and $\int_{\partial V_k} g\mu_{p,q} = \int_{b_k} \mu_{p,q}$. The integral $\frac{1}{2\pi i} \int_{\partial B_p \cup B_q} = g(p) - g(q)$ is the sum of the residues of $g\mu_{p,q}$ at p and q. Moreover,

$$\int_{b_k} \mu_{p,q} = \int_{\partial(B_p \cup B_q)} g\,\mu_{p,q} = 2\pi i(g(p) - g(q)) = 2\pi i \int_{p}^{q} \omega_k$$

and

$$\int_{b_k} \mu_{p,q;a} = 2\pi i \int_{p}^{q} \omega_k + 2\pi i \sum_{l=1}^{g} m_{p,q;l}.$$

Here, $m_{p,q;l}$ are the integers $\frac{1}{2\pi i} \int_{a_l} \mu_{p,q}$.

Define $\mu_{D_1,D_2;a} = \sum_{e=1}^{E} \mu_{p_e,q_e;a}$ and

$$\mu_{D_1,D_2;a,b} = \mu_{D_1,D_2;a} - 2\pi i \sum_{e=1}^{E} \sum_{l=1}^{g} m_{p_e,q_e,l}\omega_l.$$

The assumption $u(D_1 - D_2) = 0$ implies by the definition of u the vanishing of $\sum_{e=1}^{E} \int_{p_e}^{q_e} \omega_k$, $k = 1, \ldots, g$. It now follows that all a- and b-periods of $\mu_{D_1,D_2;a,b}$ are integral multiples of $2\pi i$. Hence for any multi-valued primitive $M_{D_1,D_2;a,b}$ the expression $e^{M_{D_1,D_2;a,b}}$ defines a meromorphic function f on S. One checks $(f) = D_1 - D_2$. $\qquad\square$

Theorem 18.7. *The map $u_b \colon S \to \mathrm{Jac}(S, J)$ is a holomorphic embedding.*

Proof. We have seen that u_b is a holomorphic immersion. If $u_b(p) = u_b(q)$ with $p \neq q$ then by Abel's Theorem the divisor $D = p - q$ is principal, which implies the existence of a meromorphic function with exactly one simple pole. $\qquad\square$

The Abel–Jacobi map $u \colon \mathrm{Div}(S, J) \to \mathrm{Jac}(S, J)$ induces a map $v \colon \mathrm{Pic}(S, J) \to \mathrm{Jac}(S, J)$ by putting $v(L) = u(s)$ for a line bundle $L \in \mathrm{Pic}(S, J)$ and a meromorphic section s of L. Clearly, $v(L_1 \otimes L_2) = v(L_1) + v(L_2)$, so v is a group homomorphism. Remember that the Picard group $\text{Pic}(S, J)$ is homeomorphic to $(S^1)^g \times \mathbb{Z}$. Abel's Theorem 18.6 implies the following.

Theorem 18.8. *The restriction of* v *to* $\text{Pic}_0(S, J)$ *is an isomorphism of complex tori.* $\square$

The image $u_b(S)$ in $\text{Jac}(S, J)$ is a holomorphic, hence by Chow's Theorem 17.1, a projective sub-variety, which is a copy of the surface (S, J) in its Jacobian variety. This copy is, up to translation, independent of the base point b.

Denote by $\text{Div}_{\text{eff}}^k(S)$ the set of effective divisors of degree k. This set can also be understood as the k-fold symmetric product $S^{[k]}$. So it becomes a topological space, even a projective variety. The images W_b^k of similar maps $u_b^k : \text{Div}_{\text{eff}}^k(S) \to \text{Jac}(S, J)$ form an increasing sequence of irreducible projective sub-varieties.

Theorem 18.9. *The map* $u_b^g : \text{Div}_{\text{eff}}^g(S) \to \text{Jac}(S, J)$ *is surjective, so* $W_b^g = \text{Jac}(S, J)$. *The image* $W_b^{g-1} = u_b^{g-1}(\text{Div}_{\text{eff}}^{g-1}(S))$ *is of codimension* 1.

Proof. If $W_b^k \neq W_b^{k+1}$ then $\text{Dim}(W_b^k) < \text{Dim}(W_b^{k+1})$ by irreducibility. If $W_b^k = W_b^{k+1}$ then $\cup_h W_b^h = W_b^k$ and moreover W_b^k is a subgroup of $\text{Jac}(S, J)$. From $v(\text{Pic}_0(S, J)) = \text{Jac}(S, J)$ follows $\cup_h W_b^h = \text{Jac}(S, J)$. We conclude that $u(S) = W_b^1 \neq W_b^2 \neq \cdots \neq W_b^{g-1} \neq W_b^g = \text{Jac}(S, J)$ since $\text{Dim}(\text{Jac}(S, J)) = g$. $\square$

In his study of the multiplicities of values taken by an integral quadratic form $\langle n, Q(n) \rangle$ on $\mathbb{Z}^d$, Jacobi, following Euler, introduced ϑ-sequences $\sum_{n \in \mathbb{Z}^d} e^{\pi i \langle n, Q(n) \rangle t}$. For his study of elliptic integrals Jacobi introduced the functions ϑ_τ. The chain of sub-varieties W_b^k were studied by Riemann and he discovered that the divisor W_b^{g-1} in the Jacobian variety and the divisor Θ of his ϑ-functions $\vartheta(z, \Omega)$ with g periods and g weak periods are translates of each other.

Chapter 19
Special Metrics on J-Surfaces

19.1 The Bergman Metric

Let $S = (S, J)$ be a Riemann surface of genus $g \geq 2$. The Uniformization Theorem states that there exists a unique volume form ω_J with $\omega_J(u, J(u)) \geq 0$, $u \in TS$, and such that the metric $\mu_J(u, v) = \omega_J(u, J(v))$ has constant curvature $k_{\mu_J} = -1$.

The Bergman metric[1] $\beta = \beta_J$ is also a Riemannian metric depending on the J-field. The complex vector space $\mathrm{Hol}(S, J)$ of holomorphic differentials has dimension g. The space $\mathrm{Hol}(S, J)$ is endowed with a canonical hermitian form ρ, namely

$$\rho(\alpha', \alpha) = \frac{1}{-2i} \int_S \alpha' \wedge \bar{\alpha}, \ \alpha', \alpha \in \mathrm{Hol}(S, J).$$

The real part $\mathrm{Re}\rho$ defines a scalar product. The canonical Bergman metric β is the Riemannian metric on the surface S defined by

$$\beta_p(u)^2 = \sum_{k=1}^{g} |\alpha_{k,p}(u)|^2, \ u \in T_p S,$$

where α_k, $k = 1, \ldots, g$, is a complex orthonormal basis in $\mathrm{Hol}(S, J)$. It follows from the Riemann–Roch Theorem that $l(K_S - p) = g - 1$ for every $p \in S$. Indeed, $l(K_S - p) - l(p) = 2g - 3 - g + 1$ and $l(p) = 1$. So, for every $p \in S$ not all $\alpha_{k,p}$ vanish. It follows that β is a Riemannian metric on S. A direct computation shows that β does not depend on the chosen orthonormal basis in $\mathrm{Hol}(S, J)$. Moreover the metric β is J-invariant.

The Uniformization Theorem implies that there exists a unique real function u on S such that the metric $e^{2u}\beta$ has constant curvature -1.

The metric β and the function u deserve to be called *canonical* since clearly

Theorem 19.1. *The group of automorphisms of the Riemann surface (S, J) leaves the Bergman metric β and the function u invariant.* □

---

[1] Named after Stefan Bergman (1895–1977), Polish-born American mathematician.

© The Author(s), under exclusive license to Springer Nature Switzerland AG 2021
N. A'Campo, *Topological, Differential and Conformal Geometry of Surfaces*, Universitext,
https://doi.org/10.1007/978-3-030-89032-2_19

Two functions u, v on S are called *level-equivalent* if there exists a $\phi \in \text{Diff}_0(S)$ that maps each connected component of a level of u to a connected component of a level of v. Level-equivalence is an equivalence relation on functions. Two function u, v are level-equivalent if and only if the induced (singular) foliations are isotopic.

Interestingly, the notion of level-equivalence enables us to define an equivalence relation on the Teichmüller space $\mathbb{T}_g$. Two points $p, q \in \mathbb{T}(S_g)$ are called *Bergman-level-equivalent* if there exists a path $t \in [0, 1] \mapsto J_t \in J(S)$ connecting $p = (S, J_0)$ and $q = (S, J_1)$ such that the canonical functions u_{S,J_t} on S are all level-equivalent, more precisely the level-equivalence from u_{S,J_t} to $u_{S,p}$ is realized by an isotopy ϕ_t.

The Bergman-level-equivalence relation on $\mathbb{T}(S_g)$ is invariant under the action of the mapping class group.

The complete trivalent graph appears as a $\text{PSL}(2, \mathbb{Z})$-invariant spine in the Teichmüller space $T_1 = \mathbb{H}^2$. The following is an attempt to construct mapping class group-invariant spines in $\mathbb{T}_g$, $g \geq 2$:

Question. Is the decomposition of $\mathbb{T}(S_g)$ by the Bergman-level equivalence classes a cell decomposition?

Question. Is the union $B_g \subset \mathbb{T}(S_g)$ of all the Bergman-level-equivalence classes that have a compact closure in $\mathbb{T}(S_g)$ a spine for $\mathbb{T}(S_g)$?

The Bergman energy functional $E_\beta(S, J)$ of a Riemann surface is defined as the energy of its Bergman metric. Remember $E_\beta(S, J) = \int_S k_\beta K_\beta = \int_S k_\beta^2 \text{dVol}_\beta$. The critical point set B of the map $E_\beta \colon \mathbb{T}(S_g) \to \mathbb{R}$ is of particular interest.

Question. Is B the vertex set of the expected spine B_g of the previous question?

19.2 Special Metrics and Covering Spaces

If the "special" metric class is characterized by a local property such as a curvature property, then this class is preserved by coverings. This is the case for curvature 0 and -1 metrics. We recall:

Theorem 19.2. *Let S be a Riemann surface of genus g. The universal covering space $\bar{S}$ of S is naturally a Riemann surface, which is conformal to $\mathbb{P}^1(\mathbb{C})$ if $g = 0$, to $\mathbb{C}$ if $g = 1$, and to $\mathbb{C}_+$ if $g \geq 2$.* $\square$

Proof. If $g = 0$ the Riemann surface S is conformal to the simply connected Riemann sphere $\mathbb{C} \cup \{\infty\} = \mathbb{P}^1(\mathbb{C})$, hence $\bar{S} = \mathbb{P}^1(\mathbb{C})$. If $g = 1$ the Riemann surface S is conformal to $\mathbb{C}/\Gamma$ for a lattice $\Gamma \subset \mathbb{C}$ that acts by translations. Hence $\bar{S}$ is conformal to $\mathbb{C}$. If $g \geq 2$ the Riemann surface S carries a conformal Riemannian metric of constant curvature -1. Hence $\bar{S}$ is a complete simply connected oriented hyperbolic surface, thus it is conformal to any model of the oriented hyperbolic plane, so for instance to $\mathbb{C}_+$, $\mathbb{D}$, $\mathbb{H}_I$, $\mathbb{H}_J$. The group of conformal automorphisms of $\mathbb{C}_+$ is the group of fractional transformations $z \in \mathbb{C}_+ \mapsto \frac{az+b}{cz+d} \in \mathbb{C}_+$, $\left(\begin{smallmatrix} a & b \\ c & d \end{smallmatrix}\right) \in \text{PSL}(2, \mathbb{R})$, that also acts by isometries on the hyperbolic plane $\mathbb{C}_+$. Therefore the surface S is metrically and conformally isomorphic to $\mathbb{C}_+/\Gamma$, where Γ is a lattice in $\text{PSL}(2, \mathbb{R})$. $\square$

Surfaces of genus 0 do not have connected coverings. However, it is remarkable, as we saw earlier in Section 10.4, that the space of metrics on (S^2, J) that are conformal to a given J-field and have constant curvature $+1$ is a model for the hyperbolic three space $\mathbb{H}^3$.

The Bergman metric on surfaces of genus ≥ 2 is not characterized by a local property. Let $\pi: S' \to S$ be a covering map above a Riemann surface S of genus $g \geq 2$ and degree $k \geq 2$. Let $\alpha_1, \ldots, \alpha_g$ be an orthonormal basis for the space $\text{Hol}(S, J)$. Then the pull back forms $\frac{1}{\sqrt{k}} \overset{*}{\alpha_i}$, $1 \leq i \leq g$, can be completed by forms α_i, $g < i \leq kg - k + 1$, to an orthonormal basis for the space $\text{Hol}(S', J)$. The pull back $\frac{1}{\sqrt{k}} \pi^* \beta_S$ of the Bergman metric of S to S' will differ from the Bergman metric $\beta_{S'}$ since $\pi^* \beta_S$ only takes into account the first g forms $\frac{1}{\sqrt{k}} \overset{*}{\alpha_i}$, $1 \leq i \leq g$, of the completed orthonormal basis for $\text{Hol}(S', J)$.

19.3 The Energy of Canonical Embeddings

Let S be a Riemann surface of genus $g \geq 2$ and let $e_S: S \to (\mathbb{P}((K_S^{\otimes 3}))^* = \mathbb{P}^N(\mathbb{C})$, $N = 5g - 6$, be its tri-canonical embedding. Let μ be a normalized Fubini–Study metric on $\mathbb{P}^N(\mathbb{C})$. Let $\mu_S = e_S^* \mu$ be the pull back metric on S. Let $E_S(\mu) = \int_S k_{\mu_S}^2 \, d\text{Vol}_{\mu_S}$ be the energy of the Riemannian surface (S, μ_S).

The *tri-canonical energy functional* of the Riemann surface S is the map

$$E_S: \mu \in FS^N \mapsto E_S(\mu) \in \mathbb{R}$$

defined on the space of normalized Fubini–Study metrics FS^N. The space FS^N is parametrized by the homogeneous space $\text{PSL}(N + 1, \mathbb{C})/\text{PU}(N + 1)$, which has the very rich structure of a symmetric space. The quantity $E_S(\mu_S) + 2\pi\chi(S)$ measures how far the metric μ_S is away from a metric with constant curvature.

A positive answer to the following question would perhaps help to answer the question posed by David Kazhdan below.

Question. Is the restriction of $E_S: \mu \mapsto E_S(\mu)$ to geodesics in the space $\text{PSL}(N + 1, \mathbb{C})/\text{PU}(N + 1)$ convex? Does E_S have a unique minimum?

Let $e_{S,k}: S \to \mathbb{P}^N(\mathbb{C})$, $N = (2k - 1)g - 2k$, be the k-canonical embedding and let $E_{S,k}$ be the corresponding energy functional. Let $ME_{S,k}$ be the minimal value of $E_{S,k}$.

Question. (David Kazhdan) Does $\liminf_{k \to \infty} ME_{S,k} = 2\pi\chi(S)$ hold? Let $M\mu_{S,k}$ be a minimizer of $E_{S,k}$. For a subsequence of $(M\mu_{S,k})$, is the limit of $(S, M\mu_{S,k})$ the uniformizing Poincaré metric?

Chapter 20
The Fundamental Group and Coverings

20.1 Simply Connected Riemann Surfaces and the Universal Uniformization Theorem

We have encountered only three simply connected Riemann surfaces, namely: the Riemann sphere or complex projective line $\mathbb{C} \cup \{\infty\} = \mathbb{P}^1(\mathbb{C})$, the Gaussian plane $\mathbb{C}$ and the upper half plane $\mathbb{C}_+$. The first, being compact, is topologically distinct from the last two. The last two are homeomorphic, even diffeomorphic. As Riemann surfaces, the surfaces $\mathbb{C}_+$ and $\mathbb{C}$ are not isomorphic. Both are homogeneous, but the groups of holomorphic automorphism $\mathrm{Aut}(\mathbb{C}_+)$ and $\mathrm{Aut}(\mathbb{C})$ are not isomorphic: the first group contains as subgroups all finitely generated free groups, but in the second group any pair of commutators commute!

The following theorem is a key result in mathematics. We state it without proof (we advise the reader to consult the book [30] for an interesting discussion of many aspects of this theorem):

Theorem 20.1 (Universal Uniformization Theorem, Henri Poincaré and Paul Köbe). *Any connected, simply connected Riemann surface S is isomorphic to one of the three surfaces* $\mathbb{P}^1(\mathbb{C})$, $\mathbb{C}$, $\mathbb{C}_+$. □

See for instance the books of Albert Pfluger [94] or of Simon Donaldson [34] for a proof. The first steps of the proof show that the topology of S admits a countable basis, a theorem due to T. Radó,[1] and hence the topology is metrizable. This first step is essential in view of the surface constructed by H. Prüfer[2] [98].

A classical weaker version of above theorem deals with simply connected open subsets of $\mathbb{C}$. For a proof, see the book of Henri Cartan [23].

Theorem 20.2. *Every simply connected non-empty proper open subset $U \subset \mathbb{C}$ is biholomorphically isomorphic to the upper half plane* $\mathbb{C}_+$. □

It is interesting to see how these three spaces differ. All three are 2-dimensional manifolds. Only $\mathbb{P}^1(\mathbb{C})$ is compact and the manifolds $\mathbb{C}$, $\mathbb{C}_+$ are diffeomorphic. The

---

[1] Tibor Radó (1895–1965), Hungarian-American mathematician.

[2] Heinz Prüfer (1896–1934), German mathematician.

© The Author(s), under exclusive license to Springer Nature Switzerland AG 2021
N. A'Campo, *Topological, Differential and Conformal Geometry of Surfaces*, Universitext,
https://doi.org/10.1007/978-3-030-89032-2_20

J-surfaces $\mathbb{C}$, $\mathbb{C}_+$ are not isomorphic, since any holomorphic map $\phi\colon \mathbb{C} \to \mathbb{C}_+$ composed with $u \in \mathbb{C}_+ \mapsto \frac{1-u}{1+u} \in \mathbb{D}$ is constant by Rouché's theorem.

The groups of holomorphic automorphisms are the group of projective transformations $\mathrm{PGL}(2, \mathbb{C})$ for $\mathbb{P}^1(\mathbb{C})$, the group of linear affine transformations $\mathrm{Aff}(\mathbb{C}) = \{z \in \mathbb{C} \mapsto \lambda z + t \mid t \in \mathbb{C}, \lambda \in \mathbb{C}^*\}$ for $\mathbb{C}$ and the group of fractional transformations $\mathrm{PSL}(2, \mathbb{R})$ for $\mathbb{C}_+$. These groups are real Lie groups of real dimensions 6, 4 and 3, respectively.

These groups are also pairwise not isomorphic as abstract groups. For instance, in the group $\mathrm{Aff}(\mathbb{C})$ any pair of commutators $[a, b]$, $[c, d]$ commutes, i.e. satisfies $[[a, b], [c, d]] = \mathrm{Id}_{\mathbb{C}}$, and the group $\mathrm{PSL}(2, \mathbb{R})$ has the non-abelian free group F_2 on two generators as a subgroup.

Exercise 20.3. Show that the abstract groups $\mathrm{PSL}(2, \mathbb{R})$ and $\mathrm{PGL}(2, \mathbb{C})$ are not isomorphic. Hint: the alternating group A_5 is a subgroup of $\mathrm{PGL}(2, \mathbb{C})$ and not of $\mathrm{PSL}(2, \mathbb{R})$.

20.2 The Universal Cover and Uniformization of Riemann Surfaces

Let (S, J) be a connected Riemann surface and let $\phi\colon \bar{S} \to S$ be a continuous differentiable structure on $\bar{S}$ such that ϕ is the universal covering map. Let $\bar{J} = \phi_*^*(J)$ be the pull back J-field on $\bar{S}$.

The resulting Riemann surface $(\bar{S}, \bar{J})$ is simply connected. So the above Universal Uniformization Theorem applies. In particular, the fundamental group $\pi_1(S)$ will be realized as a subgroup by deck transformations in $\mathrm{Aut}(\bar{S}, \bar{J})$. This is a very strong restriction!

For a compact connected Riemann surface S all three possibilities for the universal cover appear.

In case $g = 0$ the surface S is simply connected, so $S = \bar{S} = \mathbb{P}^1(\mathbb{C})$.

In case $g \geq 2$ the fundamental group $\pi_1(S)$ is infinite, so the universal cover $\bar{S}$ is not compact. Moreover, two commutators in $\pi_1(S)$ do not commute in general, so $\bar{S}$ is isomorphic to $\mathbb{C}_+$ as a Riemann surface.

In case $g = 1$ the fundamental group $\pi_1(S)$ is abelian, and infinite, moreover of rank 2, hence the only possibility is that $\bar{S}$ is isomorphic to $\mathbb{C}$ as a Riemann surface.

These remarks show that the Universal Uniformization Theorem implies the uniformization of compact Riemann surfaces of genus ≥ 2. Indeed, let S be a compact Riemann surfaces of genus ≥ 2. The Universal Uniformization Theorem shows the existence of a covering map $\phi\colon \mathbb{C}_+ \to S$. The fundamental group $\pi_1(S)$ acts as deck transformations on $\mathbb{C}_+$. This deck transformation action $\rho\colon \pi_1(S) \times \mathbb{C}_+ \to \mathbb{C}_+$ has the following properties:

(i) The action is holomorphic: for every $\gamma \in \pi_1(S)$ the map $\rho(\gamma)\colon \mathbb{C}_+ \to \mathbb{C}_+$ is biholomorphic, hence $z \in \mathbb{C}_+ \mapsto \rho(\gamma)(z) = \frac{a_\gamma z + b_\gamma}{c_\gamma z + d_\gamma} \in \mathbb{C}_+$. It follows that the action ρ is also isometric for the hyperbolic metric on $\mathbb{C}_+$.

(ii) The action is free: for all $\gamma \in \pi_1(S)$, $\gamma \neq 1_{\pi_1(S)}$ and for all $z \in \mathbb{C}_+$ one has $\rho(\gamma)(z) \neq z$.

The Riemann surface S is the orbit space of the action ρ. So, the hyperbolic metric on $\mathbb{C}_+$ induces a hyperbolic metric on S, since the free action ρ is by isometries.

This quotient construction applies for many more Riemann surfaces.

Theorem 20.4. *Let S be a connected Riemann surface with a non-abelian fundamental group. The Riemann surface S admits a hyperbolic metric in its conformal class.*

Proof. The universal cover $\bar{S}$ of S cannot be biholomorphic to $\mathbb{C}$. Indeed, the action ρ by deck transformations would be non-abelian, so would contain a biholomorphic map $z \mapsto \lambda z + t$ that is not a translation, hence with $\lambda \neq 1$. This map then has a fixed point and cannot be a deck transformation.

For the same reason, $\bar{S}$ is not biholomorphic to $\mathbb{P}^1(\mathbb{C})$, since all biholomorphic automorphisms of $\mathbb{P}^1(\mathbb{C})$ have a fixed point.

It follows by the Universal Uniformization Theorem 20.1 that $\bar{S}$ is biholomorphic to $\mathbb{C}_+$. One argues as above. $\square$

Many more Riemann surfaces S with hyperbolic metrics in their conformal class appear. The simplest ones are the so-called surfaces of *finite type*: non-compact surfaces obtained by deleting a non-empty finite set from a compact connected surface. In order for the fundamental group to be non-abelian one has to delete at least three points if the compact surface is of genus 0. Those surfaces admit a finite pant decomposition and the construction of a hyperbolic metric could also be done at the price of many new difficulties, as in the compact case of genus ≥ 2.

For more general surfaces a direct construction, without using the universal cover surface, seems to be difficult. Interesting examples of Riemann surfaces that are not of finite type are obtained by deleting a Cantor set from a connected compact surface.

Instead of pairs of pants we work with viruses. A *k-virus* is a 2-sphere from which are removed k open discs that have disjoint closures. Moreover the virus is called *marked* if each boundary component is marked by a point. For example, a 3-virus is a pair of pants. A *hyperbolic* virus is a virus with a hyperbolic metric on it, such that the boundary components are totally geodesic and of length 1.

Exercise 20.5. Construct for $k \in \mathbb{N}$, $k \geq 3$, a hyperbolic k-virus.

Let Γ be an infinite graph with all vertices of finite valency ≥ 3.

The graph Γ can be thickened to a hyperbolic surface $S(\Gamma)$ as follows: choose for each $k \geq 3$ a marked hyperbolic k-virus and replace each vertex of valency k by the chosen marked k-virus, and glue the boundaries according to the graph and markings. This provides many examples of hyperbolic surfaces that are not of finite type.

As a graph one can choose the Cayley graph $\Gamma(G, \Sigma)$ of a group G together with a finite generating system Σ. All vertices will be of equal valency and we need only to choose one k-virus.

The following question pops up: How much group-theoretic information about G is hidden in the hyperbolic surface $S(\Gamma(G, \Sigma))$? As an example, the random walk on $S(\Gamma(\mathbb{Z}^2, \{(\pm 1, 0), (0, \pm 1)\}))$ is recurrent, the random walk on

$$S(\Gamma(\mathbb{Z}^3, \{(\pm 1, 0, 0), (0, \pm 1, 0), (0, 0, \pm 1)\}))$$

is transient.

Appendix A
Reminder: Topology

A.1 Topological Properties

A topology T on a set X is defined by a set T of subsets in X that satisfies three axioms. A pair (X, T) consisting of a set X and a set T of subsets of X is called a topological space if the set T satisfies the following axioms:

(i) The empty subset $\emptyset$ and the subset X are elements in T.
(ii) For every family $(U_\alpha)_{\alpha \in A}$ of elements in T the union $\bigcup_{\alpha \in A} U_\alpha$ is an element in T.
(iii) For every finite family $(U_\alpha)_{\alpha \in A}$ of elements in T the intersection $\bigcap_{\alpha \in A} U_\alpha$ is an element in T.

The notion of topological space is central in mathematics: no branch of mathematics is free from using topological spaces, and every branch of mathematics provides methods for the study of topological spaces. Let (X, T) be a topological space.

A subset $A \subset X$ is called *open in the space* X if $A \in T$. A subset $A \subset X$ is called *closed in the space* X if its complement $X \setminus A$ is open in the space X. A *neighborhood of a point* $p \in X$ is a subset $N \subset X$ for which there exists an open subset U with $p \in U \subset N$.

Let A be a subset of a topological space (X, T). The set of traces $\{U \cap A \mid U \in T\}$ is a topology on A. A subset A of X endowed with the induced topology is called a *(topological) subspace of* X.

Extreme examples of topologies T on a set X are the topology $T = \{\emptyset, X\}$ with smallest possible set T and the topology $T = P(X)$ with biggest possible set T, namely the power set $P(X)$ of all subsets in X. The topology $P(X)$ is called the *discrete topology* on X.

We have already encountered many topological spaces, such as $\mathbb{R}$, $\mathbb{C}$, metric spaces (X, d), surfaces S, normed vector spaces $(V, \| \cdot \|_V)$, Banach and Fréchet spaces like $C^0(S, \mathbb{R})$, $C^\infty(S, \mathbb{R})$, ...

There are many properties that a topological space or a map between topological spaces can satisfy or not satisfy. We list and comment on some of the most important properties.

© The Author(s), under exclusive license to Springer Nature Switzerland AG 2021
N. A'Campo, *Topological, Differential and Conformal Geometry of Surfaces*, Universitext,
https://doi.org/10.1007/978-3-030-89032-2

Let (X, T) be a topological space. For short, we say "X is a topological space" or even "X is a space" without explicitly mentioning its topology T.

A (topological) space X is *connected* if the only subsets $A \subset X$ that are open and closed in X are the subsets $A = \emptyset$ and $A = X$.

A space X is *separated* if for every pair of distinct points $p, q \in X$ there exist open subsets U, V in X with $p \in U$, $q \in V$, $U \cap V = \emptyset$. A separated space is also called a Hausdorff space.

A space X is *compact* if X is separated and if for every family $(U_\alpha)_{\alpha \in A}$ of open subsets in X that covers X there exists a finite sub-family $(U_\alpha)_{\alpha \in B}$, $B \subset A$, B *finite*, that covers X. (A family $(X_i)_{i \in I}$ of subsets of X *covers* X if $\bigcup_{i \in I} X_i = X$ holds).

A space X is *metrizable* if its topology is the topology that underlies a metric on X.

Most basic property for a map $f \colon X \to Y$ between topological spaces $X = (X, T_X)$ and $Y = (Y, T_Y)$ is the property of continuity. A map f is *continuous* if for every open subset U of Y the preimage $f^{-1}(U) \subset X$ is open in X. Clearly from the definition of closed set it follows that a map is continuous if the preimages of all closed sets are closed. The composition of continuous maps is again a continuous map.

A map $f \colon X \to Y$ is *open/closed* if for every open/closed subset A of X the image $f(A) \subset Y$ is open/closed in Y.

A map $f \colon X \to Y$ is a *homeomorphism* if the map f is bijective, continuous and open. It follows then that f^{-1} is continuous too. Two spaces X, Y are called *homeomorphic* if there exists a homeomorphism $f \colon X \to Y$. Two homeomorphic spaces share all topological properties.

A space X is *path connected* if for every pair of points $p, q \in X$ there exists a path, i.e. a continuous map $\gamma \colon [0, 1] \to X$, which connects p and q, i.e. with $\gamma(0) = p$, $\gamma(1) = q$.

Exercise A.1. Show that the union of the Y-axis and $\{p \in \mathbb{R}^2 \mid x(p) \neq 0,\ y(p) = \sin(1/x(p))\}$ is a subspace of the plane that is connected, but not path connected.

The *path connected components* in a space X are the path connected subsets that are maximal in the set of all path connected subsets of X, with respect to inclusion.

Given a set X and a set S of subsets of X, there exist two extreme topologies on X related to the set S. One is the topology S^+ with smallest possible set of open sets such that $S \subset S^+$. The other is the topology S^- with biggest possible set of open sets such that $S^- \subset S$.

Given a space X and a subset A of X, there exists on A an *induced topology* with smallest possible set of open sets such that the inclusion map $i_A \colon A \to X$ is continuous. In fact the induced topology on A is the set of traces of open subsets of X on A, already introduced previously. A subset A of a topological space together with the induced topology from X is again a topological space. In short, A is a subspace of X.

Given a family $(X_\alpha, T_\alpha)_{\alpha \in A}$ of topological spaces, there exists a *product topology* T on the product set $\prod_\alpha X_\alpha$ which has the smallest possible set of open subsets such that each projection map $\pi_\alpha \colon \prod_\alpha X_\alpha \to X_\alpha$ is continuous.

Given a space X and an equivalence $\sim$-relation on X, the *quotient topology* on $X/\sim$ is the topology with biggest possible set of open sets such that the quotient map $X \to X/\sim$ is continuous.

Since a topology T on a set X is a subset of the power set $P(X)$, topologies on X can be compared by the inclusion relation. If two topologies T, T' on X satisfy $T \subset T'$ one says that T' is *finer* than T, or that T is *coarser* than T'. Making a topology finer or coarser can preserve or destroy a property. For instance if the space (X, T) is *separated* and T' is finer than T, the space (X, T') will also be separated. In contrast, if T' is coarser than T, the space (X, T') will not necessary be separated. The property of being separated asks for the existence of open sets: more precisely, given two distinct points $p, p' \in X$, separation requires the existence of two open disjoint sets U, U' in X with $p \in U$, $p' \in U'$. A similar, but opposite behavior, happens with respect to the *covering property* appearing in the definition of compactness. More precisely, the covering property for a space X requires that for every family $(U_\alpha)_{\alpha \in A}$ of open sets in X that covers X, i.e. with $\cup_{\alpha \in A} = X$, there exist a finite subset $B \subset A$ with $\cup_{\alpha \in B} = X$.

The definition of compactness is an example of a sharp compromise, which are rare in real life. A space X is compact if the separation property and the covering property hold. We state:

Theorem A.2. *A compact topology T on a set X is doubly extremal: if T' is finer than T, then (X, T') loses the covering property, if T' is coarser than T, then the space (X, T') is no longer separated.* $\square$

This double extremality of compact topologies explains why they have so many (useful) properties.

Exercise A.3. Show that a bounded map $f : \mathbb{R}^n \to \mathbb{R}^m$ with closed graph in $\mathbb{R}^n \times \mathbb{R}^m$ is continuous. Find a discontinuous map $f : \mathbb{R} \to \mathbb{R}$ with closed graph.

Given two topological spaces X, Y it is interesting to construct a homeomorphism $f : X \to Y$, if any exists, or to show that the spaces X, Y are not homeomorphic.

Rather elementary observations can show that two spaces are not homeomorphic. Let X be the subspace of $\mathbb{R}^2$ that looks like the letter "X" and is defined by one equality and one inequality

$$X = \{p \in \mathbb{R}^2 \mid x(p)y(p) = 0, \ x(p)^2 + y(p)^2 \le 1\}.$$

The space X is compact and path connected. There exist four points $a \in X$ such that the space $X \setminus \{a\}$ is connected. This shows that the space X is not homeomorphic to a compact, path connected subspace of the plane that looks like the letter "Y".

Exercise A.4. Show that the group G of self homeomorphisms of the space X (the letter "X") acts on X with 3 orbits. Describe the quotient topology of X/G.

More generally one can show that two graphs X, Y, both without vertices of valency 2, are homeomorphic if and only if the graphs are combinatorially equivalent. Here we define a *graph* as follows. Let $(I_\alpha)_{\alpha \in A}$ a family of copies of the interval $[0, 1]$ and let the topological space X be the (disjoint) union $X = \bigcup_\alpha I_\alpha$ equipped with the finest topology such that all inclusions $I_\alpha \subset X$ are continuous. Let R be an equivalence relation on X such that all equivalence classes with 2 or more elements consist only of boundary points of intervals. A graph is the space X/R equipped with the quotient topology. The images of the intervals I_α are the *edges* of the graph X/R. A *vertex* $v \in X/R$ is an equivalence class containing only boundary points of intervals. The *degree of a vertex* v is the cardinality of the equivalence class v.

Two continuous maps $f, g : X \to Y$ are called *homotopic* if there exists a continuous map $H : X \times [0, 1] \to Y$ such that the restriction of H to $X \times \{0\}$ equals f and the restriction of H to $X \times \{1\}$ equals g.

A continuous map $\phi : X \to Y$ is a *homotopy equivalence* if there exists a continuous map $\psi : Y \to X$ such that the composition $\psi \circ \phi$ is homotopic to the identity map $\mathrm{Id}_X : X \to X$ and the composition $\phi \circ \psi$ is homotopic to the identity map Id_Y of the space Y.

Two spaces X, Y are called homotopically equivalent if there exists a homotopy equivalence between X and Y.

A space X is called contractible if the constant map $c : X \to \{pt\}$ to a space with one point is a homotopy equivalence. Open or closed balls in normed vector spaces are examples of contractible spaces.

A graph is contractible if and only if it is connected and has no cycles. A connected graph without cycles is called a *tree*. A finite connected graph is a tree if and only if the number of vertices exceeds by one the number of edges.

What about compact connected Riemann surfaces? Can one detect the genus homotopically, topologically, or by using differential or holomorphic methods?

The most elementary and historical approach is as follows. Let S be a surface. A simply closed curve Λ in S is by definition the image of a continuous map $\gamma : [0, 1] \to S$ such that $\gamma(t) \neq \gamma(s)$ for $0 \leq t < s < 1$ and $\gamma(0) = \gamma(1)$. The following elementary observation, but with a rather non-elementary proof, was made by Riemann and later almost proved by Jordan (see Thomas Hales' article [48] for a defence of Jordan's original proof). It was finally proved by Oswald Veblen:[1]

Theorem A.5 (Jordan Curve Theorem). *Let Λ be a simply closed curve in the two-sphere S^2. Then the complement $S^2 \setminus \Lambda$ is not connected and consists of two connected components.* □

Two simply closed curves Λ, Λ' on a surface S are called *parallel* if they are disjoint and they are the boundary components of a cylinder on S, a cylinder C on S being the image of an injective continuous map $c : S^1 \times [0, 1] \to S$. The frontier of C is the difference set $C \setminus C'$, where C' is the interior of C.

Riemann's topological definition of the genus of a compact connected surface S is the maximal number g of pairwise non-parallel simply closed curves $\Lambda_1, \Lambda_2, \ldots, \Lambda_g$ on S such that the space $S \setminus \bigcup_{i=1,2,\ldots,g} \Lambda_i$ is connected.

The Jordan Curve Theorem becomes more elementary if stated for regular smooth simply closed curves. Riemann's topological definition then becomes a smooth definition of the genus.

Exercise A.6. Let Λ be polygonal simply closed curve in $\mathbb{R}^2$. Show for $p \in \mathbb{R}^2 \setminus \Lambda$ that there exist half rays R that start from p and that intersect Λ in finitely many points. Show that the parity modulo 2 of the number $\#R \cap \Lambda$ only depends on the starting point of the half ray R. Show that $\mathbb{R}^2 \setminus \Lambda$ has two connected components.

---

[1] Oswald Veblen (1880–1960), American mathematician.

Exercise A.7. A second proof of the Jordan Curve Theorem for polygonal curves. Let $A_1, A_2, \ldots, A_n$, $n > 3$, be the vertices of a simple polygonal closed curve Γ in $\mathbb{R}^2$. The vertices are labeled such that the segments

$$[A_1, A_2], [A_2, A_3], \ldots, [A_{n-1}, A_n], [A_n, A_1]$$

are the edges of the polygon Γ. Then there exist three consecutive vertices P, Q, R of Γ such that the intersection of Γ with the convex hull of P, Q, R is the union of the edges $[P, Q]$ and $[Q, R]$. Prove the polygonal Jordan Curve Theorem by induction on the number of vertices.

Exercise A.8. Prove the Jordan Curve Theorem for smooth curves. Hint: Use a polygonal approximation and describe the curve edge-wise as the graph of a function mapping the edge to its normal.

Recall that a pant decomposition of a compact connected Riemann surface S of genus $g \geq 2$ is given by a maximal system of $3(g - 1)$ pairwise non-parallel and disjoint simply closed curves on S.

Using de Rham cohomology, one gets for a compact connected Riemann surface that twice the genus is the dimension of the first de Rham cohomology group. This is in fact a differentiable definition of the genus.

Riemann gave for compact connected Riemann surfaces S a holomorphic definition of the genus as the dimension of the space of holomorphic differentials. In his proof Riemann already anticipated in the special case of surfaces the theorem of de Rham. In this definition $g(S) = l(K_S)$, the canonical divisor K_S appears.

From the Riemann–Roch Theorem one deduces that the genus realizes the extremal value

$$g(S) - \text{Max}_D \, (\deg(D) + 1 - l(D))$$

of the right-hand side if D runs over all divisors on the Riemann surface S.

In [123] Weil proved that for a proper algebraic curve defined over a field k the expression $\deg(D) + 1 - l(D)$ is bounded and uses as definition for the genus $g(C)$ of the algebraic curve C the upper extremal value of the expression $\deg(D) + 1 - l(D)$.

The answer to the question of how to detect the genus by homotopy is given in Theorem A.13.

A.2 The Fundamental Group

Let X be topological space and let $b \in X$ be a distinguished point, also called a base point. Let $\Lambda_b(X)$ be the topological space of all continuous paths $\gamma \colon [0, 1] \to X$ with $\gamma(0) = \gamma(1) = b$. Such paths are called loops in X based at b. The topology on $\Lambda_b(X)$ is the coarsest topology such all subsets $\beta(K, U) \subset \Lambda_b(X)$ are open. Here, for $K \subset [0, 1]$ compact and $U \subset X$ open in X, define $\beta(K, U) = \{\gamma \in \Lambda_b(X) \mid \gamma(K) \subset U\}$. A continuous path $s \in [0, 1] \mapsto \gamma_s \in \Lambda_b(X)$ connecting γ_0 and γ_1 corresponds to a based homotopy from γ_0 to γ_1. A *based homotopy* from γ_0 to γ_1 is a continuous map $\Gamma \colon [0, 1] \times [0, 1] \to X$ such that the restriction of Γ to the boundary of the square $[0, 1] \times [0, 1]$ satisfies $\Gamma(t, 0) = \gamma_0(t)$, $\Gamma(t, 1) = \gamma_1(t)$, $\Gamma(0, s) = \Gamma(1, s) = b$.

The space $\Lambda_b(X)$ carries an internal law of composition

$$*\colon \Lambda_b(X) \times \Lambda_b(X) \to \Lambda_b(X)$$

given by catenation $(\gamma, \gamma') \mapsto \gamma * \gamma'$, where $\gamma * \gamma' \colon [0, 1] \to X$ is the path with $\gamma * \gamma'(t) = \gamma(2t)$, $t \le \frac{1}{2}$, and $\gamma * \gamma'(t) = \gamma(2t - 1)$, $t \ge \frac{1}{2}$.

The set $\Lambda_b(X)$ carries the equivalence relation BasHom that has as equivalence classes the path connected components of the topological space $\Lambda_b(X)$. Two elements γ_0 and γ_1 are in the same BasHom-equivalence class if and only if they are based homotopic. We denote by $[\gamma]_{\text{BasHom}}$ the equivalence class of γ for the relation $\sim_{\text{BasHom}}$.

Theorem A.9. *The law of internal composition $*$ induces on the quotient set $\Lambda_b(X)/\!\sim_{\text{BasHom}}$ a group law $*/\!\sim_{\text{BasHom}}$.*

The group $(\Lambda_b(X)/H, */\!\sim_{\text{BasHom}})$ is called the *fundamental group of the space* X *with base point* b. The commonly used notation is $\pi_1(X, b)$.

A *simply connected* space X is a path connected space X such that for a point $b \in X$ the fundamental group $\pi_1(X, b)$ is trivial.

A *locally simply connected* space X is a space X such that for every point $p \in X$ and for an open subset U containing p there exists an open subset U' with $p \in U' \subset U$ such that (U', p) is simply connected.

The fundamental group of a topological space is a strong invariant, especially for those spaces that are locally simply connected. Manifolds are locally simply connected, since the fundamental group of a ball B in a normed vector space V is trivial. Indeed, by the homotopy $\Gamma(t, s) = \gamma(st)$ any loop $\gamma \in \Lambda_0(B)$ is homotopic to the constant loop at $0 \in B$.

Proof. Catenation and homotopy are compatible, which means: If loops α, α' and β, β' are based homotopic, then $\alpha * \beta$ and $\alpha' * \beta'$ are based homotopic too. Indeed, if $\Gamma_1(t, s)$ and $\Gamma_2(t, s)$ are homotopies from α to α' and from β to β' then

$$\Gamma(t, s) = \begin{cases} \Gamma_1(t, s), \ t \le \frac{1}{2}, \\ \Gamma_2(t, s), \ t \ge \frac{1}{2}, \end{cases}$$

is a based homotopy from $\alpha * \beta$ to $\alpha' * \beta'$. So, catenation induces a binary operation on based homotopy classes of loops. Moreover, if catenations of loops satisfies up to based homotopy the axioms of a group law, the induced law on based homotopy classes of loops will be a group law.

The binary internal composition of catenation $*$ on $\Lambda_b(X)$ satisfies up to based homotopy the axioms of a group law.

The class of the constant loop $e \colon [0, 1] \to X$, $e(t) = b$, will be the neutral element in $\pi_1(X, b)$.

The inverse of the class of a loop γ is the class of the loop $\gamma^{-1}(t) = \gamma(1 - t)$. Indeed,

$$\Gamma(t, s) = \begin{cases} \gamma(2st), & t \le \frac{1}{2}, \\ \Gamma(t, s) = \gamma(s - 2st), \ t \ge \frac{1}{2}, \end{cases}$$

is a based homotopy from e to the catenation $\gamma * \gamma^{-1}$.

The class of e is a right inverse. Indeed,

$$\Gamma(t, s) = \begin{cases} \gamma(\frac{2t}{s+1}), & t \le \frac{s+1}{2}, \\ \Gamma(t, s) = e(t), \ t \ge \frac{s+1}{2}, \end{cases}$$

is a based homotopy from the catenation $\gamma * e$ to γ.

The composition by catenation is associative. Indeed, the catenations $\alpha * (\beta * \gamma)$ and $(\alpha * \beta) * \gamma$ are loops that differ only with respect to their parametrizations. The first has breakpoints at $\frac{1}{2}$ and $\frac{3}{4}$, the second at $\frac{1}{4}$ and $\frac{1}{2}$. By gliding the breakpoints one creates a based homotopy between the two catenations as follows

$$\Gamma(t, s) = \begin{cases} \alpha(\frac{4t}{2-s}), & t \le \frac{2-s}{4}, \\ \beta(4t - 2 + s), & \frac{2-s}{4} \le t \le \frac{3-s}{4}, \\ \gamma(\frac{4t}{1+s}), & \frac{3-s}{4} \le t. \end{cases}$$

The group law up to based homotopy $*$ induces a group law on $\pi_1(X, b)$. □

Let (X, b), (Y, c) be pointed spaces. A based continuous map $f : X \to Y$, i.e. a mapping with $f(b) = c$, induces a map $\Lambda f : \Lambda_b(X) \to \Lambda_c(Y)$ that commutes with catenation and moreover a group homomorphism $f_\# : \pi_1(X, b) \to \pi_1(Y, c)$. If two based continuous maps $f, g : X \to Y$ are based homotopic, the induced homomorphism on the fundamental groups are equal. It follows that the induced map by a based homotopy equivalence on fundamental groups is an isomorphism of groups.

The graph B_k having one vertex v and k edges is sometimes called a bouquet. Let Γ be a finite graph and let T be a maximal spanning tree in Γ. One can shrink the tree T inside Γ to a point and obtain a bouquet $B_k = \Gamma/T$. The edges of B_k correspond to the edges of Γ that do not belong to the subtree T. In fact, one obtains a map $\Gamma \to B_k = \Gamma/T$ which is a based homotopy equivalence, if one uses a point $p \in T$ as base point for Γ. The fundamental groups $\pi_1(\Gamma, p)$ and $\pi_1(B_k, v)$ are isomorphic. Knowing that the fundamental group of a bouquet B_k is the free group F_k on k generators (see Exercise A.11 below), and that $k + 1$ equals the Euler characteristics $\chi(\Gamma) = \chi(B_k)$ one concludes

Theorem A.10. *The fundamental group of a finite connected graph Γ is isomorphic to the free group on $\chi(\Gamma) - 1$ generators.* □

Exercise A.11. Prove that $\pi_1(B_k, v)$ is isomorphic to the free group on k generators. In particular the fundamental group of the circle is isomorphic to $\mathbb{Z}$. Hint (written for the case of two edges): Parametrize each edge of B_k by $[0, 1]$. Call the half edges of the first edge a and A, and those of the second b and B. Let T_k be the infinite tree without terminal vertices, every vertex of degree $2k = 4$. Think of the tree as the Cayley graph of the free group on the generators a, b. Label at each vertex the half edges by a, A, b, B such that each edge carries a, A or b, B. Let $E : T_k \to B_k$ be the continuous map that maps vertices to v, that is linear on each edge and that preserves the labelings of the half edges. Choose a vertex $V \in T_k$. The best choice for V is the neutral element in the Cayley graph of F_2. Let $\gamma : [0, 1] \to B_k$ be a based loop. Construct a path $p_\gamma : [0, 1] \to T_k$ with $p_\gamma(0) = V$ and $E \circ p_\gamma = \gamma$. Show that $p_\gamma(1)$

only depends upon the based homotopy class of the loop γ. The map E has locally on small open subsets $U \subset B_k$ inverses, like the exponential $t \in \mathbb{R} \mapsto e^{2\pi i t} \in S^1 = B_1$ and locally defined logarithms.

Exercise A.12. The surface S_g of genus $g \geq 1$ can be obtained by identifying pairwise edges of a $4g$-gon. Show that the image of the edges of the $4g$-gon in S_g is a subset homeomorphic to the bouquet B_{2g} with vertex v being the image of the vertices of the $4g$-gon. The inclusion $B_{2g} \subset S_g$ induces a surjective group homomorphism of the fundamental groups. Determine the kernel. Find a group presentation with generators and relations of the fundamental group of S_g.

The commutator subgroup of a group G is the smallest normal subgroup H of G that contains all commutators of G. The group G/H is abelian. Moreover every group homomorphism $G \to A$ from G to an abelian group A factors over the quotient group homomorphism $G \to G/H$. The group G/H is called the *abelianization* of G. The abelianization of the free group F_k is the free abelian group $\mathbb{Z}^k$ of free rank k.

Theorem A.13. *The abelianization of the fundamental group of a surface S_g of genus g is isomorphic to the abelian group $\mathbb{Z}^{2g}$ of free rank $2g$. Consequently, the genus of a compact connected orientable surface is a homotopic invariant.*

Proof. The presentation of $\pi_1(S_g, b)$, see Exercise A.12 above, is with $2g$ generators $a_1, b_1, a_2, b_2, \ldots, a_g, b_g$ and the product of the g commutators $a_g b_g a_g^{-1} b_g^{-1}$. Hence the abelianization of $\pi_1(S_g, b)$ is the free abelian group generated by the classes of $a_1, b_1, a_2, b_2, \ldots, a_g, b_g$. $\qquad\square$

A.3 Covering Spaces

Let (X, b) be a connected, locally path connected, topological space, pointed by $b \in X$. The aim is to construct spaces Y together with continuous maps $\pi: Y \to X$ which are onto such that for every point $p \in X$ there exists a path connected open subset $U_p \subset X$ and a system $(V_q)_{q \in \pi^{-1}(p)}$ of open subsets V_q in Y such that the following for $p \in X, q, q' \in \pi^{-1}(p)$ hold:

(i) $q \in V_q$ and $V_q \cap V_{q'} = \emptyset, q \neq q'$,
(ii) $\pi(V_q) = U_p$,
(iii) the restriction $\pi_{q,p}$ of π to V_q and U_p is a homeomorphism,
(iv) $\pi^{-1}(U_p) = \cup_{q \in \pi^{-1}(p)} V_q$.

A map $\pi: Y \to X$ as above is a *covering map* and the space Y is a *covering space* of the space X.

Covering maps $\pi: Y \to X$ are continuous, locally injective and open. Not all continuous, open and locally injective maps are covering maps.

One main interest of covering spaces is that local properties of a space X are inherited by the covering spaces of X. Typical such properties are being a manifold, or being a manifold with extra local data, such as a field. A fundamental example is that a covering space of a Riemann surface is again a Riemann surface.

The following theorem is an important tool for the study of covering maps and spaces. A *lifting* of a continuous map $f: A \to X$ is a continuous map $F: A \to Y$ with $f = \pi \circ F$.

Theorem A.14 (Path and Homotopy lifting). *Let (X, b) be a path connected topological space and let $\pi: Y \to X$ be a covering map. For each path $\gamma: [0, 1] \to X$ and for each $q \in \pi^{-1}(b)$ there exists a unique path $\Gamma: [0, 1] \to Y$ with $\Gamma(0) = q$ and $\pi \circ \Gamma = \gamma$. For each continuous map $\sigma: [0, 1] \times [0, 1]: \to X$ with $\sigma(0, 0) = b$ and $q \in \pi^{-1}(b)$, there exists a unique map $\Sigma: [0, 1] \times [0, 1]: \to Y$ with $\Sigma(0, 0) = q$ and $\pi \circ \Sigma = \sigma$.*

Proof. For $p \in X$, $q \in Y$, $p \in \pi^{-1}(p)$, let $U_p \subset X, V_q \subset Y$, be open sets satisfying the properties as above expressing that π is a covering map. Let $\sigma_{p,q}: U_p \to V_q$ be the inverse of the restriction of π to V_q.

Let $\gamma: [0, 1] \to X$ be a continuous path with $\gamma(0) = b$. For each $t \in [0, 1]$ let $I_t \subset [0, 1]$ be an interval, open in the space $[0, 1]$ with $t \in I_t$ and $\gamma(I_t) \subset U_{\gamma(t)}$. Since $[0, 1]$ is compact, there exists finitely many $t_1, t_2, \ldots, t_k$ such that the intervals I_{t_i} cover $[0, 1]$. Number the intervals I_i such that $0 \in I_1, I_{i+1} \not\subset \cup_{j=1}^{i} I_j$ and $I_{i+1} \cap \cup_{j=1}^{i} I_j \neq \emptyset$ hold. Choose $q_1 \in \pi^{-1}(\gamma(t_1))$ such that $q \in V_{q_1}$ holds. Define Γ on I_1 by $\Gamma(s) = \pi^{-1}_{q_1, \gamma(t_1)}(\gamma(s))$. Choose $q_i \in \pi^{-1}(\gamma(t_i))$ such that $V_{q_i} \cap V_{q_{i-1}} \neq \emptyset$, $i = 2, \ldots, k$. Extend Γ on I_i by $\Gamma(s) = \pi^{-1}_{q_i, \gamma(t_i)}(\gamma(s))$. This defines the lift Γ of γ with initial value q.

Let Γ_1, Γ_2 be two lifts of γ with equal initial values. The set $I' = \{t \in [0, 1] \mid \Gamma_1(t) = \Gamma_2(t)\}$ is non-empty, closed, and open in $[0, 1]$. It follows that $I' = [0, 1]$ since $[0, 1]$ is connected. This proves the uniqueness of path liftings.

The construction of homotopy liftings uses path liftings. For $e \in [0, 1] \times [0, 1]$ let γ_e be the path $\gamma_e(t) = \sigma(te)$. Let Γ_e be the lifting of γ_e with $\Gamma_e(0, 0) = q$. Define the lifting of σ by $\Sigma(e) = \Gamma_e(1)$. The covering property of π implies the continuity and uniqueness of Σ. ☐

A first corollary is that the cardinality of the fiber $\pi^{-1}(p)$ does not depend upon p if X is path connected. It is called the *degree* of the covering map $\pi: Y \to X$. A stronger corollary is:

Theorem A.15 (Monodromy by bijections and permutations). *Let (X, b) be a path connected topological space and let $\pi: Y \to X$ be a covering map. Each path $\gamma: [0, 1] \to X$, $b = \gamma(0)$, $p = \gamma(1)$, defines a bijection $m(\gamma): \pi^{-1}(b) \to \pi^{-1}(p)$. For a loop $\lambda \in \Lambda_b(X)$ the bijection $m(\lambda): \pi^{-1}(b) \to \pi^{-1}(b)$ depends only on the class $[\lambda]_{BasHom}$, hence defines a representation, called a* monodromy *representation, of the fundamental group by permutations of the fiber $m: \pi_1(X, b) \to \mathrm{Perm}(\pi^{-1}(b))$.*

Proof. The liftings of a path γ in X that connects the base point b to a point p transport the fiber of b to the fiber of p. The bijection between the fibers is, by the homotopy lifting property, determined by the homotopy class of the path with respect to homotopies that fix both end points of the path. Pre-catenation of paths in $\Lambda_b(X)$ induce a representation of $\pi_1(X, b)$ by permutations of the fiber over b. ☐

Two covering maps $\pi\colon Y \to X$ and $\pi'\colon Y' \to X$ are isomorphic if there exists a homeomorphism $\phi\colon Y \to Y'$ such that π factors $\pi = \pi' \circ \phi$.

Exercise A.16. The map $E\colon T_k \to B_k$ of the previous section (Exercise A.11) is an example of a covering map.

In order for the following construction of covering maps to work, the space X has to be locally simply connected. This is the case for manifolds, graphs and many more topological spaces.

Let X be a path connected and locally simply connected space. Let $b \in X$ be a base point. The *path space* of (X, b) is the space $P_b(X)$ of continuous maps $\gamma\colon [0, 1] \to X$ with $\gamma(0) = b$ and endowed with the compact-open topology. The map $\pi\colon P_b(X) \to X, \gamma \mapsto \gamma(1)$, is continuous and surjective. For two path $\gamma_1, \gamma_2 \in P_b(X)$ with $\gamma_1(1) = \gamma_2(1)$ the catenation $\gamma_1 * \gamma_2^{-1}$ defines an element in the loop space $\Lambda_b(X)$.

Each subgroup $H \subset \pi_1(X, b)$ defines by the following an equivalence relation $\sim_{P,H}$ on the path space $P_b(X)$:

$$\gamma_1 \sim_{P,H} \gamma_2 \Leftrightarrow \gamma_1(1) = \gamma_2(1), \ [\gamma_1 * \gamma_2^{-1}]_{\mathrm{BasHom}} \in H.$$

Denote by X_H the quotient space $P_b(X)/\sim_{P,H}$. The map $\pi\colon P_b(X) \to X$ and the equivalence relation $\sim_{P,H}$ are compatible and induce a continuous map $\pi_H\colon X_H \to X$.

The following theorems construct and classify connected covering spaces of a connected, based, and locally simply connected space X in terms of the subgroups of the fundamental group of X.

Theorem A.17. *Let X be a path connected and locally simply connected topological space based at $b \in X$. The map $\pi_H\colon X_H \to X$ is a covering map for every subgroup H in $\pi_1(X, b)$. The degree of π_H is the cardinality of the quotient space $\pi_1(X, b)/H$.*
□

Theorem A.18. *Let X be a path connected and locally simply connected topological space based at $b \in X$. For every covering map $\pi\colon Y \to X$ such that Y is connected, there exist a subgroup H of $\pi_1(X, b)$ and a homeomorphism $\phi\colon Y \to X_H$ such that π factors as $\pi = \pi_H \circ \phi$.*

Proof. Choose $b_Y \in Y$ with $\pi(b_Y) = b$. The set $\Lambda_b(X, \pi, b_Y)$ of loops in X, based at b, that lift to a closed loop in Y based at b_Y is closed under catenation and also under parameter reversal $t \mapsto 1 - t$. The based homotopy classes of loops in $\Lambda_b(X, \pi, b_Y)$ define a subgroup H in $\pi_1(X, b)$. Fix a point $b_H \in X_H$ with $\pi_H(b_H) = b$ and such that the based homotopy classes of paths in $\Lambda_b(X_H, \pi_H, b_H)$ belong to H. Such a point b_H exists, take for instance the class $[e_b, b]_H \in X_H$. We define a map $\phi\colon X_H \to Y$ as follows. For $p \in X_H$ choose a path γ_H in X_H from b_H to p. Let γ_Y be the lift of the path $\pi_H \circ \gamma$ starting at b_Y in Y. Define $q = \phi(p) = \Gamma(1) \in Y$. The map ϕ is well defined and factors as $\pi_H = \pi \circ \phi$. The construction: $p \to \gamma_H \to \gamma_Y \to q$ can be reversed. It follows that ϕ is a homeomorphism.
□

Theorem A.19. *Let X be a path connected and locally simply connected topological space based at $b \in X$. The covering maps $\pi_H: X_H \to X$ and $\pi_{H'}: X_{H'} \to X$ are isomorphic if and only if the subgroups H, H' are conjugate in $\pi_1(X, b)$.*

Proof. Assume that $\phi: X_H \to X_{H'}$ is a covering isomorphism. Choose $b_H \in X_H$, $b_{H'} \in X_{H'}$ such that the based homotopy classes of loops in

$$\Lambda_b(X_H, \pi_H, b_H), \ \Lambda_b(X_{H'}, \pi_{H'}, b_{H'})$$

are the elements of H, H' respectively. Choose a path γ in $X_{H'}$ from $b_{H'}$ to $\phi(b_H)$. Let $g \in \pi_1(X, b)$ be the based homotopy class of $\pi_{H'} \circ \gamma$. Observe $gH = H'g$, showing that H and H' are conjugate subgroups in $\pi_1(X, b)$.

Conversely, assume $gH = H'g$, where g is the based homotopy class of a loop λ. Define $\phi: X_H \to X_{H'}$ as follows. Given $p \in X_H$ represented by (γ_H, b), $\gamma_H \in P_b(X)$ let Γ be a lift starting at $b_{H'}$ of the catenation $\gamma_{H'} = \lambda * \gamma_H * \lambda^{-1}$. Define $q = \phi(p) = \Gamma(1)$. The map $\phi: X_H \to X_{H'}$ is an isomorphism of coverings. $\square$

This gives the following:

Theorem A.20. *Let X be a path connected and locally simply connected topological space based at $b \in X$. The covering maps $\pi_H: X_H \to X$ and $\pi_{H'}: X_{H'} \to X$ factor by a covering map $\phi: X_H \to X_{H'}$, $\pi_H = \pi_{H'} \circ \phi$, if and only if H is conjugate to a subgroup of H'.* $\square$

The classification of covering maps up to isomorphism is reduced to the classification of subgroups up to conjugacy. This, however, can be very complicated.

Exercise A.21. The cardinality of the set of subgroups up to conjugacy in F_2 is the cardinality of the reals. The cardinality of the set of normal subgroups of F_2 is countable. The cardinality of the set of subgroups of finite index is countable.

Exercise A.22. Construct explicitly uncountably many pairwise non-isomorphic coverings of B_2.

Exercise A.23. Construct up to isomorphism all coverings of degree 2 of B_4.

Exercise A.24. Let $P: \mathbb{C} \to \mathbb{C}$ be a degree d polynomial mapping. Let C be the roots of the derivative P', put $X = \mathbb{C} \setminus P(C)$ and $Y = P^{-1}(X)$. The restriction of P to Y is a mapping $R_P: Y \to X$. Show that R_P is a covering map.

The loop space $\Lambda_b(X)$ acts by pre-catenation on the path space $P_b(X)$

$$*: \Lambda_b(X) \times P_b(X) \to P_b(X), \ (\lambda, \gamma) \mapsto \lambda * \gamma.$$

Of special interest are those subgroups H of $\pi_b(X, b)$ for which the pre-catenation action is compatible with the equivalence relation $\sim_{P,H}$. In this case the pre-catenation action descends to an action of the fundamental group $\pi_b(X, b)$ on X_H.

Theorem A.25. *The pre-catenation action $*: \Lambda_b(X) \times P_b(X) \to P_b(X)$ is compatible with the equivalence relations $\sim_{BasHom}$ and $\sim_{P,H}$ if and only if H is a normal subgroup of $\pi_1(X, b)$.*

Proof. Assume H is normal in $\pi_1(X, b)$. Indeed, for loops $\lambda_1, \lambda_2 \in \Lambda_b(X)$ with $\lambda_1 \sim_{\text{BasHom}} \lambda_2$, and paths $\gamma_1, \gamma_2 \in P_b(X)$ with $\gamma_1 \sim_{P,H} \gamma_2$, observe

$$[(\lambda_1 * \gamma_1) * (\lambda_2 * \gamma_2)^{-1}]_{P,H} = [\lambda_1]_{\text{BasHom}} * [\gamma_1 * \gamma_2^{-1}]_{P,H} * [\lambda_2]_{\text{BasHom}}^{-1} \in H$$

since $[\gamma_1 * \gamma_2^{-1}]_{P,H} \in H$ and $[\lambda_1]_{\text{BasHom}} = [\lambda_2]_{\text{BasHom}} \in \pi_b(X, b)$. Hence the action by pre-catenation is compatible with the equivalence relations.

Assume H is not normal in $\pi_1(X, b)$. Choose $h \in H$, $g \in \pi_1(X, b)$ with $ghg^{-1} \notin H$. Choose $\lambda_1 = \lambda_2$ with $[\lambda_1]_{\text{BasHom}} = g$ and $\gamma_1 \in \Lambda_b(X) \subset P_b(X)$ with $[\gamma_1]_{\text{BasHom}} = h$. Choose $\gamma_2 = e_b$, the constant path at b. Observe that $\gamma_1 \sim_{P,H} \gamma_2$ but $\lambda_1 * \gamma_1 \sim_{P,H} \lambda_2 * \gamma_2$ does not hold. Hence, the action by pre-catenation is not compatible with the equivalence relations. $\square$

A covering map $\pi : Y \to X$ that is isomorphic to $\pi_H : X_H \to X$ with H normal in $\pi_1(X, b)$ is called a *Galois covering map*.[2]

Theorem A.26. *Let the subgroup H be normal in $\pi_b(X, b)$. Let $\pi_H : X_H \to X$ be the corresponding Galois covering map. The action of the group $\pi_b(X, b)$ on X_H factors through a free action of the group $\pi_b(X, b)/H$ on X_H that preserves the fibers of π_H and that is simply transitive on each fiber.*

Proof. Pre-catenation preserves the value of a path at 1, so the action of $\pi_b(X, b)$ preserves each fiber of π_H. For a loop $\lambda \in \Lambda_b(X)$ and path γ, observe that $\lambda * \gamma \sim_{P,H} \gamma$ holds if and only if $[\lambda]_{\text{BasHom}} \in H$, hence H acts trivially on X_H and the action factors through a free action of $\pi_b(X, b)/H$. The action of $\pi_b(X, b)/H$ is fiber-wise transitive, since the action of $\pi_b(X, b)$ is fiber-wise transitive. $\square$

Exercise A.27. Let P be an univariate polynomial map. Is it true in general, or only sometimes, that the covering map R_P from Exercise A.24 is a Galois covering map? Study $P = z^3 - z^2$.

Let (X, b) be a connected and locally simply connected space. The covering map $\pi_H : X_H \to X$ for $H = \pi_1(X, b)$ is in fact the identity map of X. The other extreme case, when H is the trivial subgroup $\{e\}$ consisting of only the neutral element, is very interesting. The covering map $\pi_H : X_H \to X$, $H = \{e\}$, is called the *universal covering map of X*. The universal covering map of a space X is denoted by $\bar{\pi} : \bar{X} \to X$. The fundamental group of (X, b) acts freely on the universal covering space $\bar{X}$.

---

[2] Named after Evariste Galois (1811–1832), French mathematician.

A.4 Tessellations and Coverings

Tessellations of the Euclidean or hyperbolic plane are useful for constructing covering spaces of surfaces. Tessellations by geometric tiles build a bridge between fundamental groups and spaces, see and study the books *Generators and Relations for Discrete Groups*, by H.S.M. Coxeter and W.O.J. Moser [26] and *Tilings and Patterns* by Branko Grünbaum and G.C. Shephard [46].

The most basic examples of tessellation are related to the tile $[0, 1] \subset \mathbb{R}$. To make a tessellation of the real line with the tile $[0, 1]$, we need to move it by a group of isometries of the real line, such that the moved tiles cover the real line and that two tiles have at most boundary points in common. The group Γ_1 of translation $u \mapsto u + n$, $n \in \mathbb{Z}$, gives a tessellation with the tile $[0, 1]$. A second group that tessellates the real line with the tile $[0, 1]$ is the group Γ_2 generated by the two reflections R_0, R_1 having 0 and 1 respectively as fixed points.

Both group actions of Γ_1 and Γ_2 induce on $\mathbb{R}$ equivalence relations whose restrictions to the interior of the tile $[0, 1]$ agree. Both quotient spaces $\mathbb{R}/\Gamma_1$ and $\mathbb{R}/\Gamma_2$ are homeomorphic to the circle. However, the quotient maps $q_1 : \mathbb{R} \to \mathbb{R}/\Gamma_1$ and $q_2 : \mathbb{R} \to \mathbb{R}/\Gamma_2$ have different topological properties. The map q_1 is a covering map, but the map q_2 is not. At points $p \in \mathbb{Z}$ the map q_2 is not locally injective, and is not open.

The subgroup Γ_2' consisting of products of the reflections R_0, R_1 with an even number of factors is a normal subgroup of index 2 in Γ_2. The group Γ_2' is generated by the translation $u \mapsto u + 2$. The group Γ_2' tessellates the real line by moving the tile $[0, 2]$. The quotient map $\mathbb{R} \to \mathbb{R}/\Gamma_2'$ is a covering map to the circle.

The most basic example of a tessellation of the Euclidean plane is with the tile $[0, 1] \times [0, 1] \subset \mathbb{R}^2$ and the group $\Gamma_1 = \mathbb{Z}^2$ of Euclidean motions generated by the two translations $u \mapsto u + e_1$ and $u \mapsto u + e_2$, e_1, e_2 being the standard basis of $\mathbb{R}^2$. This group of motions has as elements the translations by the vectors with integral coordinates. The quotient map $\mathbb{R}^2 \to \mathbb{R}^2/\mathbb{Z}^2$ is the universal covering map of the torus.

Many more groups of motions give a tessellation of the plane by moving the tile $[0, 1] \times [0, 1]$, see the remarkable book *Anschauliche Geometrie* by David Hilbert and Stephan Cohn-Vossen [51]. For instance, the group Γ_2 generated by the reflections through the side of the tile. Again the quotient $\mathbb{R}^2/\Gamma_2$ is homeomorphic to the torus, but the quotient map is not a covering map. The group Γ_3 generated by the translation $u \mapsto u + e_2$ and the map $ae_1 + be_2 \mapsto (a + 1)e_1 - be_2$ moves the tile $[0, 1] \times [0, 1]$ giving a tessellation of the Euclidean plane. The quotient map is the universal cover of the Klein bottle, which is a non-orientable surface.

The group Γ_ω of motions of the complex line $\mathbb{C}$ generated by the translations $u \mapsto u + 1$ and $u \mapsto u + \omega$ for ω in the upper-half plane $\mathbb{C}_+$ preserves the natural complex structure of the complex line. The group Γ_ω tessellates the complex line with the parallelogram $P = \{t + s\omega \mid t, s \in [0, 1]\}$ as tile. The quotient $\mathbb{C}/\Gamma_\omega$ is a Riemann surface of genus 1, as we have seen earlier.

The group Γ_ω also preserves the real affine structure and the topology of the complex line. After an adaptation, this allows us to use typical affine constructions. A subset $A \subset \mathbb{C}$ is *convex* if for all $p, q \in A$ the segment

$$[p, q] = \{tp + (1 - t)q \mid t \in [0, 1]\}$$

belongs to A.

References

1. A'Campo, Norbert, Ji, Lizhen and Papadopoulos, Athanase. On Grothendieck's construction of Teichmüller space, In *Handbook of Teichmüller Theory, Vol. VI*, ed. Athanase Papadopoulos, European Mathematical Society, Zürich (2016)
2. A'Campo, Norbert and Papadopoulos, Athanase. Notes on hyperbolic geometry, In *Strasbourg Master class on Geometry*, pp. 1–182, IRMA Lectures in Mathematics and Theoretical Physics, Vol. 18, Zürich: European Mathematical Society (EMS), 461 pages (2012)
3. A'Campo, Norbert and Papadopoulos, Athanase. Area in non-Euclidean geometry, In *Eighteen essays in non-Euclidean geometry*, ed. A. Papadopoulos, pp. 3–25, Eur. Math. Soc., Zürich (2019)
4. A'Campo-Neuen, Annette, A'Campo, Norbert, Ji, Lizhen and Papadopoulos, Athanase. A commentary on Teichmüller's paper, *Deutsche Math.* 7 (1944), 344–359, in *Handbook of Teichmüller theory*, ed. A. Papadopoulos, pp. 805–814, Eur. Math. Soc., Zürich (2014)
5. Ahlfors, Lars and Bers, Lipman. Riemann's mapping theorem for variable metrics, *Annals of Mathematics* 72, 385–404 (1960)
6. Albiac, Fernando and Kalton, Nigel J. *Topics in Banach Space Theory*, Graduate Texts in Mathematics 233, Springer, New York (2006)
7. Archimedes of Syracuse. Περὶ σφαίρας καὶ κυλίνδρου [*On the Sphere and Cylinder*], `http://openn.library.upenn.edu/Data/0014/ArchimedesPalimpsest/` (225 BC)
8. Arianrhod, Robyn. *Thomas Harriot: A Life in Science*, Oxford University Press, 376 pp. (2019)
9. Baer, Reinhold. Isotopien von Kurven auf orientierbaren, geschlossenen Flächen, *Journal für die Reine und Angewandte Mathematik* 159, 101–116 (1928)
10. Baker H.F. *Abel's theorem and the allied theory of theta functions*, Cambridge University Press (1897), Cambridge, UK (reprinted 1995)
11. Belyi, G.V. Galois extensions of a maximal cyclotomic field, *Izv. Akad. Nauk SSSR Ser. Mat.* 43, 267–276, 479 (1979)
12. Belyĭ, Gennadiĭ Vladimirovich. Translated by Neal Koblitz. Galois extensions of a maximal cyclotomic field, *Math. USSR Izv.* 14 (2), 247–256 (1980)
13. Berger, Marcel. *A Panoramic View of Riemannian Geometry*, Springer-Verlag, Berlin Heidelberg New-York (2007)
14. Bonola, Roberto. *Non-Euclidean Geometry, A critical and Historical Study* (1906). Translated by H.S. Carslaw, with a foreword by Federigo Enriques, supplemented with a translation by George Bruce Halsted of *The Theory of Parallels* by Nicholas Lobachevski, and *The Science of Absolute Space* by John Bolayai, New Dover Edition, Dover Publications, Inc., 920 Broadway, New York 10, NY (1955)
15. Bourbaki, Nicolas. *Livre III, Topologie Générale*, Chap.1–2, Hermann Paris (1961)
16. Brägger, Walter. *A Uniformisation of Weighted Maps on Compact Surfaces*, PhD Thesis, Basel and `arXiv:2001.0661` (1995)
17. Busemann, Herbert. Spaces with non-positive curvature, *Acta Math.* 80, 259–310 (1948)

18. Buser, Peter. *Geometry and Spectra of Compact Riemann Surfaces*, Modern Birkhäuser Classics, Basel (1992)
19. Cartan, Élie. Sur certaines expressions différentielles et le problème de Pfaff, *Annales scientifiques de l'École Normale Supérieure*, Sér. 3, 16, 239–332 (1899)
20. Cartan, Élie. Sur une classe remarquable d'espaces de Riemann, I., *Bulletin de la Société Mathématique de France*, 54: 214–216 (1926)
21. Cartan, Élie. Sur une classe remarquable d'espaces de Riemann, II., *Bulletin de la Société Mathématique de France*, 55: 114–134 (1927)
22. Cartan, Henri. *Théorie élémentaire des fonctions analytiques d'une ou plusieurs variables complexes*, Avec le concours de Reiji Takahashi, Hermann, Paris (1964)
23. Cartan, Henri. *Calcul différentiel*, Hermann, Paris (1967)
24. Cerf, Jean. La stratification naturelle des espaces de fonctions différentiables réelles et le théorème de la pseudo-isotopie, *Inst. Hautes Études Sci. Publ. Math. No.* 39 5–173 (1970)
25. Chow, W.-L. On Compact Complex Analytic Varieties, *American Journal of Mathematics*, Vol. 71, No. 4, 893–914 (1949)
26. Coxeter, H.S.M. and Moser, W.O.J. *Generators and Relations of Discrete Groups*, third ed., Springer, Berlin (1972)
27. Darboux, Gaston. Sur le problème de Pfaff, Bull. Sci. Math. 6, 14-36, 49-68(1882)
28. Deahna, F. Ueber die Bedingungen der Integrabilität linearer Differentialgleichungen erster Ordnung zwischen einer beliebigen Anzahl veränderlicher Größen, *J. Reine Angew. Math.* 20, 340–350 (1840)
29. de Rham, Georges, *Variétés différentiables: formes, courants, formes harmoniques* (1955) Hermann, Paris, in *Differentiable Manifolds: Forms, Currents, Harmonic Forms*, Springer, Grundlehren Math. Wiss. 266, Berlin (1984)
30. de Saint-Gervais, Paul Henri, *Uniformisation des Surfaces de Riemann. Retour sur un Théorème centenaire*, ENS Éditions, Lyon (2010)
31. de Saint-Gervais, Paul Henri. *Analysis Situs*, http://analysis-situs.math.cnrs.fr/ and sections: http://analysis-situs.math.cnrs.fr/-Introduction-a-l-Analysis-situs-par-les-surfaces-.html, http://analysis-situs.math.cnrs.fr/-Bibliographie-commentee-.html
32. Dieudonné, Jean. *Foundations of modern analysis*, Boston, MA, Academic Press (1969)
33. Dieudonné, Jean. *Exposé at the Symposium in honour of Henri Cartan's 70th birthday*, University Paris-Sud at Orsay (1974)
34. Donaldson, Simon. *Riemann Surfaces*, Oxford University Press (2011)
35. Earle, C.J. and Eells, J. The diffeomorphism group of a compact Riemann surface, *Bull. Amer. Math. Soc.*, Volume 73, Number 4, 557–559 (1967)
36. Einstein, Albert. Über die von der molekularkinetischen Theorie der Wärme geforderte Bewegung von in ruhenden Flüssigkeiten suspendierten Teilchen, *Annalen der Physik*, 322, (8) 549–560 (1905),
37. Fathi, Albert, Laudenbach, François and Poenaru, Valentin. *Thurston's Work on Surfaces*, Mathematical Notes 48, Princeton University Press, Translated by Djun Kim and Dan Margalit from the French edition (1978) *Travaux de Thurston* (2012)
38. Fischer, Arthur E. and Tromba, Anthony J. Almost complex principal fiber bundles and the complex structure on Teichmüller space, *J. Reine Angew. Math.* 352, 151–160 (1984)
39. Frobenius, Georg. Über das Pfaffsche Problem, *J. für Reine und Angew. Math.*, 82 230–315 (1877)
40. Girondo, Ernesto, González-Diez, Gabino. *Introduction to compact Riemann surfaces and dessins d'enfants*, London Mathematical Society Student Texts, 79, Cambridge University Press (2012)
41. Gramain, André. *Topology of surfaces*, translated from the French and edited by Leo F. Boron, Charles O. Christenson and Bryan A. Smith, Moscow, Idaho: BCS Associates. First published by Presses Universitaires de France, Paris (1971), under the title of *Topologie des surfaces* (1984)
42. Grothendieck, Alexandre. Résumé des resultats essentials dans la théorie des produits tensoriels topologiques et des espaces nucléaires, *Annales de l'Institut Fourier* 4, 73–112 (1952)
43. Grothendieck, Alexandre. *Produits tensoriels topologiques et espaces nucléaires*, Memoirs of the American Mathematical Society, 16 (1955)

44. Grothendieck, Alexandre. *Techniques de construction en géométrie algébrique,* Séminaire Cartan, Paris 1960–61, Exposés 1–20 (1960–61)
45. Grothendieck, Alexandre. *Esquisse d'un Programme,* (1984), in *Geometric Galois Actions, 1. Around Grothendieck's Esquisse d'un Programme,* ed. Leila Schneps and Pierre Lochak, London Mathematical Society Lecture Note Series 242, Cambridge University Press (1997)
46. Grünbaum, Branko and Shephard, G.C. *Tilings and Patterns,* W.H. Freeman and Company, New York (1986)
47. Guggenheim, H. The Jordan curve theorem and an unpublished manuscript by Max Dehn, *Archive for History of Exact Sciences* 17, 193–200 (1977)
48. Hales, Thomas. Jordan's proof of the Jordan curve theorem, *Studies in Logic, Grammar and Rhetoric* 10 (23) (2007)
49. Harnack, Axel. *Die Grundlagen der Theorie des logarithmischen Potentiales und der eindeutigen Potentialfunktion in der Ebene,* B. G. Teubner, Leipzig (1887)
50. Helgason, Sigurdur, *Differential geometry, Lie groups and symmetric spaces,* Academic Press (1978)
51. Hilbert, David and Cohn-Vossen, Stephan. *Anschauliche Geometrie,* (1932), second edition with extra Appendix, *Einfachste Grundbegriffe der Topologie,* Alexandroff, Paul, Springer Verlag (2011)
52. Hodge, W.V.D. *The theory and applications of harmonic integrals,* Cambridge Mathematical Library, Cambridge University Press (1941)
53. Huber Heinz. *Zur analytischen Theorie hyperbolischer Raumformen und Bewegungsgruppen,* Math. Annalen, Vol. 138, 1-26 (1959).
54. Huber Heinz. *Zur analytischen Theorie hyperbolischer Raumformen und Bewegungsgruppen,II,* Math. Annalen, Vol. 142, 385-398 (1960/1961), Vol. 143, 463-464 (1961).
55. Hurwitz, Adolf. *Über algebraische Gebilde mit eindeutigen Transformation in sich, Math. Ann.* 41, 408–442 (1893)
56. Ivanov, Nikolai V. A Topologist's View of the Dunford–Schwartz Proof of the Brouwer Fixed-Point Theorem, *The Mathematical Intelligencer,* Vol. **22**, Springer, New York (2000)
57. Kakutani, Shizuo. Two-dimensional Brownian motion and harmonic functions, *Proceedings of the Imperial Academy,* Volume 20, Number 10, 706–714 (1944)
58. Karamata, Jovan. Neuer Beweis und Verallgemeinerung einiger Tauberian-Sätze, *Math. Zeitschrift,* 33, 1, 294–299 (1931)
59. Katz, Nicolas M. An overview of Deligne's proof of the Riemann hypothesis for varieties over finite fields (Hilbert's problem 8), in *Mathematical Developments Arising from Hilbert Problems,* Proceedings of Symposia in Pure Mathematics XXVIII, ed. Browder, Felix E., American Mathematical Society. A collection of survey essays by experts devoted to each of the 23 problems emphasizing current developments (1976)
60. Knus, Max-Albert, Merkurjev, Alexander, Rost, Markus and Tignol, Jean-Pierre, *The Book of Involutions,* AMS (1998)
61. Kodaira, K.; Spencer D.C., *On deformation of complex analytic structures,* Ann. Math 67, 328-466 (1958)
62. Koszul, Jean-Louis. *Lectures on fibre bundles and Differential Geometry,* Tata Inst. Fund. Research, Bombay (1965)
63. Köthe, Gottfried. *Topologische lineare Räume I,* Grundlehren der mathematischen Wissenschaften in Einzeldarstellungen, Band 107: (1969, 1979), *Topological vector spaces. I. II. (Translated from German by D.J.H. Garling),* Grundlehren der mathematischen Wissenschaften, Band 159, Springer-Verlag New York (1960)
64. Kürschák, J. Über Limesbildung und allgemeine Körpertheorie, in *Proceedings of the* 5th *International Congress of Mathematicians,* Cambridge 1912, vol. 1, 285–289 (1912)
65. Kürschák, J., Über Limesbildung und allgemeine Körpertheorie, *Journal für die reine und angewandte Mathematik* 142, 211–253 (1913)
66. Ladegaillerie, Yves, *Découpes et isotopies de surfaces topologiques,* 537 Thèse de Doctorat d'État, Faculté des Sciences, Montpellier (1976)
67. Lando, Sergei K.; Zvonkin, Alexander K., *Graphs on Surfaces and Their Applications,* Encyclopaedia of Mathematical Sciences: Lower-Dimensional Topology II, 141, Berlin, New York: Springer-Verlag (2004)

68. Lévy, Paul, Le mouvement brownien plan, *Amer. Journ. of Math.*, 62 (1) 487–550 (1940)
69. Łoś, Jerzy. Quelques remarques, théorèmes et problèmes sur les classes définissables d'algèbres, in Symposium by Thoralf Skolem and Wiskundig Genootschap on *Mathematical interpretation of formal systems*, pp. 98–113. North-Holland Publishing Co., Amsterdam (1955)
70. Maehara, Ryuji. The Jordan Curve Theorem Via the Brouwer Fixed Point Theorem, *The American Mathematical Monthly*, Mathematical Association of America, 91 (10), 641–643 (1984)
71. Magnus, Wilhelm. *Noneuclidean Tesselations and Their Groups*, Academic Press, New York and London (1974)
72. Malgrange, Bernard, Sur l'intégrabilité des structures presque complexes, *Séminaire Jean Leray* N. 1 (1968–1969), 1-8 (1968)
73. Malliavin, Paul. *Géométrie différentielle intrinsèque*, Paris: Hermann (1972)
74. Maskit, Bernard. Comparison of hyperbolic and extremal lengths, *Ann. Acad. Sci. Fenn. Ser. A I Math.* 10, 381–386 (1985)
75. Milnor, John W. *Topology from the differential view point*, Based on notes by David W. Weaver, The University Press of Virginia, Charlottesville, USA (1965),
76. Milyutin Aleksei A. Isomorphism of the spaces of continuous functions over com- pact sets of the cardinality of the continuum, *Teor. Funkcii Funkcional Anal. i Prolozen*, 2, 150–156 (1966)
77. Minsky, Yair N. Extremal length estimates and product regions in Teichmüller space, *Duke Mathematical Journal*, 83, 249–286 (1996)
78. Mirzakhani, Maryam. *Growth of the number of simply closed geodesics on hyperbolic surfaces*, Annals of Mathematics, Series 2, Band 168, Nr. 1, 97–125 (2008).
79. Mischler, Bernard. Private communications, on several occasions while travelling by bicycle (c. 1975)
80. Morse, Marston. The critical points of a function of n variables, *Trans. Amer. Math. Soc.* 33, 72–91 (1931)
81. Moser, J. On the Volume Elements on a Manifold, *Transactions of the AMS*, 120(2), 286–294 (1965)
82. Mueller, W. and Wendland, K. Extremal Kaehler metrics and Ray–Singer analytic torsion, in *Geometric aspects of partial differential equations (Roskilde, 1998)*, 135–160, Contemp. math., 242 AMS, Providence, RI, USA (1999)
83. Newlander, A. and Nirenberg, L. *Complex analytic coordinates in almost complex manifolds*, Annals of Mathematics (Second Series) 65 (3), 391–404 (1957)
84. Nijenhuis, Albert. X_{n-1}-forming sets of eigenvectors, *Indag. Math.* 13, 200–212 (1951)
85. Nijenhuis, Albert and Woolf, William. Some integration problems in almost-complex and complex manifolds, *Ann. Math.* 77, 424–489 (1963)
86. Osgood, B, Phillips, R. and Sarnak, P. Extremals of determinants of Laplacians, *J. Funct. Anal.* 80, 148–211 (1988)
87. Ostrowski, Alexander. Über einige Fragen der allgemeinen Körpertheorie, *J. fur Reine Angew. Math.* 143, 255–284 (1913)
88. Ostrowski, Alexander. Untersuchungen zur arthmetischen Theorie der Körper, *Math. Z.* 39, 1, 269–320 and *Math. Z.* 39, 1, 321–404 (1935),
89. Penner, Robert. *Decorated Teichmüller theory*, with a foreword by Yuri I. Manin, QGM Master Class Series, European Mathematical Society, Zürich, xviii+360pp (2012)
90. Penner, Robert. Moduli spaces and macromolecules, *Bull. Amer. Math. Soc.* 53 217–268 (2016)
91. Penrose, Roger. *The Road to Reality. A Complete Guide to the Laws of the Universe,* Jonathan Cape, Random House, London (2004)
92. Perron, Oskar. Eine neue Behandlung der ersten Randwertaufgabe $\Delta u = 0$, *Math. Z.* 18, 42–54 (1923)
93. Petersson, Hans. Über eine Metrisierung der automorphen Formen und die Theorie der Poincaréschen Reihen, *Math. Ann.* 117, 453–537 (1940)
94. Pfluger, Albert. *Theorie der Riemannschen Flächen*, Springer-Verlag, Berlin Göttingen Heidelberg (1957)

95. Phelps, Robert R. *Lectures on Choquet's Theorem*, Second Edition, Lectures Notes in Mathematics, Springer, Heidelberg, Berlin, New York (2001); First Edition van Nostrand, Princeton, NJ (1966)

96. Polyakov, A.M. Quantum geometry of bosonic strings, *Phys. Lett.* 103B, 211–213 (1981)

97. Popescu-Pampu, Patrick. *What is the Genus?*, Lecture Notes in Mathematics, Springer (2016)

98. Radó, Tibor. Über den Begriff der Riemannschen Fläche, *Acta Szeged*, 2 (2), 101–121 (1925)

99. Radó, T. *Subharmonic Functions*, Erg. Math. u. i. Grenzgebiete, Berlin, Julius Springer (1937)

100. Rashed, Roshdi and Papadopoulos, Athanase. *Menelaus' Spherics*. Early Translation and al-Māhānī / al-Harawī's Version. De Gruyter, Scientia Graeco-Arabica 21 (2017)

101. Revuz, Daniel and Yor, Marc. *Continuous martingales and Brownian motion*, Springer Verlag (1999)

102. Riemann, Bernhard. *Grundlagen für eine allgemeinen Theorie der Funktionen einer veränderlichen complexen Grösse,* Inaugural dissertation, Göttingen, (1851)

103. Riemann, Bernhard. Theorie der Abelschen Functionen, *J. Reine Angew. Math.* 54, 115–155 (1857)

104. Riemann, Bernhard. Ueber das Verschwinden der Theta-Functionen, *J. Reine Angew. Math.* 65, 161–172 (1866)

105. Riemann, Bernhard. *Gesammelte Mathematische Werke und wissenschaftlicher Nachlass*, Herausgegeben unter Mitwirkung von R. Dedekind von H. Weber, Leipzig, Druck und Verlag von B.G. Teubner (1876)

106. Розенфельд, Б.А. История неевклидовоуи геометрии. Развитие понятия о геометрическом пространстве. Izdat. Nauka, Moscow (1976)

107. Rosenfeld, B.A. *A history of non-euclidean geometry. Evolution of the concept of a geometric space*, Translated from the Russian by Abe Shenitzer. Studies in the History of Mathematics and Physical Sciences, 12. Springer-Verlag, New York-Berlin (1988)

108. Samelson, Hans. Differential Forms, the Early Days; or the Stories of Deahna's Theorem and of Volterra's Theorem, *American Mathematical Monthly*, Vol. 108, No. 6, 522–530 (2001)

109. Sard, Arthur. The measure of critical values of differentiable maps, *Bull. Amer. Math. Soc.* 48, 883–890 (1942)

110. Schoutens, Hans. *The Use of Ultraproducts in Commutative Algebra*, Lecture Notes in Mathematics, Vol. 1999, Springer, Heidelberg-Dordrecht-London-New York (2010)

111. Schwartz, Laurent. *Théorie des distributions*, Publications de l'Institut de Mathématique de l'Université de Strasbourg, No. IX-X. Nouvelle édition, entiérement corrigée, refondue et augmentée. Hermann (1950),

112. Serre, Jean-Pierre. Un théorème de dualité, *Comment. Math. Helv.* 29, 9–26 (1955)

113. Serre, Jean-Pierre. Géométrie algébrique et géométrie analytique, *Annales de l'Institut Fourier*, Vol. 6, 1–42 (1956)

114. Steinitz, Ernst. Algebraische Theorie der Körper, *J. Reine Angew. Math.* 137, 167–309 (1910)

115. Steinitz, Ernst. Algebraische Theorie der Körper, New publication with permission by Ernst Steinitz and with the addition: *Abriss der Galoisschen Theorie,* by Reinhold Baer and Helmut Hasse, W. de Gruyter and Co., Berlin (1930)

116. Sullivan, Dennis. On the ergodic theory at infinity of an arbitrary group of hyperbolic motions, in *Riemann surfaces and related topics*, Proceedings of the 1978 Stony Brook Conference, edited by I. Kra and B. Maskit, Ann. of Math. Studies 97, Princeton University Press, Princeton NJ, 465–496 (1981)

117. Teichmüller, Oswald. Veränderliche Riemannsche Flächen, *Deutsche Math.* 7 (1944), English translation: Variable Riemann surfaces, translated by Annette A'Campo-Neuen. In: *Handbook of Teichmüller theory*, ed. A. Papadopoulos, Vol. IV, pp. 787–803, Eur. Math. Soc., Zürich (2014)

118. Thom, René. Quelques propriétés globales des variétés différentiables, Commentarii Mathematici Helvetici, 28, 178-6, (1954).

119. Thurston, William. A norm for the homology of 3-manifolds, *Memoirs of the American Mathematical Society*, 59 (33), 99–130 (1986)

120. Tits, Jacques. Sur les analogues algébriques des groupes semi-simples complexes, in *Colloque d'algèbre supérieure*, tenu à Bruxelles du 19 au 22 décembre 1956, Centre Belge de Recherches Mathématiques, Établissements Ceuterick, Louvain, Paris: Librairie Gauthier-Villars, pp. 261–289 (1957)

121. Tromba, Anthony J. On a natural algebraic affine connection on the space of almost complex structures and the curvature of Teichmüller space with respect to its Weil–Petersson metric, *Manuscripta Math.* 56, 475–497 (1986)

122. Voisin, Claire. *Hodge Theory and Complex Algebraic Geometry*, Cambridge University Press (2007), *Théorie de Hodge et géométrie algébrique complexe*, Cours Spécialisés 10, SMF (2002)

123. Weil, André. *Sur les courbes algébriques et variétés qui s'en déduisent*, Publications de l'Institut de Mathématiques de l'Université de Strasbourg, Hermann and Cie, 6, rue de la Sorbonne, Paris (1948)

124. Weyl, Hermann. Das asymptotische Verteilungsgesetz der Eigenwerte linearer partieller Differentialgleichungen (mit einer Anwendung auf die Theorie der Hohlraumstrahlung), *Mathematische Annalen* 71(4), 441–479 (1912),

125. Weyl, Hermann. *Die Idee der Riemannschen Fläche*, Teubner, Leipzig (1913)

126. Wiener, Norbert. Generalized harmonic analysis, *Acta Math.* 55(1), 117–258 (1930)

127. Wolpert, Scott. On the Weil–Petersson Geometry of the Moduli Space of Curves, *American Journal of Mathematics*, Vol. 107, 4, 969–997 (1985)

128. Wolpert, Scott A. Chern forms and the Riemann tensor for the moduli space of curves, *Invent. Math.* 85, no. 1, 119–145 (1986)

129. Yamada, Sumio. Weil–Petersson convexity of the energy functional on classical and universal Teichmüller space, *Journal of Differential Geometry*, 51, 35–96 (1999)

130. Yamada, Sumio. Local and global aspects of Weil–Petersson geometry, in *Handbook of Teichmüller Theory*, Volume IV, ed. A. Papadopoulos, pp. 44–107 European Mathematical Society, Zürich (2014)

131. Yor, Marc. *Some Aspects of Brownian Motion. Part I, Some Special Functionals*, Birkhäuser (1992)

132. Yor, Marc. *Some Aspects of Brownian Motion. Part II, Some Recent Martingale Problems*, Birkhäuser (1997)

Index

abstract field of complex numbers, 23
Additivity of Euler characteristic,
      215
almost complex structure, 48
Archimedes of Syracuse, 89
Arzelà–Ascoli theorem, 210
Arzelà, Cesare, 210
Ascoli, Giulio, 210

Baker, Henry Frederick, 181
Banach Fixed Point Theorem, 21
Banach manifold, 125
Banach–Schauder Theorem, 187
Banach space, 32
Banach, Stefan, 21
basic smooth rigidities, 29
Belyi, Gennadii Vladimirovich, 185
Belyi map, 89
Bendixson, Ivar Otto, 30
Berger, Marcel, 130
Bergman metric, 156, 253
Bergman, Stefan, 253
Blij, Frederik van der, 6
Bolyai, János, 63, 76
Borel, Edouard Justin Émile, 210
Brägger metric, 161
Brägger, Walter, 161
Brouwer Fixed Point Theorem, 18
Brouwer, Luitzen Egbertus Jan, 19
Brownian motion, 127
Brown, Robert, 127
Busemann Herbert, 86

Cartan, Élie, 18
Cartan formula, 31
Cartan, Henri, 18
Čech cohomology, 204
Čech, Eduard, 203
Chasles, Michel, 61
Chow, Wei-Liang, 237
Chow's theorem, 237
Christoffel, Elwin Bruno, 142
Christoffel functions Γ_{ij}^{k}, 142
Clairaut, Alexis Claude, 14
Clebsch, Rudolf Friedrich Alfred, 50
closed form, 16
co-rank Morse lemma, 39
covariant derivation, 140
cross-ratio, 62

Deahna, Heinrich Wilhelm Feodor,
      50
Dehn, Max, 15
Deligne, Pierre, 197
de Rham cohomology group, 100
de Rham, Georges, 99
differentiable manifold, 42
differential of a function/mapping, 3
differential form
      1-differential form, 2
      of degree k, 47
Dirac, Paul Adrien Maurice, 130
Dirichlet, Peter Gustav Lejeune, 127
distance function, 1
division algebra, 22

© The Author(s), under exclusive license to Springer Nature Switzerland AG 2021
N. A'Campo, *Topological, Differential and Conformal Geometry of Surfaces*, Universitext,
https://doi.org/10.1007/978-3-030-89032-2

Printed in the United States
by Baker & Taylor Publisher Services